#72

Progress in HPLC-HPCE Volume 5

Progress in HPLC-HPCE

Series Editors: H. Parvez and S. Parvez

Previously published in this series:

Progress in HPLC-HPCE Volume 5

Capillary electrophoresis in biotechnology and environmental analysis

Editors

H. Parvez, P. Caudy, S. Parvez
and P. Roland-Gosselin

Utrecht, The Netherlands, 1997

VSP BV
P.O. Box 346
3700 AH Zeist
The Netherlands

© VSP BV 1997

First published in 1997

ISBN 90-6764-213-4

Printed in The Netherlands by Ridderprint bv, Ridderkerk.

This book is dedicated to

Claude REISS
a great scientist, a friend, a teacher and a true humanist

and to

Simone, Natacha, Alexandre and Salima

SpectraPHORESIS ULTRA

Contents

Preface

The collection of papers presented in Volume 5 of PROGRESS IN HPLC-HPCE tries to evaluate the true impact of high performance capillary electrophoresis (HPCE) today on analytical biotechnology and environmental analysis. This analytical method of investigation, though new, has already given a promising application to peptide mapping, molecular biology vectors, proteins and carbohydrate analysis. The application of HPCE using negative UV absorption is a tremendous advance for the detection of anions and cations. This ion analysis by HPCE adds a new dimension to the control of environmental and industrial pollution. Ion analysis by HPCE is truly a novel concept, its use is highly economical and reliable.

The first section of the present volume provides a complete survey of all present innovations in instrument design, interface to Mass Spectrometry and adaptation of laser fluorescence detection for the achievement of higher sensitivity essential for biomedical analysis. Different methods of preconcentration techniques such as column stacking and use of variable surfactants have been extensively treated to give a comprehensive outlook for the improvment of separations at a higher sensitivity. Practical application of HPCE to protein and peptide analysis, a technique widely used in biotechnology and pharmaceutical industry, represents a major part of the book and each related field is treated by a specialist. The chapters in this section are dedicated to the analysis of protein folding, determination of phosphorylated histone H1 variants, process monitoring in biotechnology and peptide mapping of monoclonal antibodies and biologically active peptides.

A novel application of HPCE is thoroughly investigated for the screening of drugs of abuse. This new use of HPCE may lead to considerable benefit in biomedical analysis. The determination of protein-drug binding by capillary-zone electrophoresis as well as assay of catecholamines and other transmitters are further additions to the utility of this new analytical tool. The analysis of DNA analogs and antisense therapeutics are further proofs for the future adaptation of HPCE to clinical applications. The possibility to employ HPCE at very low UV wavelengths carries a tremendous advantage over HPLC for the analysis of carbohydrates, chiral separation of oligosaccharides and enantiomers. Finally, negative UV absorption for the determination of anions and cations offers alternative methods for the control of environmental and industrial pollution at a very economical cost.

This compilation summarizes today's status of the possibilities of HPCE in biotechnology and environmental analysis and provides promising openings for its interface with other superior physical methods. The reader of this volume will be able to appreciate that, with better knowledge of different HPCE methods, one can find a capillary electrophoresis solution to fulfil selectivity as well as sensitivity requirements for each particular analytical problem.

The editors express their sincere gratitude to Dr Peter Jandik of TSP, USA, as well as to all colleagues in TSP, France, for their kind dedication and help to advance the final accomplishment of this volume. Our thanks are also due to all staff at VSP International Science Publishers, for their continued support and contribution towards the realization of this book.

Hasan Parvez
Pierre Caudy
Simone Parvez
Pierre Roland-Gosselin
Editors Progress in HPLC-HPCE volume 5

TECHNOLOGICAL INNOVATIONS

Progress in HPLC-HPCE, Vol. 5, pp. 3–18
H. Parvez *et al.* (Eds)
© VSP 1997.

Capillary electrophoresis/mass spectrometry for the determination of bioactive proteins and peptides

KENNETH B. TOMER, LEESA J. DETERDING and CAROL E. PARKER

Laboratory of Molecular Biophysics, National Institute of Environmental Health Sciences, PO Box 12233, Research Triangle Park, North Carolina 27709, USA

INTRODUCTION

This chapter is intended to provide a solid experimental foundation for interfacing capillary electrophoresis with two mass spectrometric ionization techniques that are widely used for the analysis of peptides and proteins: continuous flow fast atom bombardment and electrospray ionization. The specific details of interfacing CE to electrospray ionization using a sheath flow interface on a quadrupole mass spectrometer and a coaxial interface with continuous flow fast atom bombardment, including modification of the CE system, buffer requirements, sample concentration and column derivatization as well as constraints put on the combination because of the requirements of the mass spectrometric ionization techniques, will be discussed. In particular, the application of CE/MS separation to the analysis of a mixture of bioactive peptides and a complex mixture of proteins, both at the femtomole/low picomole per component level, is detailed.

Capillary electrophoresis, as a field, has grown tremendously over the past decade, and is becoming more and more accepted as a practical analytical tool. Several obstacles still remain to its widespread acceptance, however. Among these are detection sensitivity and identification of analytes. The latter actually exists as a two-part problem. One part is reproducible migration times, so that an analyte in a complex mixture can be identified reproducibly and the second is structural assignment of analytes. Extensive and successful efforts have been made in instrument design to address the question of reproducible migration times, and a number of detection devices and methodologies have been developed to provide enhanced sensitivity and selectivity (Guzman, 1993). One detection system which combines good sensitivity and structural information about the analyte (thus, making it suitable for unknown mixtures) while being of general applicability is mass spectrometry. In addition, the use of a mass spectrometer as a detector for CE also reduces the need for precisely reproducible migration times, because analytes can be identified by a combination of the appearance of compound specific ions and approximate migration times.

Mass spectrometrists have also been interested in developing the combination of CE with MS because CE can deliver a low amount of analyte to the mass spectrometer in a short time period, i.e. high analyte flux, which extends the MS sensitivity. The bulk fluid flow rates associated with CE are also more compatible with most MS source vacuum and pumping requirements than are many other flowing liquid introduction techniques.

Not surprisingly, therefore, there has been a great deal of interest in the use of MS as a detector for CE (or alternatively, in the use of CE as an introduction system for MS). The initial report of the successful coupling of CE with MS was made by Smith and co-workers in 1987 (Olivares *et al.*, 1987), and over the next two years, a number of successful reports were made by other groups (Lee *et al.*, 1988; Minard *et al.*, 1988; Moseley *et al.*, 1989a, b; Caprioli *et al.*, 1989; Reinhoud *et al.*, 1989). It has been only within the last two years, however, that the number of groups working with CE/MS has really expanded, with at least 28 groups actively involved (as determined from abstracts at the 1993 and 1994 High Performance Capillary Electrophoresis and Mass Spectrometry conferences). As the number of commercially available CE systems compatible with interfacing to MS increases and as the availability of low cost mass spectrometers increases, the number of groups combining the two will also rapidly increase.

MASS SPECTROMETRY INTERFACED WITH CE

There are two main ionization techniques in mass spectrometry that have been most often coupled with CE, continuous flow FAB (CF–FAB) (Caprioli, 1990) and electrospray ionization (ESI) (Fenn *et al.*, 1990). Both techniques yield protonated (positive ion conditions) or deprotonated (negative ion conditions) molecule ion species and both are considered to be soft ionization techniques in that little fragmentation of analytes is observed. The most significant difference between the two as it relates to the mechanics of CE interfacing is that, in CF–FAB, ions are produced at high vacuum, while in ESI, the ions are produced at atmospheric pressure.

CF–FAB

In CF–FAB, the analyte is introduced into the ion source in an aqueous glycerol (5–25%) solution at a flow rate of *ca.* 3–5 µl/min. The analyte solution is bombarded by a beam of high energy atoms, typically xenon, which desorbs ions into the gas phase. CE has been interfaced with CF–FAB using a liquid junction in which the end of the CE capillary is external to the ion source (Caprioli *et al.*, 1989) and using a coaxial approach in which the CE capillary extends into the ion source (Moseley *et al.*, 1989a, b). A coaxial approach is desirable when the end of the CE capillary is in the source, because the total flow necessary for stable operation is greater than the liquid flow from the CE capillary which necessitates supplemental fluid delivery (make-up flow) and because of the requirements for a matrix fluid (e.g. glycerol). The outer column, the sheath column, carries the make-up flow and the matrix.

ESI

In an ESI source (and the closely related atmospheric pressure ionization source (API)), a flowing liquid solution (typically 5–10 μl/min) passes through a needle electrode. There is usually a 3 to 5 kV voltage potential between the electrode and the first elements of the source. This causes charge formation and disruption of the stream into charged liquid droplets (often pneumatically-assisted as in ion spray). These droplets undergo repetitive desolvation and coulombic explosions until analyte ions are desorbed into the gas phase. Both liquid junction (Lee *et al.*, 1988) and sheath flow interfaces (Smith *et al.*, 1988) have been developed for coupling CE with ESI. The sheath flow carries the bulk of the fluid and is typically acidified for positive ion conditions.

CF–FAB – ESI differences

In addition to the source pressure, there are other significant differences between ESI and CF–FAB which may be of significance for specific analyses. One is that ESI can produce multiply protonated species. That is, a single analyte can be produced with a distribution of charges due to differences in the number of protons added. Because mass spectrometers measure mass-to-charge, m/z, rather than just mass, this means that an ion whose mass is out of the mass range of the mass spectrometer may be observed if it is produced with a sufficient number of added protons. For example, a protein of $M_r = 30\,000$ will be observed on an instrument with a 1000 dalton mass range, if the protein ion is produced with 30 or more charges. This obviously greatly extends the capabilities of mass spectrometry for compound detection and identification.

A second difference is in the ability to separate analytes as negative ions with detection as positive ions. Under these conditions, there must be sufficient mixing of the sheath fluid with the analyte so that the analyte undergoes efficient protonation. This mixing is quite efficient in CF–FAB (Moseley *et al.*, 1991), but is less efficient in ESI (Perkins *et al.*, 1992; Moseley *et al.*, 1992; Smith and Udseth, 1993). This can lead to reduced sensitivity or necessitate alteration of separation conditions under ESI.

To illustrate the experimental procedures for coupling CE to both CF–FAB and ESI, detailed methodology will be provided in this chapter for the separation and mass spectrometric detection of a suite of bioactive peptides by coaxial

CF–FAB and of a mixture of proteins by ESI. Prior to discussing the individual interfaces, however, some procedures will be described that are applicable to both interface types.

CE/MS INTERFACES

Most commercial CE instruments are not designed for use with mass spectrometry, and the following discussion is primarily directed to those interfacing such instruments to a mass spectrometer. (If the CE instrumentation is designed for interfacing with your mass spectrometer, follow the manufacturer's instructions.)

1) Many commercial instruments have a current continuity check to verify electrical continuity between the cathode and anode buffer compartments. This circuitry must be altered when the outlet buffer compartment is inside the mass spectrometer. This should only be done by qualified personnel.

2) A hole must usually be placed in the wall of the CE instrument to serve as an outlet for the capillary to pass through. Placing a piece of Teflon tubing into the hole through which the CE column extends helps prevent scratching the polyimide coating on the column and, thus, reduces the chance of electrical breakdown.

3) The inlet end of the capillary and end of the CE/MS probe must be at the same height.

4) Buffers: Only volatile buffers should be routinely used. Non-volatile buffers will rapidly coat the source with consequent MS problems. Non-volatile buffers also usually contain alkali metal ions which can decrease analyte sensitivity through formation of numerous metalated ions (Moseley *et al.*, 1989b). Surface active agents are usually avoided, because they also can cause severe source contamination and/or inhibit analyte ion formation (but see Varghese and Cole (1993) and Kirby *et al.* (1994) for some examples of MECC/ESI/MS). Although it is possible to run with buffer concentrations above *ca.* 10 mM, this can often lead to excessive current, changing the voltage on the probe (and, thus, detuning the source) (Perkins *et al.*, 1992) or excessive ion formation from the buffer.

5) Although electromigration of analytes onto the capillary column can be used, the electromagnetic pulse created by the relays involved has been known to cause problems, such as locking-up or automatic rebooting, with some computers. Sample introduction by electromigration is also prone to discrimination effects due to differential mobilities of the analytes.

6) Sheath flow designs help to maintain temperature control within the CE column. In general we have not found it necessary to add cooling gas to the probe.

7) All solutions should be made with high-purity, filtered water (e.g. from a Milli-Rho/Milli-Q system from Millipore) and HPLC grade solvents. All solutions should be degassed prior to use.

8) Use fresh buffer solutions and check the pH often. Changes in the pH of buffers sufficient to affect migration times have been noted when the buffer has sat for several days, or has been repeatedly sonicated.

Sample concentration

A major concern about the utility of CE, especially when interfaced to MS, is that the sample concentration required to obtain reasonable MS sensitivity can be quite high, often millimolar (which can create sample solubility problems). This is due to the combination of MS sensitivity and CE injection volume. There are several viable techniques for concentrating the sample inside the CE capillary.

Sample stacking. The simplest method for sample concentration on the CE column is sample stacking. In sample stacking, the analyte is injected onto the column in deionized water which does not contain any buffers. The analyte ions migrate rapidly to the boundary between the high conductivity zone (column buffer) and the low conductivity zone (analyte in water). At this boundary, the ions slow down and concentrate at the level of the leading buffer. Burgi and Chien (1991) have noted a ten-fold increase in sample concentration using this technique over injecting the sample in buffer.

Transient isotachophoresis/CE. Thompson and co-workers (1992) have reported on the combination of transient isotachophoretic focusing followed by CE separation (ITP/CE). In ITP/CE, the column is filled with a background electrolyte (BGE). A leading electrolyte (LE) band with a higher mobility than the BGE is injected, followed by an injection of the sample. The column is placed into a trailing buffer solution, and high voltage is applied. The sample ions will adapt their concentration to that of the LE, forming concentrated ITP zones, and then the LE will move through the BGE leaving the concentrated analyte ITP zones in the BGE. At this point, a normal CE separation can begin. In one example, the authors filled over 33% of the column volume with sample and demonstrated good CE efficiency and a 24-fold improvement in sensitivity over that obtained by sample stacking.

Sample injection

In addition to the standard injection procedures compatible with our home-built system – electromigration, and hydrodynamic – we have developed a simple technique for pressurized injection of fairly large sample amounts for use in 'scouting' runs and for use with smaller i.d. capillaries, e.g. 13 µm i.d. columns. This consists of a pressurizable stainless steel container which is drilled out to hold a 1 ml conical vial (or other desired size) (Deterding *et al.*, 1989; Caprioli and Tomer, 1990) (Fig. 1). The CE column is placed through the fitting at the top of the vessel into the analyte solution and pressure is applied for a specified

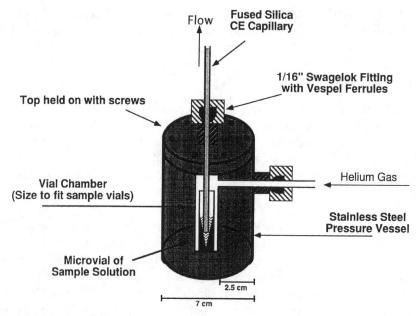

Figure 1. Pressurized injection vessel. (Reprinted from Caprioli and Tomer, 1990, in *Continuous Flow Fast Atom Bombardment*, with permission.)

time. After sample injection, the column is removed and returned to the CE buffer compartment. As with other pressure based injection techniques, analyte discrimination, as found with electromigration, is not observed. This injection system is also useful for flushing the capillary, which is routinely done for 5 min between analyses or for column equilibration.

Capillary columns

Prior to use, bare fused silica columns were flushed with 1 M KOH for 20 min and then with 0.1 M KOH for 45 min and then with water as per the method of Lauer and McManigill (1986). After flushing with water, the capillaries were then flushed with running buffer for 15 min.

Although most CE/MS analyses reported in the literature have been carried out with underivatized fused-silica, derivatized columns are becoming increasingly popular. Garcia and Henion (1992) have also demonstrated that gel-filled capillaries can be successfully used with ESI/MS. Because the end of the capillary is at atmospheric pressure, the gel is not extruded into the source. Commercially available, covalently-modified CE columns can be used routinely in CE/MS, but the use of sieving and/or polymer containing buffers in CE/MS may contaminate the source.

The impetus for use of coated capillaries has come, to a great extent, from adsorption problems encountered in protein separations. Basic proteins can be adsorbed onto the negatively charged capillary walls at pHs below the PI of the protein. As most protein separations are performed under acidic conditions, especially in conjunction with ESI/MS, this adsorption leads to peak broadening and to reduced sensitivity. We have been preparing our own columns in which the free silanol groups have been converted to aminopropyldimethoxysilane groups. These columns, called APS columns, have a positively charged surface at a buffer pH less than 10.6, and will repel basic proteins.

Preparation of APS columns (Moseley *et al.*, 1991)

1) Rinse the column with 6 M HCl for approximately 4 h.
2) Dehydrate the column by installing in a capillary GC oven and purging with dry helium at 300°C for 12–15 h.
3) Rinse with toluene for 30 min.
4) Pump a 5% solution of 3-aminopropyltrimethoxysilane in toluene through the column for 1 h, followed by a 30 min toluene rinse.
5) Dry the column by purging with helium for 30 min.
6) Condition the column with CE buffer for one hour prior to use.
7) The column can be regenerated if migration times begin to increase by flushing with 0.01 M HCl.
8) Because the column wall is now positively charged, electroosmotic flow is reversed and the high voltage power supply is run at a negative potential so that the end of the capillary is at negative potential relative to the CE/MS interface.

CE/MS EXPERIMENTAL

Electrospray ionization

Probe design. The probe illustrated in this section (Fig. 2) was modified from the probe designed for use with a Vestec electrospray source, but the details are transferable to other probe designs (Parker, 1992; Perkins, 1992).

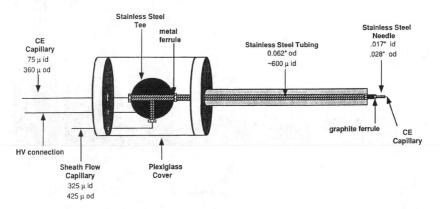

Figure 2. CE/ESI probe. (Adapted from Parker *et al.*, 1992, with permission from *J. Am. Soc. Mass Spectrom.*)

As mentioned above, the ESI interface is of a coaxial design in which the outer column is for the sheath fluid and the inner column is the separations capillary. The CE capillary used is 75 μm i.d. × 360 μm o.d. and *ca.* 1 m in length. Use of thinner-walled tubing leads to less reliable operation due to electrodrilling of holes through the capillary wall. The 1 m length of capillary is sufficiently long to reach from the buffer reservoir to the end of the probe with some room for manipulation. The electrospray needle is cut from 0.017″ i.d. × 0.028″ o.d. stainless steel tubing (22 gauge). A length of 1/16″ o.d., 600 μm i.d. stainless steel tubing is used as the outer column (sheath column). The capillary column is threaded through one arm of a stainless steel tee into the sheath column which is connected to the opposite arm. The sheath fluid is delivered through a 325 μm i.d. × 425 μm o.d. fused silica column which is connected to the third leg of the tee. The high voltage lead for the ESI needle is also connected to the tee (or ground connection, if the needle is at ground potential). In the original Vestec design, the tee was enclosed in a metal can. This was replaced by a Plexiglas cylinder to prevent arcing, and both metal endplates of the probe handle are grounded. We found that optimum sensitivity and stability of the spray was obtained when the CE column protruded slightly from the end of the needle (*ca.* 1 mm). The sheath fluid is delivered using a Harvard syringe pump. The sheath fluid, in addition to forming the spray, also provides electrical contact between the end of the CE column and 'ground'.

CE/MS analysis of a complex protein mixture using ESI. The CE/MS analysis of the venom of the snake *Dendroaspis polylepis polylepis*, the black mamba, is used as an example of a separation of a complex protein mixture (Perkins *et al.*,

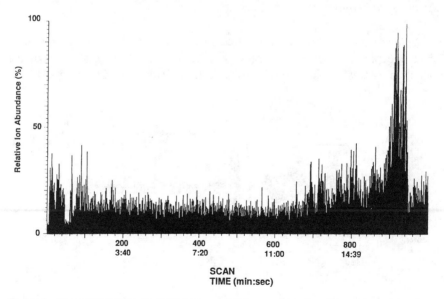

Figure 3. Reconstructed ion electropherogram of the separation of snake venom proteins from *Dendroaspis polylepis polylepis*.

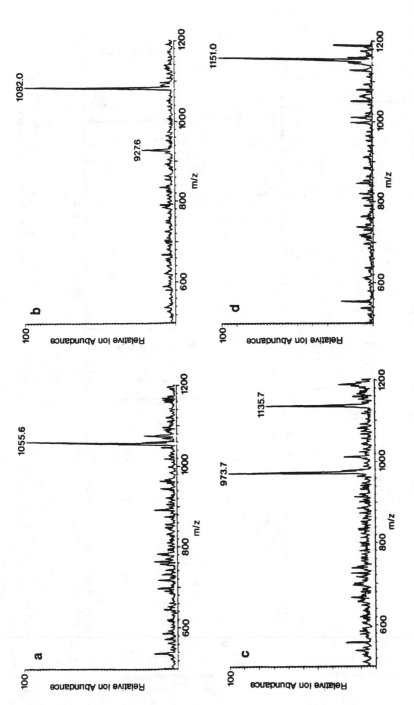

Figure 4. The on-line full scan CE/ESI/MS spectra of: a) Protein of M_r 7379; b) Protein of M_r 6484; c) Toxin C, M_r 6809; d) Toxin α, M_r 6907. (Reprinted from Perkins *et al.*, 1993, with permission from *Electrophoresis*.)

K. B. *Tomer* et al.

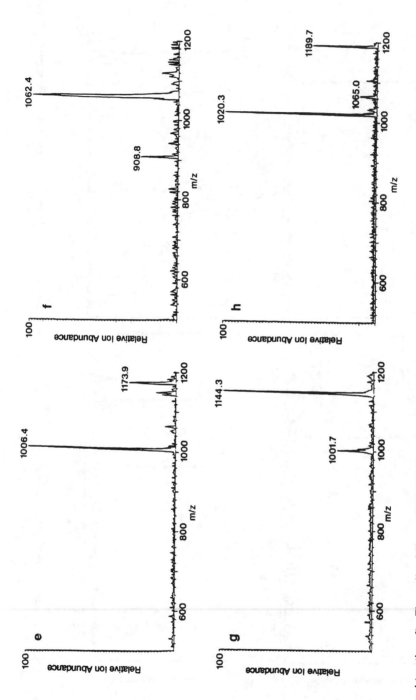

Figure 4 (continued). The on-line full scan CE/ESI/MS spectra of: e) Toxin β, M_r 6470; f) Protein of M_r 6360; g) Protein of M_r 8005; h) Toxin I, M_r 7132, with Protein CM-3, M_r 7336, and peptides yielding multiply-charged ions of *m/z* 1026, 1031 and 1065. (Reprinted from Perkins *et al.*, 1993, with permission from *Electrophoresis*.)

1993). At least 28 proteins have been previously purified from this venom, and our CE/MS analysis demonstrated the existence of at least 70 proteins. For these experiments a VG 12-250 quadrupole equipped with a Vestec ESI source and the modified interface probe described above was used. The CE capillary was an APS derivatized fused silica column. The sheath fluid was a 50 : 50 methanol : 3% aq. acetic acid solution. The mass spectrometer was scanned from m/z 500 to its maximum m/z value, 1200, at a scan rate of 1 s per scan. Because an APS column is used, the CE voltage was set at −33 kV during the analysis while the electrospray needle was held at +3 kV for a total voltage drop of 36 kV. The buffer used in the experiment was 0.01 M acetic acid at pH 3.5. The column was flushed with buffer solution for 10 min prior to sample analysis using the pressurized injection vessel described above. The snake venom (*Dendroaspis polylepis polylepis*, Sigma Chemical Co., St. Louis, MO, USA) was dissolved in water at a concentration of 1 mg/ml. The buffer solution in the pressurized injection vessel was replaced with the snake venom solution. The vessel was pressurized at 10 psi for 2 s. This was sufficient to inject *ca.* 50 nl of the analyte solution onto the column. The column was removed from the injection vessel and placed into the buffer compartment (with the high voltage off). The high voltage was turned on and the separation commenced. After the separation was completed, the high voltage turned off. The column was removed from the buffer compartment and placed into a vial of running buffer in the injection vessel to flush the system. The reconstructed total ion electropherogram is shown in Fig. 3 and some full scan spectra obtained from the separation are shown in Fig. 4.

Coaxial continuous flow FAB

Probe design (Moseley *et al.*, 1989b, 1991). As indicated above, the end of the CF–FAB probe, and thus the end of the CE capillary, is under high vacuum. To prevent excessive vacuum-induced flow in the separation column,

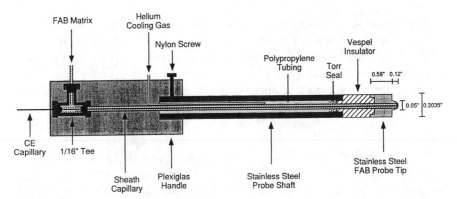

Figure 5. Schematic of a CE/coaxial continuous flow fast atom bombardment probe interface. (Reprinted from Moseley *et al.*, 1991, with permission from *Anal. Chem.*)

a column with a maximum i.d. of 13 μm is used. A schematic of the overall coaxial CE/CF–FAB interface is shown in Fig. 5. The probe illustrated was built in the laboratory, but a standard CF–FAB probe can be used with minor modifications. The design shown here allows incorporation of a helium cooling gas line. We have found, however, that the matrix flow surrounding the CE capillary is normally sufficient to remove excess heat. The sheath column should fit tightly into the hole in the probe tip to prevent backflow of the column effluent and/or FAB matrix into the probe tip. Teflon tape, a septum or Torr Seal can be used to minimize backflow. A length of 160 μm i.d. × 365 μm o.d. fused silica is used as the outer column (sheath column). The separation column enters the sheath column through one arm of a stainless steel tee. The sheath fluid is delivered through a 3 m long 25 μm i.d. fused silica column which is connected to the third leg of the tee. Vespel ferrules are used, because graphite ferrules can lead to column plugging. The CE column terminates approximately 1–2 mm from the probe tip. This helps prevent vacuum induced flow. The column is always threaded from the tip of the probe rather than from the tee. In the latter case, we have noted more problems with high voltage arcing. As with the ESI probe, the tee is mounted in a Plexiglas handle to prevent electrical contact between the tee and the source housing and/or operator.

Because the source is under vacuum, the flow of matrix solution must balance the rate of evaporation from the probe tip. Too little flow will lead to a dry tip and no matrix. Too much flow will lead to drop formation which causes unstable source operation and peak-broadening. We have found that using a 150 μm o.d. column with a 160 μm i.d. sheath column permits us to easily control this balance with relatively little sheath flow – less than 1 μl/min. The sheath fluid is a 25/75 glycerol/water mixture containing 0.5 mM heptafluorobutyric acid which is thoroughly degassed prior to use. The acid serves two purposes. It helps to promote protonation of the analytes, and it ensures electrical continuity between the CE effluent and the probe tip. The flow of matrix is controlled by an Isco syringe pump.

CE/MS analysis of chemotactic peptides using coaxial CF–FAB MS. This experiment illustrates both the technique of CE coaxial CF–FAB and the ability to carry out separations under conditions in which the analytes are negatively charged followed by analysis as positive ions (Moseley *et al.*, 1991). The running buffer in these experiments is 0.005 M ammonium acetate adjusted to pH 8.5 with ammonium hydroxide. The pH of the FAB matrix was 3.5 which was sufficient to rapidly neutralize and then protonate the negatively charged analytes.

The mass spectrometer used in this experiment was a VG ZAB 4F tandem mass spectrometer with a source accelerating voltage of +8 kV. This potential is applied to the interface probe tip which means that 'ground' potential is at +8 kV. Thus, the CE buffer (positive electrode) is at +38 kV to provide a voltage drop of 30 kV (300 V/cm for a 1 m long column). Xenon (8 kV

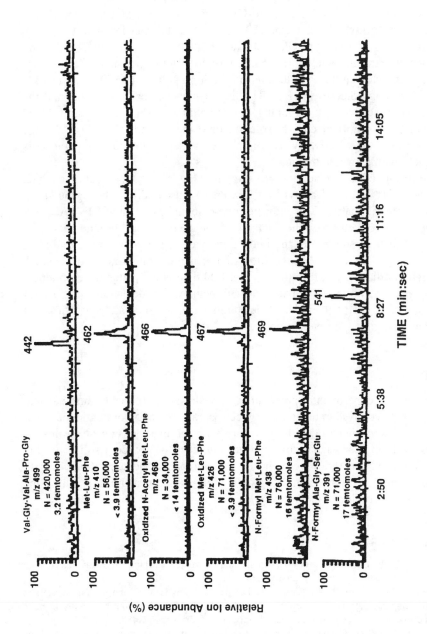

Figure 6. Single ion electropherograms of the protonated molecular ions of a mixture of chemotactic peptides. (Reprinted from Moseley *et al.*, 1991, with permission from *Anal. Chem.*)

at 1 mA) was used as the FAB gas. The mass spectrometer was scanned exponentially at a rate of 3 s/decade over the mass range of 520 dalton to 370 dalton. This experiment also uses bare fused-silica capillaries.

A solution of Val-Gly-Val-Ala-Pro-Gly (2.5×10^{-5} M), Met-Leu-Phe (2.5×10^{-5} M), *N*-formyl Met-Leu-Phe (10^{-4} M), *N*-acetyl Met-Leu-Phe (10^{-4} M), and *N*-formyl Ala-Gly-Ser-Glu (10^{-4} M) is prepared in water. Samples are injected onto the column by electromigration by placing the end of the capillary into the analyte solution and turning on the high voltage to 150 V/cm (15 kV for a 1 m long column) for 2 s. The high voltage is then turned off and the column is placed into the running buffer solution. This introduces 0.16 nl (4 fmol to 16 fmol) of analyte solution onto the column. The high voltage is then turned on to +38 kV to commence the separation. The separation is complete within twelve minutes.

The results of this separation are shown in Fig. 6. At this level of analyte, the $(M + H)^+$ ion of *N*-acetyl Met-Leu-Phe is obscured by a background ion arising from the matrix. This indicates one potential problem of CF–FAB, that is not as severe with ESI – the intensity of the chemical background. It should be noted that, in addition to the known analytes in the solution, a number of impurities arising from the oxidation of the methionine were also detected. Separation efficiencies of up to $N = 420\,000$ theoretical plates were obtained. When this separation was carried out with negative ion detection using a FAB matrix with a pH of 9, limits of detection were approximately one order of magnitude higher (poorer).

SUMMARY

These sets of experimental procedures were designed to enable the reader to set up a CE/MS system using either electrospray ionization or continuous flow FAB. The experimental details may vary with the exact design of the instruments in the reader's laboratory, but are intended to provide a solid working foundation. The analytes were chosen to demonstrate the separations of both a peptide mixture and a complex protein/polypeptide mixture, illustrating the power of the combined techniques.

REFERENCES

Burgi, D. S. and Chien, R.-L. (1991). Application of sample stacking to gravity injection in capillary electrophoresis. *J. Microcol. Sep.* **3**, 199–202.

Caprioli, R. M., Moore, W. T., Martin, M., DaGue, B. B., Wilson, K. and Moring, S. (1989). Coupling capillary zone electrophoresis and continuous-flow fast atom bombardment mass spectrometry for the analysis of peptide mixtures. *J. Chromatogr.* **480**, 247–257.

Caprioli, R. M. (1990). Continuous-flow fast atom bombardment mass spectrometry. *Anal. Chem.* **62**, 477A–485A.

Caprioli, R. M. and Tomer, K. B. (1990). Liquid chromatography/mass spectrometry. In: *Continuous-Flow Fast Atom Bombardment Mass Spectrometry*, Caprioli, R. M. (Ed.). John Wiley, Chichester, pp. 107–117.

Deterding, L. J., Moseley, M. A., Tomer, K. B. and Jorgenson, J. W. (1989). Coaxial continuous flow fast atom bombardment in conjunction with tandem mass spectrometry for the analysis of biomolecules. *Anal. Chem.* **61**, 2504–2511.

Fenn, J. B., Mann, M., Meng, C. K. and Wong, S. F. (1990). Electrospray ionization – principles and practice. *Mass Spectrom. Rev.* **9**, 37–70.

Garcia, F. and Henion, J. D. (1992). Gel-filled capillary electrophoresis/mass spectrometry using a liquid junction-ion spray interface. *Anal. Chem.* **64**, 985–990.

Guzman, N. A. (1993). *Capillary Electrophoresis Technology*. Marcel Dekker, New York.

Kirby, D., Greve, K. F., Foret, F., Karger, B. L. and Nashabeh, W. (1994). Capillary electrophoresis-electrospray ionization mass spectrometry utilizing background electrolytes containing surfactants. In: *Proc. 42nd ASMS Conf. on Mass Spectrometry and Allied Topics.* Chicago, p. 1014.

Lauer, H. H. and McManigill, D. (1986). Capillary zone electrophoresis of proteins in untreated fused silica tubing. *Anal. Chem.* **58**, 166–170.

Lee, E. D., Mueck, W. M., Covey, T. R. and Henion, J. D. (1988). *J. Chromatogr.* **458**, 313–321.

Minard, R. D., Chin-Fatt, D., Curry, P. and Ewing, A. G. (1988). Capillary electrophoresis/flow FAB MS. In: *Proc. of the 36th Annual Conf. on Mass Spectrometry and Allied Topics*, 950–951.

Moseley, M. A., Deterding, L. J., Tomer, K. B. and Jorgenson, J. W. (1989a). Capillary zone electrophoresis/fast atom bombardment mass spectrometry: Design of an on-line coaxial continuous-flow interface. *Rapid Commun. Mass Spectrom.* **3**, 87–93.

Moseley, M. A., Deterding, L. J., Tomer, K. B. and Jorgenson, J. W. (1989b). Coupling of capillary zone electrophoresis and capillary liquid chromatography with coaxial continuous-flow fast atom bombardment tandem sector mass spectrometry. *J. Chromatogr.* **480**, 197–209.

Moseley, M. A., Deterding, L. J., de Wit, J. S. M., Tomer, K. B. and Jorgenson, J. W. (1989c). Optimization of a coaxial continuous flow fast atom bombardment interface between sector mass spectrometry for the analysis of biomolecules. *Anal. Chem.* **61**, 1577–1584.

Moseley, M. A., Deterding, L. J., Tomer, K. B. and Jorgenson, J. W. (1991). Determination of bioactive peptides using capillary zone electrophoresis/mass spectrometry. *Anal. Chem.* **63**, 109–144.

Moseley, M. A., Jorgenson, J. W., Shabanowitz, J., Hunt, D. F. and Tomer, K. B. (1992). Optimization of capillary zone electrophoresis/electrospray ionization parameters for the mass spectrometry and tandem mass spectrometry analysis of peptides. *J. Am. Soc. Mass Spectrom.* **3**, 289–300.

Parker, C. E., Perkins, J. R., Tomer, K. B., Shida, Y., O'Hara, K. and Kono, M. (1992). Application of capillary zone electrophoresis and nanoscale packed capillary liquid chromatography/electrospray ionization mass spectrometry to the analysis of macrolide antibiotics. *J. Am. Soc. Mass Spectrom.* **3**, 563–574.

Perkins, J. R., Parker, C. E. and Tomer, K. B. (1992). Nanoscale separations combined with electrospray ionization mass spectrometry: sulfonamide analysis. *J. Am. Soc. Mass Spectrom.* **3**, 139–149.

Perkins, J. R., Parker, C. E. and Tomer, K. B. (1993). Characterization of snake venoms using capillary electrophoresis in conjunction with electrospray mass spectrometry: Black mambas. *Electrophoresis* **14**, 458–468.

Olivares, J. A., Nguyen, N. T., Yonker, C. R. and Smith, R. D. (1987). On-line mass spectrometric detection for capillary zone electrophoresis. *Anal. Chem.* **59**, 1230–1232.

Reinhoud, N. J., Niessen, W. M. A., Tjaden, U. R., Gramber, L. G., Verheij, E. R. and van der Greef, J. (1989). Performance of a liquid-junction interface for capillary electrophoresis mass spectrometry using continuous flow fast atom bombardment. *Rapid Commun. Mass Spectrom.* **3**, 348–351.

Smith, R. D., Barinaga, C. J. and Udseth, H. R. (1988). Improved electrospray ionization interface for capillary zone electrophoresis-mass spectrometry. *Anal. Chem.* **60**, 1948–1952.

Smith, R. D. and Udseth, H. R. (1993). Mass spectrometric detection for capillary electrophoresis. In: *Capillary Electrophoresis Technology*, Guzman, N. A. (Ed.). Marcell Dekker, New York, pp. 525–567.

Thompson, T. J., Vouros, P., Foret, F. and Karger, B. L. (1992). Capillary electrophoresis/electrospray ionization mass spectrometry: improvement of protein detection limits using on-column transient isotachophoretic sample preconcentration. *Anal. Chem.* **65**, 900–906.

Varghese, J. and Cole, R. B. (1993). Cetyltrimethylammonium chloride as a surfactant buffer additive for reversed-polarity capillary electrophoresis-electrospray mass spectrometry. *J. Chromatogr.* **A652**, 369–376.

Progress in HPLC-HPCE, Vol. 5, pp. 19–48
H. Parvez *et al.* (Eds)
© VSP 1997.

Preconcentration techniques in high performance capillary electrophoresis

PIERRE ROLAND-GOSSELIN,[1] PIERRE CAUDY,[1] HASAN PARVEZ[2] and SIMONE PARVEZ[3]

[1] *Thermo Separation Products, France, Hightec Sud, 12 Avenue des Tropiques, Z. A. de Courtabeuf, BP 141, 91944 Les Ulis Cedex, France*
[2] *CNRS, 91400 Orsay, France*
[3] *Neuroendocrinologie et Neuropharmacologie du Développement, Université de Reims-UFR Sciences, BP 347, 51062 Reims Cedex, France*

INTRODUCTION

Detection limit is a very important criterion in the choice of an analytical technique. As in capillary electrophoresis, the detection has to be performed through a 50 or 75 mm internal diameter capillary, it is generally admitted that the detection limit is within the range of 10^{-5} to 10^{-6} with a UV detector. In order to make capillary electrophoresis more attractive as a separation technique, it is necessary to improve this limit of detection as far as possible. This can be achieved by several processes:

- exchanging the mode of detection from UV either by fluorescence mediated by laser or not (Gassmann *et al.*, 1985; Burton and Sepaniak, 1986; Toulas and Hernandez, 1992), by electrochemical detection (Lu *et al.*, 1993) or by conductimetry (Avdalovic *et al.*, 1993),
- modifying the optical pathway of the capillary,
- using different ways of sample introduction (hydrodynamic or electrokinetic process) classically offered by commercial instrumentation.

Our main aim here is not to discuss the utilisation of different possible modes of detection of increased optical pathway: briefly, the latest can be done utilising either a bubble capillary (Gordon, 1991) or a Z-shaped cell (Albin *et al.*, 1993; Moring *et al.*, 1993). Our concern is to examine the possibilities of optimising the final yield in signal/background ratio. These possibilities are linked to sample introduction techniques.

THEORY AND RESULTS

In most of the commercial systems two injection modes are available:
  1) hydrodynamic mode,
  2) electrokinetic mode.

HYDRODYNAMIC MODE

The capillary is subjected to a pressure difference between its extremities. This presssure difference can be realised by gravity (Fig. 1A), by overpressure applied inside the sample vial (Fig. 1B), or by aspiration of the sample through the capillary tube (Fig. 1C).

Whatever is the chosen mode, the total volume of sample introduced in the capillary tube follows the Poiseuille law

$$V = \frac{\Delta P D^4 \pi T}{128 \eta L_t}, \tag{1}$$

where $\Delta P = \rho g \Delta h$ when injection mode with gravity is used or ΔP is the difference in the pressure, D is the diameter of the capillaty tube, η the fluid

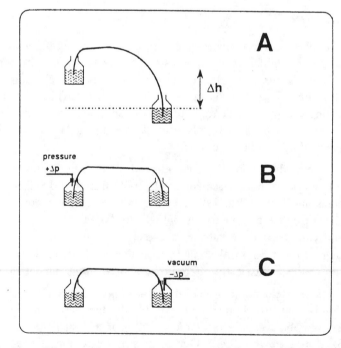

Figure 1. Three different modes of injection by pressure difference in High Performance Capillary Electrophoresis (HPCE). A: by gravity; B: by overpressure applied inside the sample vial; C: by aspiration of the sample through the capillary tube. Reproduced with permission.

viscosity, L_t the total length of the capillaxy tube, ρ the density of the sample solution, T the time during which the difference of pressure is applied and $g = 9.81 \text{ m s}^{-2}$.

According to (1), the injected volume is proportional to the difference of pressure and to the duration of injection. Table 1 shows the collection made by Weinberger (1993) on volumes introduced in the capillary tube with the second injection according to the three possible modes of injection as proposed in Fig. 1.

One can wonder about the influence of the volume of injection on the area and the shape of the peak, on the resolution and on the number of theoretical plates?

Elution peaks can be considered as Gauss curves as used in chromatography in agreement with the theory of plates.

The total variance of the elution peak, σ_t^2, is the resultant of several parameters: diffusion, convection, Joule effect, length of the sample zone, electroendoosmosis, electrokinetic dispersion, interactions (adsorption) with capillary wall, conductivity differences, pH differences, detection window, etc.

Since we are considering only the different parameters involved during injection mode, and as all these parameters are additive, the total variance of a peak can be expressed according to the following equation (Foret and Bocek, 1988; Jones *et al.*, 1990).

$$\sigma_t^2 = \sigma_{\text{dif}}^2 + \sigma_{\text{long}}^2 + \sigma_{\text{niv}}^2 + \sum_i \sigma_i^2, \tag{2}$$

where 'dif' represent the influence of diffusion, 'long' the influence of the length of the capillary, 'niv' the influence of the difference in the liquid level (buffer or sample) in the bottle situated at both ends of the capillary which induces *de facto* a displacement of liquid by gravity inside the capillary and 'i' the influence of all other parameters.

The influence of σ_{niv}^2 can be minimised or even suppressed if one takes care about setting the same amount of liquid at the same level in each bottle.

The influence of

$$\sigma_{\text{long}}^2 \quad \text{is} \quad l_{\text{inj}}/12 \tag{3}$$

for a rectangular profile of injection (Hjertén, 1990).

Table 1.
Hydrodynamic injection mode and volume of injection

Sampling method Internal diameter	Overpressure $\Delta P = 0.5$ psi	Aspiration $\Delta P = 12.7$ cm Hg	Gravity $\Delta H = 10$ cm
50 μm	0.6 nl/s	2.9 nl/s	0.17 nl/s
75 μm	3 nl/s	1.5 nl/s	0.87 nl/s

[a]Capillary length: 1 m, temperature: 30°C, viscosity: 0.801 gm cm^{-2} s^{-1}, density: 0.997 g ml^{-1} (Reproduced with permission: Weinberger, 1993).

Table 2.

Length and maximal volume for a capillary of 50 μm × 50 cm (Reproduced with permission: Otsuka and Terabe, 1989)

Variation in the number of theoretical plates	Number of theoretical plates (a)	100 000	500 000	1, 000 000
	Peak width in mm $t_m = 10$ min	4.47	2.00	1.41
5%	l_{inj} in mm	1.24	0.56	0.39
	inj in nl	2.43	1.09	0.77
10%	l_{inj} in mm	1.77	0.79	0.56
	inj in nl	3.48	1.55	1.10

The term $\sigma_{dif}^2 = 2Dt$ is the equation of Einstein in which D is the diffusion constant and t the duration during which the solute has run a certain length (Hjertén, 1990).

Otsuka and Terabe (1989) studied the influence of the injection volume on the number of theoretical plates for a given sample disolved in the working electrolyte. The results are provided in Table 2. One can see that with a given starting base of 500 000 plates, it is possible to inject only up to 1.09 nl if one does not want to decrease by more than 5% the number of plates in order to increase by 5% the area of the peak. This observation shows that in pre-fixed experimental conditions (e.g. sample already dissolved in buffer), one has limited choice as far as the time of injection is concerned (cf. Table 1) if one does not want to deteriorate dramatically the number of theoretical plates or the symmetry of the peak and therefore the separation performance. According to the mode of sampling (1A, 1B or 1C) the time of injection should be within the limits of 0.5 to 10 s. It is then necessary to modulate the sample concentration which is neither an easy task nor the best way to improve the deteetion limit.

Enrichment using the buffer conductivity differences

We have taken into consideration the hypothesis till now that the sample is already disolved in the working buffer, a normal attitude for someone familiar with HPLC. We should also consider another situation if the sample is solubilized in water or in diluted working buffer.

Mikkers et al. (1979) showed important differences in behavior of the same sample according to its dilution in a variable matrix keeping the analytical conditions constant. Figure 2a shows as related to a sample dissolved in pure water. One can observe that the peaks are narrower and higher than in Fig. 2b which corresponds to the same sample dissolved in the working buffer. What can be the mechanism responsible for this distortion?

When the sample is dissolved in the buffer, the total ionic concentration of the sample zone can be estimated overall similar to that of the buffer zone;

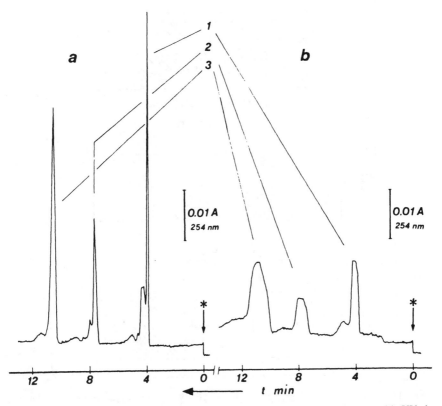

Figure 2. Zone electrophoretic separation of three anionic sample constituents with UV detection, using operational system 3. (1) Sulphanilic acid, 70×10^{-12} mole; (2) 5-bromo-2,4-dihydroxybenzoic acid, 140×10^{-12} mole; (3) adenosine-5'-monophosphoric acid, 35×10^{-12} mole. t (min) = time of analysis. a) Constituents dissolved in water; b) constituents dissolved in carrier electrolyte. Reproduced with permission.

therefore, the conductivity and the electric field will be identical in each zone. However, when the sample is dissolved in water the total ionic concentration in the sample zone is very low, and the electric resistance of the sample zone is therefore very high as compared to that of the buffer zone (Fig. 3A).

At the time when the electric current is applied (Fig. 3B) the electrical field E_1 will be much higher in the sample zone 1 than in the buffer zone 2 ($E_1 \gg E_2$) and the translation speed index of a solute will be higher since this speed is proportional to the electric field as indicated by (4):

$$V_i = \mu_i E, \tag{4}$$

where μ_i the apparent mobility of the solute i.

The solutes i will move very fast in the sample zone 1 until they reach the border sample/buffer 2 (Fig. 3B). As soon as they have passed this border (Fig. 3C), their translation speed will slow down extremely when the electric field E_2 in that zone 2 is much weaker than in zone 1. According to Chien

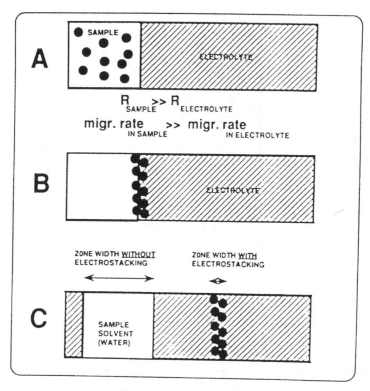

Figure 3. Electrostacking in High Performance Capillary Electrophoresis (HPCE) for high sensitivity yield. Reproduced with permission.

and Burgi (1992a) if one neglects the diffusion factor, the solutes will be concentrated in a very narrow band and can be estimated as:

$$X_i = X_{\text{inj}}/\gamma,$$

where X_{inj} is the length of the sample zone taken immediately after the introduction of the sample in the capillary tube, γ the augmentation factor of concentration specified as below. When the capillary tube has a constant diameter, the differences in the speed of electroosmotic flux can be neglected between zone 1 and 2.

$$\gamma = \frac{E_1}{E_2} = \frac{r_1}{r_2} = \frac{C_2}{C_1}, \tag{5}$$

where C_1 and C_2 are the buffer concentrations, respectively, in zones 1 and 2. Theoretically, this preconcentration factor γ is proportional to the yield of the two electric fields E_1 and E_2. The greater the yields of the concentrations C_1 and C_2, the more important will be the preconcentration factor.

With extrapolation, it is possible to think about the introduction of a reduced 'sample zone' prepared from a previous long sample zone once all the solutes

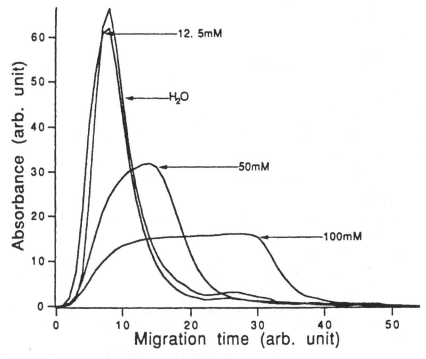

Figure 4. Graphs of PTH-arginine peak shapes as a function of the buffer concentration in the sample plug. All sample plugs are 5-min injections with sample buffer concentrations ranging from 100 mM to pure water. Reproduced with permission.

have passed through highly concentrated buffer. The discrepancy induced by the length of the sample in the sampling becomes

$$\sigma^2_{\text{long}} = \frac{1}{12}\left\{\frac{X_{\text{inj}}}{\gamma}\right\}^2.$$

Figure 4 illustrates the impact of sample buffer concentration on the shape of the peak of PTH-arginine. One can notice that contrary to what can be expected from the theory, the shape and the amplitude of the peak are equivalent for a sample buffer which is either water or 12.5 mM MES/HIS (12.5 mM 2-(N-morpholino)-ethane sulfonic acid adjusted to pH 6.13 with 12.5 mM histidine).

In the employed protocol, a laminar flux is created in the capillary tubing, induced by the discrepancy between the electrophoretical flux and the translation speed of the buffer in the sample zone 1 and buffer zone 2. This laminar flux, well described by Chien and Helmer (1991), results in an enlargment of the sample width: the greater the difference between the two concentrations, the higher the value of this laminar flux working in the opposite way to the pre-concentration phenomenon.

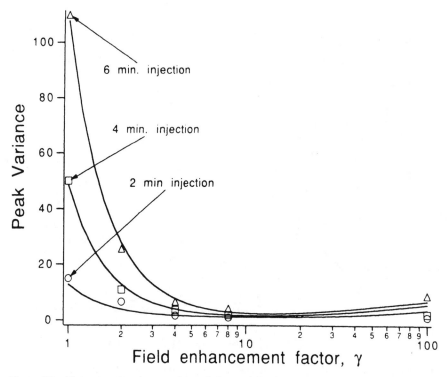

Figure 5A. Plot comparing the experimental data and the theoretical calculated total peak variance of PTH-arg vs the field enhancement factor as a function of the injection time. The circles, squares, and triangles are for the 2-, 4-, and 6-min injections, respectively. The theoretical calculations are the solid lines. $\gamma = 1$ corresponds to the sample buffer concentration equal to the separation buffer concentration. Reproduced with permission.

Figures 5A, 5B from Burgi and Chien (1991), shows these effects on the peak variance, and demonstrates that the results of the pre-concentration process reach optimal value for a γ factor between 10 and 20. When γ is small, the variance σ^2 is proportional to the injection duration. For high γ values, the laminar flux becomes significant and the peak symmetry is therefore altered.

Chien and Burgi (1992) have shown that the injection volume should represent 1% of the total capillary volume (cf. Table 3). If one compares these values to that of Table 1, these volumes give the operator more latitude.

The wise use of this ionic strength induced preconcentration method allows a 10 to 20 factor increase in the sensibility.

Use of a very high volume

The continual search for a greater sensibility has led to an attempt to introduce greater volumes: however, the introduction of greater volumes in the capillary has an important and negative impact on the peak symmetry. Therefore, two attempts have been made by Chien and Burgi (1991a, 1992b) to overcome this

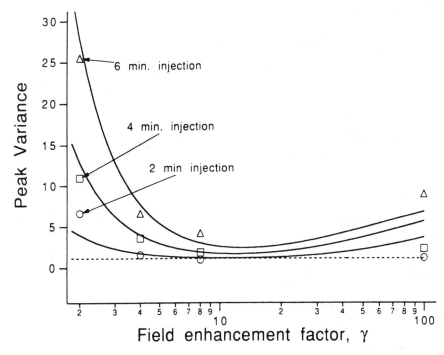

Figure 5B. Expanded plot of the calculated optimum field enlargement factor in Fig. 5A. The dashed horizontal line is the location of the peak variance which is 10% greater than the diffusion-limited case. Reproduced with permission.

Table 3.
Injection volume (nl) equal to 1% of total capillary volume

Internal capillary diameter	Internal capillary volume per length in cm	Total volume for a 44 cm capillary length	Corresponding injection volume, 1% precision for $L = 44$ cm	Total volume for a 70 cm capillary length	Corresponding injection volume, 1% precision for $L = 70$ cm
50	19.6	862.4	8.62	1372	13.72
75 μm	44.1	1940.4	19.40	3087	30.87

difficulty, which means injection of a greater sample volume, keeping at the same time a correct peak symmetry.

Using a time inversion of the polarity with electronegatively charged solutes. When the wall of the capillary is electronegatively charged and when simultaneously a positive tension is applied to the injection side, the buffer moves towards the detector by the electroosmotic flux. The positively charged solutes are concentrated in front of the sample whereas the electronegatively charged solutes concentrate at the end of the sample as soon as the tension is applied (Fig. 6).

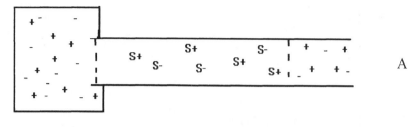

Buffer reservoir   A

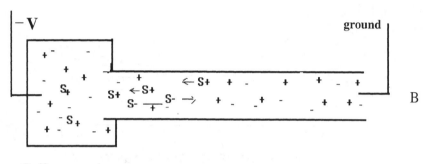

Buffer reservoir   B

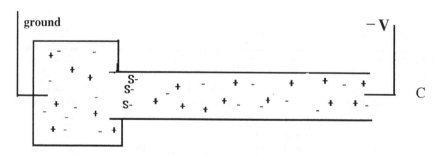

Buffer reservoir   C

Figure 6. A: A long plug of sample prepared in water is introduced hydrodynamically into the column. B: High voltage with reversed polarity is applied to the column to push water out while the column retains the analytes. C: When the current reaches about 95% of the original value, the polarity of the electrodes is switched back to normal configuration and separation is performed. Reproduced with permission.

If one applies a negative tension on the injection side as soon as the capillary has been filled with the sample, the electroosmotic flow is directed towards the injection side and the positively charged solutes are the first to leave the capillary. They are followed by the buffer molecules. At the same time the electronegatively charged solutes migrate rapidly towards the positive electrode and concentrate at the border zone sample-buffer, where their migration speed becomes almost zero. Simultaneously, working buffer is slowly follows the other end of the capillary. During all this period, the total intensity of the electric current increases progressively to a maximum, attesting that all the sample solvent has been evacuated from the capillary. Immediately, before reaching this maximum, one should reverse the applied electric intensity. The real separation will then develop its normal course as the electroosmotic flux takes its 'normal' direction.

Two examples have been provided in Figs 7 and 8. Figure 7 reports the separation of aspartic-acid-PTH and glutamic acid-PTH in the absence of commutation of the polarity (curve A) and with commutation of the polarity (curve B). The latter process allows an improvement of the densitivity and a better resolution between the peaks. Figure 8 illustrates HPCE separation of hirudin (Geyer, 1994). Curve A is related to an injection of 0.1 mg/ml of hirudin solute. Curve B is representative of an injection of the same solution under similar conditions but the polarity of the electrodes has been reversed after 2 min.

Without inversion of the polarity with time. The obligation to commute the current polarity in the course of the separation can bring problems in repeatability if the intensity variation is not checked carefully and if the matrix is

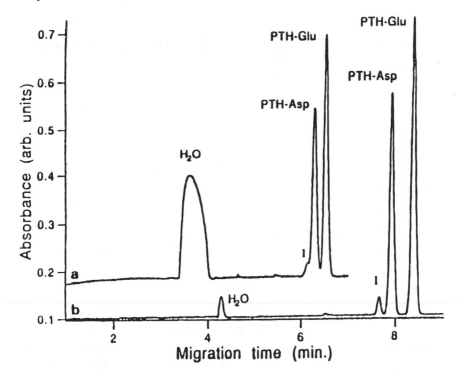

Figure 7. Large volume stacking with water removed in HPCE. Reproduced with permission.

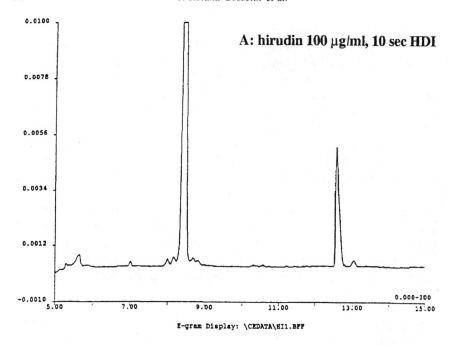

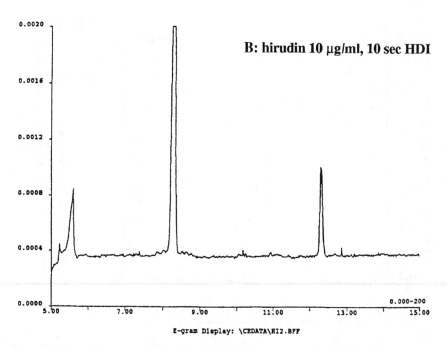

Figure 8. Electropherograms of hirudin under different conditions. Curve A: represents an injection of 100 µg/ml of hirudin solute. Curve B: shows an injection of the same solution but of a concentration of 10 µg/ml, under similar conditions but after 2 min, the polarity of the electrodes is reversed.

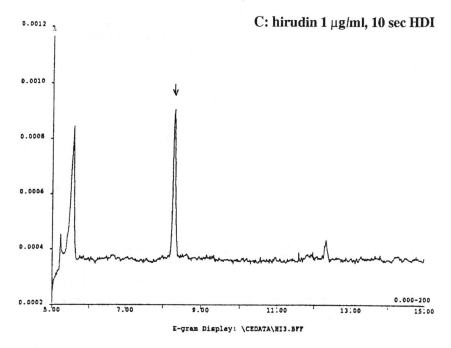

C: hirudin 1 μg/ml, 10 sec HDI

E-gram Display: \CEDATA\HI3.BFF

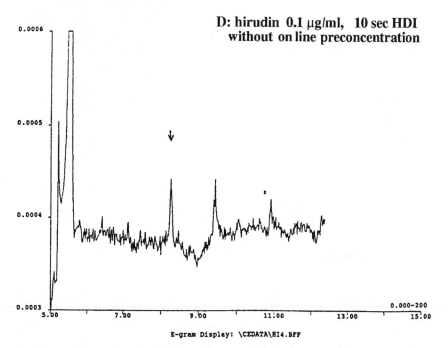

D: hirudin 0.1 μg/ml, 10 sec HDI
without on line preconcentration

E-gram Display: \CEDATA\HI4.BFF

Figure 8 (**continued**). Curves C and D represent, respectively, hirudin concentrations of 1 μg/ml and 0.1 μg/ml.

P. Roland-Gosselin et al.

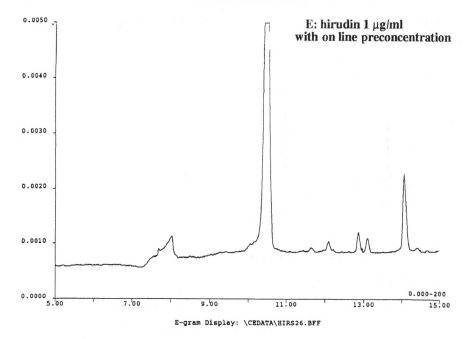

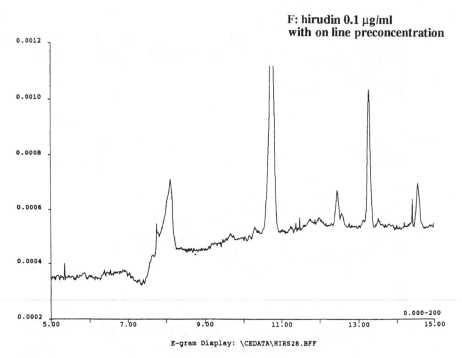

Figure 8 (**continued**). Separation by HPCE of hirudin, 1 μg/ml (E) and hirudin 0.1 μg/ml (F) with on line concentration. The experimental are the same as described in legend of Figs 8(A–D). Reproduced with permission.

different from one sample to the other. Therefore, Chien and Burgi (1992b), Burgi (1993) have explored the possibility to obtain the same type of enrichment without inversion of the polarity.

Separation of electropositively charged solutes

A similar process as above can be used for positively charged solutes if one adds to the working electrolyte some organic modifier in order to reverse the electroosmotic flux. Figure 9 reports the separation of arginine-PTH and histidine-PTH. Trace (a) corresponds to a sample length of 0.7 cm for a TTAB free sample, trace (b) to a sample length of 7 cm for a 1 mM TTAB added sample whereas trace (c) is representative of a 35 cm length of a 1 mM TTAB added sample. The solvent has been eliminated out of the column using the electroosmotic flux for traces (b) and (c). As the values of the electroosmotic flux are different in the sample compartment and the electrolyte, one can see that the so-created laminar flux induces similar effects as previously resulting in broadening of the peaks.

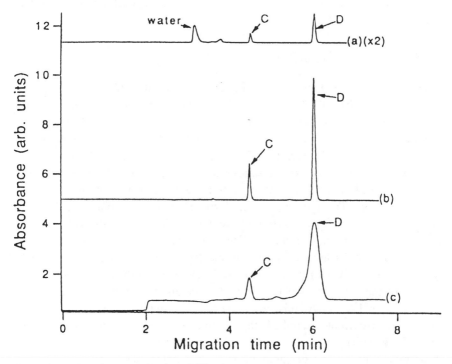

Figure 9. Positive ion separation by HPCE. The electropherogram represents a 0.7 cm injection without the sample buffer removed. Running conditions are: 1 mM TTAB in support buffer, water in sample buffer, −30 kV, 25 µA. PTH-his (peak C); PTH-arg (peak D). Panel (b) is a 7 cm injection with the sample buffer removed with similar running conditions as in panel (a). Panel (c) represents a 35 cm injection with the sample buffer removed. There is 1 mM TTAB included in the sample buffer. Reproduced with permission.

Separation of electronegatively charged solutes (27)

Whenever the working electrolyte contains EDTA (1 mM), the electroosmotic flux within the capillary is extremely reduced as the neutral marker cannot be detected after a delay of 3 h. If the capillary is filled with a EDTA-free sample, the EDTA lining the capillary wall becomes dissolved and the electroosmotic flux increases thereafter. With the action of the increased electroosmotic flux, the solvent sample is evacuated from the column whereas the electronegatively charged solutes concentrate on the border zones (Fig. 10). The EDTA added working buffer is simultaneously pumped by the other end of the capillary, the electroosmotic flux is again at zero value in the capillary zone previously occupied by sample. Figure 11 shows 2 electropherograms illustrating the process described above. The sensitivity is increased 10-fold. Figure 12 reports an electropherogram of a 10 µg/l solution. It also shows the trace of current intensity demonstrating that 1.5 min is required to eliminate sample solvent from the capillary.

These processes of preconcentration are extremely powerful since they can improve sensitivity 100-fold. They can also be employed for fraction collection.

pH dependent preconcentration

Whenever the required molecules to be separated are of amphoteric character, it is possible to utilize pH differences between the sample solvent and the working electrolyte. Aebersold and Morrison (1990) have analyzed synthetized peptides solubilized in the working electrolyte. The obtained electropherograms are presented in Fig. 13. The trace C shows that it is impossible to work with a solution at concentrations below 6.25 ng/µl. However, the detection of HPLC collected fractions implies a 10-fold increase in sensitivity of the detector. The introduction of a greater volume of injection does not improve the sensitivity as described previously. Therefore, the authors have tried a preconcentration process. The sample was dissolved in a buffer of pH 10, a pH much higher than the isoelectric point (pI) of the peptide analyzed. Immediately after the injection of the sample into the capillary (Fig. 14), the end of the capillary is placed in a vial containing low pH electrolyte (citrate buffer, pH 2.5). At this border, a pH gradient will form. As soon as the electrical current is applied, the peptide will migrate towards the anode since these molecules are negatively charged and concentrate at the level of the sample/buffer. The pH gradient rapidly disintegrates and the peptide already migrated in the citrate buffer (pH 2.5) get positively charged, eventually migrating towards the cathode. The outcome of this concentration process is illustrated in Fig. 15 which shows the possibility to detect concentrations in the range of ng/µl which corresponds to the concentrations of the fractions collected after HPLC.

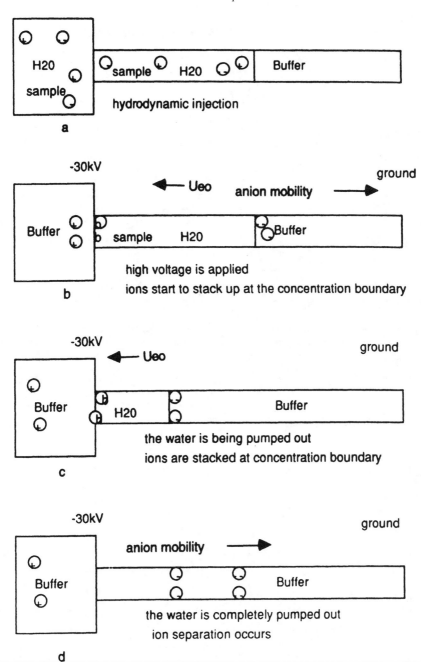

Figure 10. The mechanism of how the water pumps itself out of the column in HPCE: a) once the sample is injected into the column, the DETA on the capillary wall dissolves into the water of the sample plug increasing the potential in the same region; b) and c) after the voltage is applied, the ions stay at the boundary until the water plug is pumped out of the column; d) the water is completely pumped out of the column, and the ions separate under normal CE conditions. Reproduced with permission.

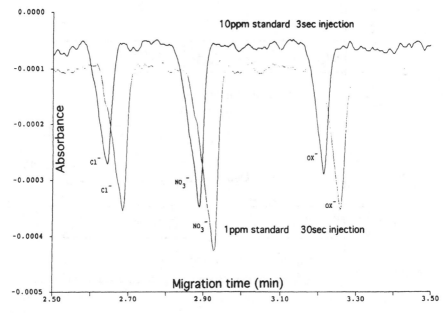

Figure 11. Electropherogram showing a comparison of a 3 s hydrodynamic injection of 10 ppm of anions with a 30 s injection of 1 ppm of anions in a 50 μm-i.d. column. Reproduced with permission.

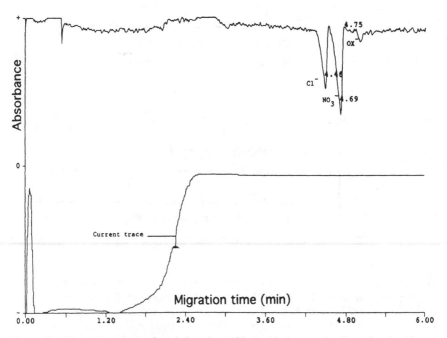

Figure 12. Electropherogram of a whole column filled with the sample. Reproduced with permission.

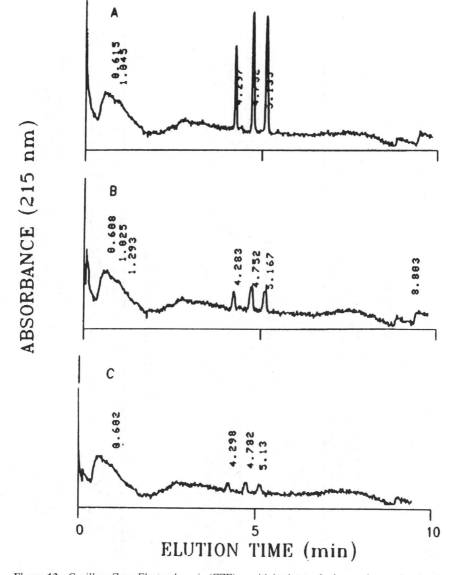

Figure 13. Capillary Zone Electrophoresis (CZE) sensitivity in standard operating mode. A calibrated mixture of 3 synthetic peptides of standard evaluation mixture was dissolved in 10 mM citrate buffer, pH 2.5. Samples were loaded to the capillary by applying reduced pressure to the cathode end of the capillary for 5 s. Electrophoresis carried out at 30 kV at the detector setting of 0.005 a.u.f.s. at 215 nm. Sample concentration (A) 25 ng/μl; (B) 12.5 ng/μl; (C) 6.25 ng/μl. Reproduced with permission.

CZE OF DILUTE PEPTIDE SAMPLES

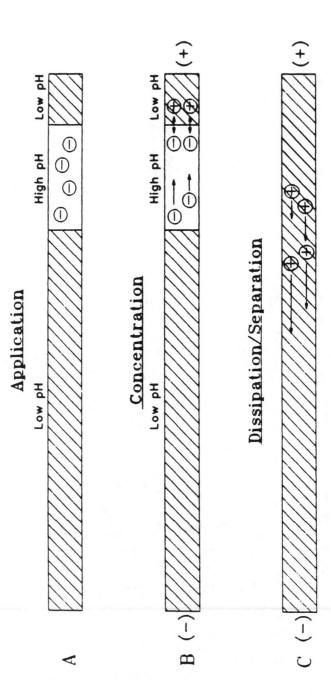

Figure 14. Schematic illustration of electrophoretic stacking. A: Samples are applied to the capillary in a solution with a pH value higher than the pI of the peptides, resulting in deprotonated, negatively charged peptides. B: Application of a potential gradient over the capillary leads to concentration of the peptides at the electrophoresis buffer-sample interface. C: After dissipation of the pH-step gradient, peptides resume mobility towards the cathode and are separated in the electric field. Reproduced with permission.

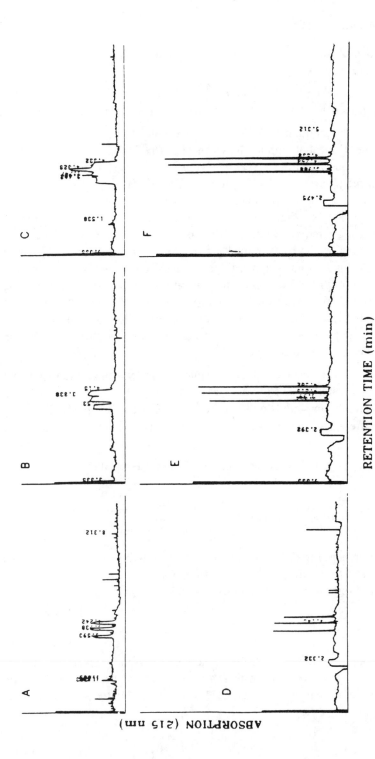

Figure 15. Effect of the concentration procedure. Performance evaluation standard was dissolved at 10 ng/µl in either 10 mM citrate buffer, pH 2.5 (A, B, C) or in 10 mM NH₄OH (D, E, F). Samples were loaded to the capillary by applying reduced pressure at the cathode end of the capillary for 10 s (A, D) or 30 s (C, F). Electrophoresis was performed at 30 kV. Peptides were detected at 215 nm at detector setting of 0.005 a.u.f.s. Reproduced with permission.

ELECTROKINETIC MODE

Another mode of sample injection into the capillary is the electrokinetic mode which is useful because:

- it increases the sensitivity, and
- it improves the shape of the peaks in qualitative analysis.

In the case of quantitative analysis, it is necessary to be careful as demonstrated by the electropherogram shown in Fig. 16. Jandik and Bonn (1993) have precisely described the successive steps and the description of the phenomenon in the figure as well as in the text. The end of the capillary is placed in a vial containing the sample made of two ionized species (i and ii) and one nonionized species (N) (Fig. 17). As soon as the current is employed, the ionized species start to move towards the limit of the sample/buffer. No study was performed either on the best optimal tension for sampling or on the duration during which this tension is applied, but the most commonly applied tension is in the range of 5 kV. The sign of the tension is chosen in order to allow the interesting solutes to move from the vial to the capillary and then through the border sample/buffer. The pumping of the solutes will start if the choice of the polarity is correct and the sample conductivity is below that of the buffer. Three facts have to be considered:

- the ions, with inversed charge of that of the tension applied, will not enter the capillary even with electroosmosis,
- the neutral compounds will not cross the border of sample/buffer,
- the different solutes will cross the border of sample/buffer at different speeds.

The differences in the speed of migration follow the equation:

$$t^+ = \sum_i t_i^+ = \frac{I^+}{I^+ + |I^-|} \quad \text{and} \quad t^- = \sum_i t_i^- = \frac{I^-}{I^- + |I^+|},$$

where t_i is number of transfers, I^+ and I^- are respectively anionic and cationic currents, t^+ and t^- are the number of transfers for anionic and cationic solutes, respectively. From these two equations, one can postulate that

$$t^+ + t^- = 1.$$

For anionic and cationic species, the number of transfers is given by the sum of this number for each solute which is determined by the ratio of the ionic conductance equivalent of the species (λ) to the equivalent conductance (Λ_{eq}) as

$$t_i^+ = \frac{\lambda^+}{\Lambda_{eq}} \quad \text{and} \quad t_i^- = \frac{\lambda^-}{\Lambda_{eq}}.$$

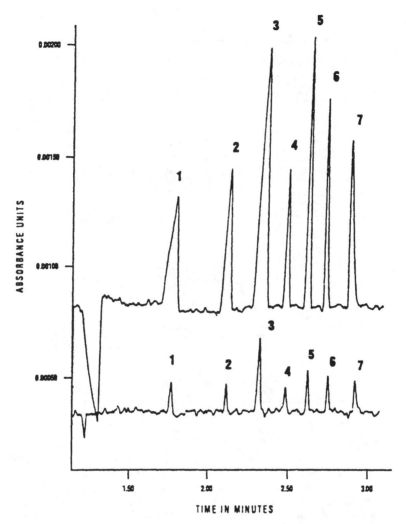

Figure 16. Separation of a standard mixture of cations: comparison of injection modes. Reproduced with permission.

[a] 1 – ammonium; 2 – potassium; 3 – sodium; 4 – calcium; 5 – magnesium; 6 – lithium; 7 – barium.

[b] Conditions. Capillary: Fused silica, 50 μm, 44 cm, 37 cm. Carrier electrolyte: 4 mM copper (II) sulphate, 4 mM formic acid, 4 mM 18-crown-6, the TSP Cation Buffer Kit P/N 4659-010. Voltage: +20 kV. Injection: 2 s electrokinetic at 10 kV (upper trace), or hydrodynamic for 2 s (lower trace). Detection: 215 nm, indirect. Temperature: 15°C. Sample concentration: 2.5 ppm ammonium, 2.5 ppm potassium, 0.125 ppm sodium, 1.25 ppm calcium, 0.125 ppm magnesium, 0.05 ppm lithium and 5 ppm barium as the chloride salts in water, stored in plastic vials. Instrument: SpectraPHORESIS® 1000.

[c] Cations in low ppm concentrations are easily separated using the TSP Cation Buffer Kit. The figure shows the focusing effect obtained by using electrokinetic injection. Peaks are much larger, with concomitantly lower detection limits. For quantitation, however, hydrodynamic injection is recommended because it offers much wider range of dynamic response. The electrokinetic injections usually show a good dynamic response at low concentrations, while at higher levels the response becomes essentially independent of analyte concentration.

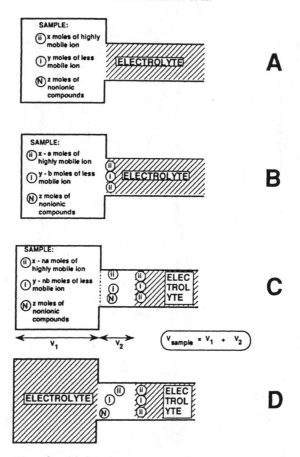

Figure 17. Four stages of sample introduction by electromigration. A: Sampling end of capillary and the corresponding electrodes are removed from the electrolyte vessel and placed in sample. B: First time segment after application of sampling potential, electrophoretic sampling prevails. C: Electrophoretic sampling continues, electroosmotic contribution becomes apparent. D: The sampling potential is turned off. The sampling end of capillary, along with the electrode, is removed from the sample and placed in the carrier electrolyte. (Reproduced with permission: Jandik and Bonn, 1993.)

The electrophoretic mobility being proportional to the equivalent conductance can be expressed according to the following equation:

$$\mu_{\text{sampling}} = \frac{\lambda_i}{\Lambda_{\text{eq}} F} \quad \text{and} \quad \nu_{\text{sampling}} = \frac{\lambda_i E}{\Lambda_{\text{eq}} F}. \tag{6}$$

This equation shows that each chemical species is collected at a different speed which is not dependent on equivalent conductance of each species only but also on the total conductance of the sample solution. The problem arises from the fact that neither the exact ionic composition of each sample nor the equivalent conductance of the sample (Λ_{eq}) are known. It is always possible to find practical solutions. Yet another factor has to be taken into consideration for sample processing: in most cases, there exists an electroosmotic flux. This flux starts as soon as the electrical current is applied and allows the elimination of low mobility ions present in some matrixes and therefore, decreases the total conductivity favourably. Steps B and C of Fig. 17 are observed only when the electrophoretic speed is superior to the electroosmotic flux, which is certainly the case for low molecular weight ions. Heavier molecules such as proteins, peptides and DNA fragments showing comparable or lower migratory speeds will enter the capillary more by electroosmosis than by their own displacement. As a result, step B does not take place. The different ionic species (i and ii) are introduced with a speed proportional to their sample concentration which thwarts the mechanism of pure electromigration. In the course of the step D, an electrostacking phenomenon takes place for the second time for two reasons: the electroosmotic flux has introduced neutral sample molecules of the sample in the capillary, and oppositely charged molecules, the proper mobility of which being in absolute value lower than that of the electroosmotic flux. If the concentration of ions a and b in the electrolyte is stated as n_a and n_b, the ratio a/b is directly proportional to the ratio of the number of transfers. The value n takes into consideration elapsed time between steps B and C. The concentration of the two ions and neutral species should be approximately equivalent in the volume $V1$ and $V2$. The transfer of the ions from volume $V2$ towards the electrolyte is proportional to the transfer between $V1$ and $V2$. When the sampling current is turned off, the end of the capillary is replaced in a vial containing working electrolyte and then the required current of separation is applied. At the beginning of this last step, on condition that the sample conductivity is below that of the working buffer, the preconcentration phenomenon comes into effect and induces the migration of all the ions initiated by electroosmosis to the border of sample/working buffer.

To summarize the process, there are two extreme cases which are defined according to the relative value of the electrophoretic mobility of a species and of the electroosmotic mobility. The impact of the C step will be negligible when the electrophoretic mobility is much higher than electroosmotic mobility. Indeed, the V step will be negligible. Generally, the total quantity (mole) introduced into the capillary by electromigration satisfies the following equation:

$$Q \text{ (moles)} = (\mu_{\text{sampling}} + \mu_{\text{eo}})C\pi r^2 t_i \frac{V_i}{L_t}, \tag{7}$$

where μ_{sampling} and μ_{eo} respectively represent the electrophoretic mobility of a solute and the electroosmotic mobility, r is capillary radius and V is applied current, L_t is the total length of the capillary, C is molar concentration, and t_i, the time during which the V is applied. This equation suggests three comments:

1) this injection mode is discriminatory: only the positively or negatively ionized solutes will be introduced into the capillary depending on the sign of the applied current.

2) the quantity of a given species, i, introduced into the capillary is a function of its mobility, its concentration and of the value of electroosmotic flux. As a result, a solute showing high mobility and high concentration will be introduced into the capillary in a preponderant manner compared to a species showing low mobility and low concentration.

3) the matrix in which the sample passes through may have an important effect.

To minimize the effects related to points 2 and 3, Huang *et al.* (1988) and more recently Leube and Roeckel (1994) have developed two different analytical protocols to take into consideration the distortion induced by the difference in equivalent conductance between the species i and that of the sample. These two studies indicate that it is possible to minimize the matrix influence to obtain the repetion factor equivalent to that of the hydrodynamic mode.

Whenever the electrokinetic mode is utilized, it is possible to increase the preconcentration phenomenon by introducing some water in the capillary immediately before sample injection. This process has also the capacity to decrease the length of the sample in the capillary. It is therefore possible either to apply a longer duration of current or a higher current intensity for the injection process. Figure 18 reports a comparison between the three injection modes for the sample: hydrodynamic (a), electrokinetic (b) and electrokinetic with a water clip introduction (c).

Chien and Burgi (1991b) have succeeded in optimizing this process with a commutation of the polarity of the current at the time of sample aspiration in order to collect simultaneously positive and negative solutes. The entire spectrum of results has been presented in Table 4.

Preconcentration by isotachophoresis

Similarly to capillary zone electrophoresis, isotachophoresis separates ionized solutes according to the electrophoretic mobility. Contrary to capillary zone electrophoresis which utilizes a single electrolyte, isotachophoresis employs two electrolytes. One of them, called 'leader', has an electrophoretic mobility higher than that of the analyzed ions (one named as leader), the other one, 'terminal', demonstrating lower mobility than that of analyzed ions.

The electrophoretic speed is imposed by the ionic compound of the terminal electrolyte. The signal intensity is proportional to the width of the corresponding band of each solute in isotachophoresis whereas in capillary zone electrophoresis, the intensity is proportional to the solute concentration within that band. This concentration is directly linked to the leader electrolyte concentration for each of the solute. This effect is described by the equation derived from Kohlrausch regulatory function:

$$\frac{C_A}{C_L} = \frac{\mu_A(\mu_L + \mu_R)}{\mu_L(\mu_A + \mu_R)}, \tag{8}$$

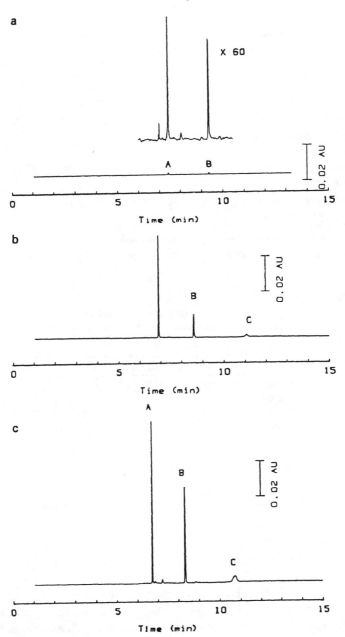

Figure 18. a) Electropherogram using conventional electro-injection. The column was filled with 100 mM MES-HIS buffer and the sample prepared in 100 mM MES-HIS buffer was injected into the column at −5 kV for 10 s. b) Electropherogram using electro-injection as (a) except the sample was prepared in water. c) Electropherogram using field amplified sample injection: a short plug of water was injected into the column first by gravity and the sample prepared in water was injected into the column at −5 kV for 10 s. All experiments were operated with a −30 kV separation voltage. Peaks A, B and C correspond to PTH-arginine, PTH-histidine and neutral marker, respectively. (Reproduced with permission: Leube and Roeckel, 1994).

P. Roland-Gosselin et al.

Table 4.

Comparison of peak heights for both positive and negative ions using various injection methods

	PTH-arginine	PTH-histidine	PTH-aspartic acid	PTH-glutamic acid
Gravity injection	1	1	1	1
Conventional electro-injection	0.311	0.225	0.025	0.022
Field amplified sample injection: positive ions only	28.04	13.44	0	0
Field amplified sample injection: negative ions only	0	0	13.66	12.58
Field amplified polarity switching injection: both positive and negative ions	32.03	9.28	2.58	1.96

All peaks normalized with respect to the gravity injection. (Reproduced with permission: Chien and Burgi, 1991b).

Table 5.

Detection limit (twice background) obtained by preconcentration in isotachophoretic mode. Reproduced with permission

Anions	Detection limit in µg/l
Chloride	0.5
Sulphate	0.3
Nitrate	0.8
Oxalate	0.6
Fluoride	0.3
Formate	0.5
Phosphate	0.3
Acetate	0.8
Propionate	0.6

where C_A and C_L are the respective concentrations of solute A and leader buffer, μ_A, μ_L and μ_R are the respective mobilities of solute A, leader electrolyte and its counter ion R. Considering the examples of extreme mobilities (hydroxy and carboxyl ions) in the field of low molecular weight ions, one can hope for a preconcentration factor in the range of 10^6 (Jandik and Bonn, 1993). These authors have achieved trace analysis of inorganic ions. They have gathered in the following table the detection limits of these ions (Table 5).

CONCLUSION

In this contribution, we have demonstrated that in capillary electrophoresis, it is possible to obtain simultaneously a useful sensitivity limit as well as an acceptable signal/background ratio to make this analytical tool attractive for biomedical and industrial analysis. The potential users can appreciate that with better knowledge of the preconcentration processes, they can find a capillary electrophoretic solution to fulfill their sensitivity requirements by investing some of their time.

ACKNOWLEDGEMENTS

The authors express their sincere gratitide to different presses or authors to allow the reproduction of some of the figures in the text.

REFERENCES

Aebersold, R. and Morrison, H. D. (1990). Analysis of dilute peptide samples by capillary zone electrophoresis. *J. Chromatogr.* **516**, 79–88.

Albin, M., Grossman, P. D. and Moring, S. E. (1993). Sensitivity and enhancement for capillary electrophoresis. *Anal. Chem.* **65**, 489A.

Avdalovic, N., Pohl, Ch. A., Rocklin, R. D. and Stillian, J. R. (1993). Determination of cations and anions by capillary electrophoresis combined with suppressed conductivity detection. *Anal. Chem.* **65**, 1470–1475.

Burgi, D. S. (1993). Large volume stacking of anions in capillary electrophoresis using electroosmotic flow modifier as a pump. *Anal. Chem.* **65**, 3726–3729.

Burgi, D. S. and Chien, R. L. (1991). Optimization in sample stacking for high performance capillary electrophoresis. *Anal. Chem.* **63**, 2042–2047.

Burgi, D. S. and Chien, R. L. (1992). Improvement in the method of sample stacking for gravity injection in capillary zone electrophoresis. *Anal. Biochem.* **202**, 306–309.

Burton, D. E., Maskarinec, M. P. and Sepaniak, M. J. (1986). High-sensitivity laser-induced fluorescence detection in capillary electrophoresis. *J. Chromatogr. Sci.* **24**, 347, 351.

Chien, R. L. and Burgi, D. S. (1991a). Field amplified sample injection in high performance capillary electrophoresis. *J. Chromatogr.* **559**, 141–152.

Chien, R. L. and Burgi, D. S. (1991b). Field-amplified polarity-switching sample injection in high performance capillary electrophoresis. *J. Chromatogr.* **559**, 153–161.

Chien, R. L. and Burgi, D. S. (1992a). On column sample concentration using field amplification in CZE. *Anal. Chem.* **64**, 489A.

Chien, R. L. and Burgi, D. S. (1992b). Sample stacking of an extremely large injection volume in high performance capillary electrophoresis. *Anal. Chem.* **64**, 1046.

Chien, L. S. and Helmer, J. C. (1991). Electroosmotic properties and peak broadening in field-amplified capillary electrophoresis. *Anal. Chem.* **63**, 1354–1361.

Everaerts, F. M., Verheggen, T. P. E. M. and Mikkers, F. E. P. (1979). Determination of substances at low concentrations in complex mixtures by isotachophoresis with column coupling. *J. Chromatogr.* **169**, 21–38.

Foret, F. and Bocek, P. (1988). Capillary zone electrophoresis: Quantitative study of the effects of some dispersive processes of the separation efficiency. *J. Chromatogr.* **452**, 601–613.

Gassmann, E., Kuo, J. E. and Zare, R. N. (1985). Electrokinetic separation of chiral compounds. *Science* **230**, 813–814.

Geyer, M. (1994). *Thermo Separation Products*. Internal Communication.

Gordon, G. B. (1991). United States Patent 5,061,361, 29th October.

Hjertén, S. (1990). Zone broadening in electrophoresis with special reference to high performance electrophoresis in capillaries: An interplay between theory and practice. *Electrophoresis* **11**, 665–690.

Huang, X., Gordon, M. J. and Zare, R. N. (1988). Bias in quantitative capillary zone electrophoresis caused by electrokinetic sample injection. *Anal. Chem.* **60**, 375–377.

Jandik, P. and Bonn, G. (1993). *Capillary Electrophoresis of Small Molecules.* VCH Press, New York.

Jones, H. K., Nguyen, N. T. and Smith, R. D. (1990). Variance contributions to band spread in capillary zone electrophoresis. *J. Chromatogr.* **504**, 1–19.

Leube, J. and Roeckel, O. (1994). Quantification in capillary zone electrophoresis for samples differing in composition from the electrophoresis buffer. *Anal. Chem.* **66**, 1090–1096.

Lu, W., Cassidy, R. M. and Baranski, A. S. (1993). End-column electromechanical detection for inorganic and organic species in high performance capillary electrophoresis. *J. Chromatogr.* **640**, 433–440.

Mikkers, F. E. P., Everaerts, F. M. and Verheggen, P. E. M. (1979). High-performance zone electrophoresis. *J. Chromatogr.* **169**, 11.

Moring, S. E., Reel, R. T. and Van Soest, R. E. J. (1993). Optical improvements of a Z-shaped cell for high sensitivity UV absorbance detection in capillary electrophoresis. *Anal. Chem.* **65**, 3454–3459.

Otsuka, K. and Terabe, S. (1989). Extra-column effects in high performance capillary electrophoresis. *J. Chromatogr.* **480**, 91–94.

Toulas, Ch. and Hernandez, L. (1992). Applications of a new laser-induced fluorescence detector for capillary electrophoresis to measure attomolar and zeptomolar amounts of compounds. LC-GC Int. 27.

Weinberger, R. (1993). *Practical Capillary Electrophoresis.* Academic Press, San Diego, USA.

Progress in HPLC-HPCE, Vol. 5, pp. 49–72
H. Parvez *et al.* (Eds)
© VSP 1997.

Capillary electrophoresis with laser-induced fluorescence detection: optical designs and applications

G. NOUADJE,[1] J. AMSELLEM,[1] B. COUDERC,[3] PH. VERDEGUER[1]
and F. COUDERC[2]

[1] *Zéta Technologie, 10 avenue de l'Europe, 31520 Ramonville, Toulouse, France*
[2] *Université P. Sabatier, LBME du CNRS, 118 Route de Narbonne,*
   31062 Toulouse, France
[3] *Centre Claudius Régaud, Laboratoire de Biochimie, Rue St Pierre,*
   31300 Toulouse, France

INTRODUCTION

Capillary electrophoresis (CE) has become a very popular analytical technique with applications in various fields including ions, proteins, sugars, and detergents (for a review see Kuhr, 1990; Kuhr and Monnig, 1992; Xu, 1993; Monnig and Kennedy, 1994). Most of these applications were developed using on-column vis-UV absorbance detectors. This mode of detection is the most common one because of its simplicity and low price. However, due to the small optical pathway of the detection cell (the capillary itself), UV detection for CE lacks in sensitivity and generally addresses samples with concentrations down to 10^{-6} M.

To overcome this problem some authors have proposed other detection modes. One of the more sensitive is electrochemical detection (Chen and Whang, 1993) which allows detection of electroactive substances at the nanomolar level. Due to the complex geometry of the electrochemical cells for CE, this kind of detector has not yet been commercialized.

Fluorescence detection and, more particularly, laser-induced fluorescence (LIF) detection is recognized to be the best way to overcome the problem of low sensitivity inherent to CE UV detection.

Fluorescence detection using an arc lamp as the excitation source, was previously reported (Kuhr, 1990). However, only a marginal gain in sensitivity over UV detection was achieved. The main problem is the focusing of a large amount of light from a divergent source onto the tiny detection region of the capillary. Only 15% of the energy can be used for excitation. In contrast to the arc lamp, lasers emit a highly coherent and powerful light, which allows better focusing onto the capillary, as well as the utilization of nearly 100% of the power. The use of lasers as exitation sources thus greatly improves the sensitivity of detection (Kuhn and Hoffstetter-Kuhn, 1993).

In the present article we discuss the different optical devices of LIF detector for CE as well as various applications of the LIF detection.

LASER-INDUCED FLUORESCENCE DETECTOR OPTICAL DESIGN

Orthogonal arrangement

When Jorgenson and Lukacs (1981a, b; 1986) published their first work on CE, they were using a fluorescence detector with a Xe arc lamp as exitation source. The design of the detector was based on an orthogonal arrangement: the excitation beam is perpendicular to the emission collector, and both emission and excitation are perpendicular to the main axis of the capillary (Fig. 1).

The authors reported a limit of concentration detection (LOCD) in the nano-molar range for dansylaminoacids.

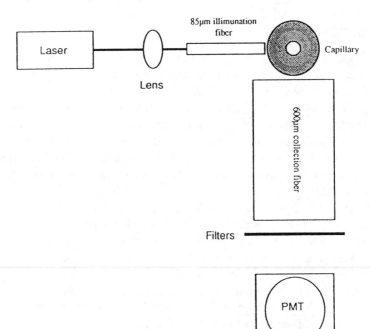

Figure 1. Diagram of orthogonal laser-induced fluorescence detection system.

Using the work of Folestad *et al.* (1982) on LIF detection for liquid capillary chromatography, the Zare group (Gassmann *et al.*, 1985; Gozel *et al.*, 1987) reported detection of dansylaminoacids at the attomole level using CE–LIF with an Helium Cadmium (He–Cd) laser 325 nm as exitation source and in orthogonal arrangement.

In order to detect four fluorophores in the same run, a modification of the orthogonal arrangement has been proposed consisting of a four photo-multiplicator tube (PMT) configuration (Smith, 1991). Each PMT detects one wavelength using an appropriate set of filters. The same result can be obtained by using a filter wheel and one PMT (Swerdlow *et al.*, 1991).

Some investigators (Cheng *et al.*, 1990; Karger *et al.*, 1991) show that the charge coupled device (CCD) detector allows multiwavelength detection and provides better analysis selectivity, than the PMT.

The orthogonal arrangement presents two main disadvantages. The first one arises from difficulty encountered in correct aligning in the system since two focusing optics are needed: one for excitation and another for emission collection. The second disadvantage is a high background noise. When a radiation strikes a cylindrical body, scattered light is emitted in all directions in a plane perpendicular to the main axis of the capillary. Background noise due to scattering generally limits detection sensitivity of orthogonal arrangements to 10^{-9} M.

One way to reduce light scattering is to use post-column sheath-flow cuvette detectors.

Sheath-flow cuvette detectors

In order to decrease background noise due to light scattering, Dovichi's group (Wu and Dovichi, 1989; Cheng and Dovichi, 1989) proposed the use of a

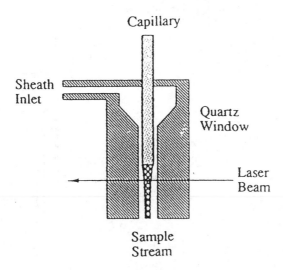

Figure 2. Diagram of sheath-flow cuvette LIF detector.

sheath-flow cuvette, commonly employed in cytometry, as post column detector. In this system, the end of the capillary is inserted into a quartz flow
chamber 250 μm × 250 μm square (Fig. 2). Sheath fluid identical to separation
buffer is introduced into the cuvette by means of a liquid chromatography pump
at a flow rate below 1 μl/min. To close the electrical circuit for electrophoresis, the stainless steel plumbing associated with the sheath stream is held at
ground potential. According to the authors, the design decreases background
noise and allows detection of fluorecein thiocarbamyl (FTC) aminoacids at the
picomolar level. By using two collecting optics it is even possible to detect six
molecules of sulforhodamine 101 (Cheng *et al.*, 1994).

Axial optical illumination

On the basis of the LIF detectors built for HPLC by Todoriki and Hirakawa
(1980), and by Sweedler *et al.* (1991), Taylor and Yeung (1992) proposed a
third type of LIF design. An optical fiber carrying the excitation radiation is
inserted into the capillary at 5 to 10 mm from the capillary end, and fluorescence
emission is collected at right angles by a PMT (Fig. 3). This design presents
a LOCD of 10^{-12} M using rhodamine 6G excited by the 488 nm ray of an
argon–ion laser.

   All systems based on the orthogonal arrangement are nevertheless difficult to
focus because two focusing optics are needed: one for excitation and another
one for collection. Hernandez *et al.* (1990, 1991) solved this the problem by
using single optics in a collinear arrangement.

Collinear arrangement

In a collinear arrangement the excitation radiation is reflected by a dichroic
mirror and focused onto the capillary by means of a microscope objective.

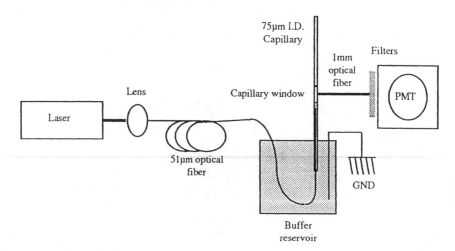

Figure 3. Diagram of axial optical illumination detector.

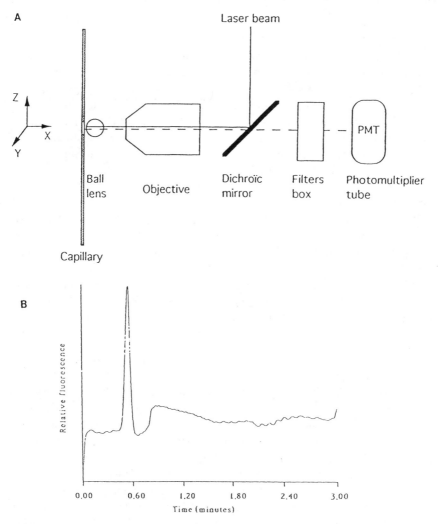

Figure 4. A: Diagram of 'ball lens' LIF detector arrangement. B: Electropherogram of 10^{-12} M rhodamine 123, 75 μm i.d. fused silica capillary 60 cm long, buffer 20 mM sodium tetraborate, pH 8.5, analysis +20 kV, 12 s 15 kV electrokinetic injection.

Fluorescence emission is collected by the same objective, crosses the dichroic mirror (as its wavelength is higher than the excitation wavelength), and is then filtered by a notch filter (to eliminate laser light), a high pass and spacial filter. The light is measured by a PMT.

One of the main advantages of the collinear design is the possibility of using a short working distance and a high numerical aperture (NA) objective in order to increase the collection efficiency (Hernandez *et al.*, 1991). Thus the authors reported a LOCD of 10^{-13} M for different FTC aminoacids.

Due to the high price of objectives with a high NA (0.85 or 0.90) due mainly to the requirement for UV transparent objectives and because the use of short working distances decreases mechanical tolerances dramatically, such a detector has hardly passed the laboratory stage.

On the basis of the Hernandez collinear arrangement our group has developed a commercial LIF detector. In this detector the high NA objective is simply replaced by two optical components: a ball lens cell (developed by Thermo Separation Product (Freemont, CA) for its UV detector) and a low NA, UV-transparent and long-working-distance objective (Fig. 4). This design decreases dramatically the price of a collinear arrangement without decreasing the performance. Moreover it presents two other important improvements. The first one is related to the design of the ball lens cell. Once the capillary is mounted on the cell, it can simply be inserted into the detector, with no other focussing procedure required. The other advantage is that both the ball lens and the low NA objective are transparent to UV light. Thus a low-UV laser whose frequencies double that of an argon–ion laser (i.e. 514 nm giving 247 nm), can be used to work with native fluorescence of organic compounds.

Using a 488 nm argon–ion laser, the performance of the ball lens design is comparable to the collinear arrangement. For rhodamine 123 we have found a LOCD of 5×10^{-13} M with a signal to noise ratio more than 3.

LASER SOURCES

The majority of applications of LIF detection have been developed using the well known argon–ion, He–Cd or He–Ne laser. Choise of the laser is generally dictated by the fluorescence reagent used to tag the substrate of interest. Six wavelengths are available using these lasers: 488, 514 nm for an argon–ion laser; 325, 442 nm for a He–Cd laser; and 543, 633 nm for a He–Ne laser. A tendency begins to appear to use diode lasers (Higashijima et al., 1992; Fuchigami et al., 1993; Jansson and Roeraade, 1993) which increase the number of available wavelengths of excitation. Another tendency is to use low-UV laser as nitrogen or deep-UV lasers (Lee and Yeung, 1992a; Nie et al., 1993). However, applications in this field are limited due to the high price of such lasers.

SOME APPLICATIONS OF CE–LIF

Nucleic acids applications

As an alternative to radiodetection, the application of fluorescence detection to DNA sequencing or to the separation and detection of polymerase chain reaction (PCR) products has been recommended by different authors for a long time (Ansorge et al., 1986; Smith et al., 1986; Prober et al., 1987). Thus the CE–LIF represents an extremely promising tool for the molecular biology scientist.

DNA sequencing. The human genome project has generated an important need for new analytical techniques to analyse sequencing reaction product (Hunkapiller *et al.*, 1991). The ultimate goal is to sequence at very high speed large DNA fragments of up to 1000 pb (for a good review, see Luckey *et al.*, 1993a). The works published to date deal mainly with the analysis of Sanger reaction products with CE–LIF using polyacrylamide gel filled capillaries (Cheng *et al.*, 1990; Cohen *et al.*, 1990; Drossman *et al.*, 1990; Luckey *et al.*, 1990; Swerdlow and Gesteland, 1990; Swerdlow *et al.*, 1990, 1991; Chen *et al.*, 1991a, 1992; Karger *et al.*, 1991a, b; Smith, 1991; Huang *et al.*, 1992a, b; Mathies and Huang, 1992; Pentoney *et al.*, 1992; Rocheleau *et al.*, 1992; Luckey and Smith, 1993a, b; Luckey *et al.*, 1993a, b; Kambara and Takahashi, 1993; Taylor and Yeung, 1993; Takahashi *et al.*, 1994), or semi-viscous solutions (Carson *et al.*, 1993; Ruiz-Martinez *et al.*, 1993) and denaturing conditions (Swerdlow *et al.*, 1992). Capillaries coated according to Hjerten are generally employed (Hjerten, 1985), or gas chromatography coated capillaries.

Due to the selectivity of LIF detection Sanger reaction products can be detected according to their dideoxynucleotide ends. Thus Smith (1991) and Drossman *et al.* (1990) analysed a mixture of the four sequencing reactions, using a single capillary. Each reaction was carried out with one particular fluorescent primer, each primer differing from the others by its maximum emission wavelength when exited by the same excitation wavelength. Using a four wavelength detection system (in an orthogonal design), each base is identified by its fluorescent intensity at a given wavelength. Due to the fact that the emission wavelengths are very close to one another, important sources of error are introduced. Swerdlow *et al.* (1991) improved the system by using two exitation wavelengths (488 nm, 543 nm) and a filter wheel emission collection system. The measure of the speed of rotation of the filter wheel allows the four bases to be distinguished using one PMT.

A new approach was introduced by using the sheath-flow cuvette with a He–Ne 543 nm laser. The Sanger reactions were carried out by using different ratios of dideoxynucleotide triphosphates (ddNTPs). The nature of the dideoxynucleotide end can be determined by the peak intensities. According to the authors this method failed to detect products containing low level of ddNTPs.

This same approach was improved by Pentoney *et al.* (1992) who used only three ddNTPs instead of four, thus attributing the missing peaks to the missing ddNTP. In order to collect a maximum of the fluorescence emission a parabolic mirror was placed behind the capillary in an orthogonal configuration. However this sequencing strategy seems not to be applicable to large DNA fragments (> 1000 pb) because the use of different ddNTP ratios do not lead to the same amounts of small, medium, and large extension products of the DNA molecules.

In order to improve the output rate of the instruments, some authors have proposed multiple capillary systems. Thus Mathies and co-workers (Huang *et al.*,

1992a, b; Mathies and Huang, 1992) have developed a 25 capillar array system using a collinear arrangement. The capillary array is mounted on an XY high speed table, with each capillary passing the detection zone at a scanning rate of 2 cm/s (Fig. 5). Migration times were not apparently reproducible from one capillary to another, even when the capillaries were prepared using the same protocol. For DNA sequencing, two different primer dyes with different emission wavelengths were employed. Using a given ratio of the two fluorotides in each Sanger reaction, each base can be identified according to its proper ratio. Since the emission wavelengths are different, identification of each base is more accurate. Starke *et al.* (1994) show that there is no significant influence on migration time in the range 50–500 bp DNA fragments, labeled with different dyes.

Taylor and Yeung (1993) proposed a capillary array multiplexed fluorescence detection with axial optical illumination (Sweedler *et al.*, 1991; Taylor and Young, 1992). Emission radiation is collected with a CCD camera. This system does not require displacement of the capillary array; each capillary is detected simultaneously. Even at very high separation speeds (> 1000 pb/h) there is no limitation due to the scanning rate of detection as there is in the Mathies system.

Different multiple capillary systems have been proposed using sheath-flow cuvette detection (Swerdlow *et al.*, 1990, 1991; Chen *et al.*, 1992). Recently Kambara and Takahashi (1993) proposed a two-excitation-wavelength, four-colour detection system. The four-colour detection system was replaced by a CCD camera.

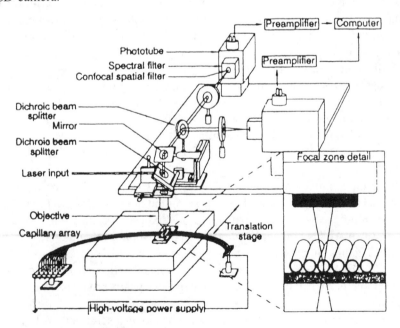

Figure 5. Diagram of LIF capillary array collinear detector.

Using the orthogonal arrangement, Karger's group (Ruiz-Martinez *et al.*, 1993) proposed a two windows scheme. Each tightly closed window is dedicated to a particular excitation wavelength. The four different fluorotides were detected using a diode array in order to measure the fluorescence spectra. The sensitivity was estimated to be in the order of 10^{-12} M, and a continuous buffer replacement system was used to decrease background noise. The preceding works deal with analysis of small DNA fragments (< 800 bp (Luckey *et al.*, 1993b)), which can be separated at a very high speed of up to 1000 bp/h (Swerdlow and Gesteland, 1990).

Oligonucleotides and PCR products analysis. Since the original publication by Kasper *et al.* (1988) on the application of CE-UV detection to the separation of restriction fragments, whole phage, viral and plasmid DNA, a number of papers have been published in this field.

Because this type of analysis requires less resolution than DNA sequencing, the use of semi-viscous solutions instead of gel filled capillaries has become more popular.

It is well known that the gel preparation of filled capillaries is a time consuming and fastidious task. The use of viscous solution buffer is an interesting alternative to overcome the problem of intercapillary reproducibility and short life time (Gelfi *et al.*, 1994).

Among the linear polymers used for the preparation of viscous buffers, the most popular are linear polyacrylamide (Carson *et al.*, 1993; Chiari *et al.*, 1993; Pariat *et al.*, 1993; Ruiz-Martinez *et al.*, 1993; Gelfi *et al.*, 1994), methyl cellulose (McGregor and Yeung, 1993; Srinivasan *et al.*, 1993), hydroxypropylmethyl cellulose (Schwartz *et al.*, 1991), hydroxyethylcellulose (McCord *et al.*, 1993) and liquid agarose (Bocek and Chrambach, 1992).

Cohen *et al.* (1990) utilised capillary gel electrophoresis with a LIF detector as a tool for sequencing synthetic oligonucleotides and single-strand phage DNA. Heiger *et al.* (1990) separated mixtures of DNA restriction fragments in gel filled capillaries and attained a sensitivity level in the attomolar range.

CE–LIF is particularly applicable to the analysis of PCR products. Using coated capillaries and viscous buffers containing 0.5% of hydroxymethyl cellulose 4000 ctp, PCR products have been identified using thiazole orange (Schwartz and Ulfelder, 1992). In this case, the sensitivity of LIF detection is 400 times that of UV detection. Using YoPro dye, in similar protocol, McCord *et al.* (1993) reached a sensitivity of 500 pg/ml.

Analysis of PCR-amplified DNA fragments for human identification has been performed (Srinivasan *et al.*, 1993). The intercalating dies Toto and Yoyo were used to visualize femtogramme per milliter amounts of DNA. This method was used to analyse different genetic loci (apolipoprotein, VNTR locus D1580, mitochondrial DNA). Very recently Zhu *et al.* (1994) studied the sensitivity of different intercalating dies. They found that, according to the dye employed, small amounts of 9-aminoacridine in viscous buffer can increase resolution.

CE–LIF separation and detection of DNA fragments with bis-intercalating dyes were also studied (Kim and Morris, 1994).

In the clinical field, identification of point DNA mutations has been performed by Khrapko *et al.* (1994) using Constant Denaturing Capillary Electrophoresis (CDCE). Four DNA duplexes of 206 bp which differed from each other by only one base pair were separated on a denaturing gel at 36 °C. Using electrokinetic injection they obtained a linear relationship between the number of molecules and peak areas in the range of 3×10^4 to 10^{11} molecules. This new approach should be applicable to identify and quantify low frequency mutations, to mutational spectrometry and to genetic screening of pooled samples for the detection of variants.

The most outstanding application of clinical gel CE–LIF to date was achieved by a French clinical laboratory (Lu *et al.*, 1994), who developed a quantitative analysis of PCR-amplified HIV-1 DNA or cDNA fragments. For the first time a linear curve was shown for the relative fluorescence function of the number of HIV DNA copies obtained by PCR with the gag primer SK145/431 and the number of HIV RNA virions obtained by reverse transcription with 20 pmol of $3'$ primer SK431. The authors show that this technique allows the quantitation of the concentration of HIV-1 virions in the sera of infected patients, and the techniques of gel CE–LIF and DNA hybridization for measuring HIV-1 virions in the sera of individuals assayed at different stages of infection. Similar results were also achieved by Butler *et al.* (1994) with coated capillary and semiviscous buffer.

Southern blot–CE was developed by Ju Wei Chen *et al.* (1991). Fluorescence-tagged oligonucleotides were used as probes for hybridization in solution with complementary DNA molecules prior to separation. Gel CE sepration was achieved with UV detection.

Because of the very low concentration detection achieved with CE–LIF, our laboratory works on the direct quantification of cellular genes using the technique of Southern or Northern blotting and capillary electrophoresis.

By gel capillary (10% T 0% C), we quantified a 28-mer template DNA ($3'$CCCTACCCTCCGGTGTTTCTCCGGTACG5$'$), which is part of the 3-hydroxy-3-methylglutarylcoenzyme A reductase (HMGCoAred) gene, and its complementary fluorescent strand. These two oligonucleotides were produced by synthesis, and latter one was $5'$-labeled with fluoroprime from Pharmacia Biochemicals. The template DNA was diluted from 10^{-6} to 10^{-10} M with a constant 10^{-5} M of fluorotide. The 488 nm wavelength of an argon–ion air-cooled laser was used for detection. The calibration curve log(intensity) vs. log(concentration) is a linear curve over the four orders of magnitude with a slope of 0.75. Poor reproducibility of different runs gave 38 to 49% errors with very dilute solutions. The sensitivity limit was found to be 1.6×10^{-10} M of template DNA.

Figure 6 shows an electropherogram of 10^{-9} M labeled DNA$_{ds}$. The amount of HMGCoAR gene expressed by PinaCoA plasmids transfected in *E. coli*

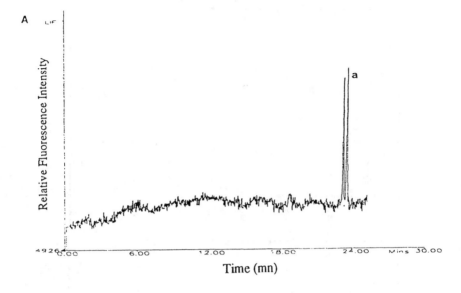

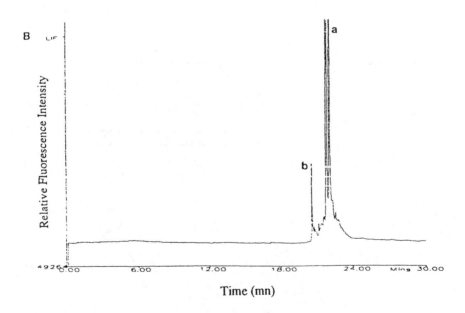

Figure 6. Fluorescent Southern pattern analysis. A: Electropherogram of 10^{-9} M HMGCoAred fluorescent probe (a). B: Electropherogram of 10^{-5} M HMGCoAred fluorescent probe and 10^{-9} M template DNA_{ss}, (b) indicates fluorescently obtained DNA_{ds}. (70 cm long 75 μm i.d. fused silica capillary filled as described in Hjerten (1985) with 10% T 0% C gel, −20 kV analysis: 20 s, −20 kV electrokinetic injection.)

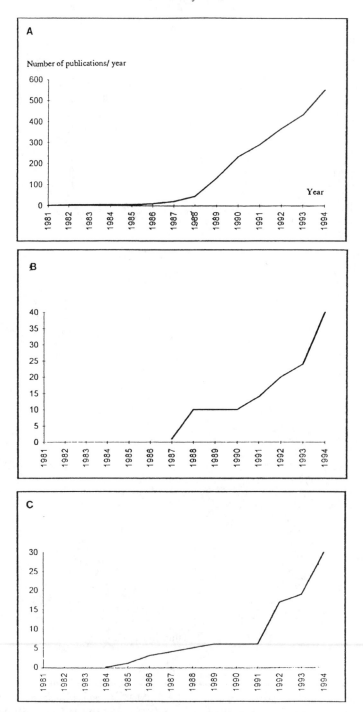

Figure 7. Evolution of CE/nucleic acids works and LIF applications during the last thirteen years. A: Number of publications per year in CE. B: Number of publications per year in CE on nucleic acids. C: Number of publications per year in CE–LIF.

Table 1.

Theoretical and observed quantities of HMGcoenzyme A reductase (HMGcoAred) in 1 μg PinacoA plasmid

HMGCoAred gene quantities obtained with CE–LIF (M) ($\times 10^9$)	Theoretical HMGCoAred gene quantities (M) ($\times 10^9$)	Culture number
2.1 ± 1.0	3.7 ± 0.3	1
0.5 ± 0.2	3.7 ± 0.3	2
3.2 ± 1.2	3.7 ± 0.3	3

was measured (Seronie Vivien *et al.*, in press). Ten microgrammes of plasmid DNA was purified from three different micropreparative cultures. After digestion with restriction enzymes, the DNA was hybridized with 0.5 ng of fluorotides and nuclease S1 treated. One tenth of each hybridization mix was subjected to CE–LIF in order to quantitate the amount of fluorescent HMG-CoAred hybridized gene. Table 1 reports these quantities.

Differences between theoretical and observed quantities are due to extensive manipulation of plasmid DNA, to difficulties in reproducing yields of extraction and poor injection reproducibility.

The number of publications dealing with the analysis of nucleic acids by CE–LIF show that this technique will become an important tool in this field of research as soon as LIF detection becomes available at a low price. As an example, Fig. 7 presents the evolution of CE–LIF works and CE-nucleic acids research over the last thirteen years. The exponential pattern of the two curves indicates that CE and CE–LIF should have the same sucess as polyacrylamide and agarose slab gels and radioactive probes twenty years ago.

Amino acids analysis

The analysis of amino acids at high sensitivity is another important field of application of CE–LIF. Jorgenson and Lukacs (1981a) first reported the separation of dansyl amino acids using arc lamp induced fluorescence detection. Subsequently, analysis conditions have been improved and chiral separations of D and L series are now possible (Gassmann *et al.*, 1985; Gozel *et al.*, 1987, 1988; Nouadje and Couderc, 1994). Using LIF detection amino acid analysis has been realised using different fluorescent reagents (Table 2). Yu and Dovichi (1988, 1989) reported the separation of 18 DABSYL (4,4-dimethylaminoazo benzene-4′-sulfonyl chloride) amino acids at the hundred attomol level using a thermo-optical absorbance detector. Native fluorescence of some amino acids is possible using low-UV lasers or double-frequency argon–ion lasers (Lee and Yeung, 1992b; Chan *et al.*, 1993).

We have measured the amount of amino acids and degradation product of histidine and tyrosine: histamine and tyramine (Joosten and Stadhouders, 1987)

Table 2.

Amino acid derivatives analysed by CE–LIF

Reagent	Laser	Excitation wavelength (nm)	Detection limit	References
Dansyl	He–Cd	325	50 attomol	(A)
FITC	Argon–ion	488	0.1 zeptomol	(B)
FTH, TRIC	Argon–ion	488	zeptomole level	(C)
NDA	Diode	424	0.9 attomol	(D)
NDA	He–Cd	442	0.8 attomol	(E)
CBQCA	He–Cd	442	attomol level	(F)
OPA	He–Cd	325	0.1 femtomol	(G)
PSE	Diode	663	0.8 attomol	(H)
DACCASE	Diode	415	0.1 femtomol	(J)

FITC: fluorescein isothiocyanate; FTH: fluoresceinthiohydantoine; TRIC: tetra-methylrhodamine thiocarbamyl; NDA: naphthalendialdehyde; CBQCA: 3-(4-carboxy-benzoyl)-2-quinolinecarboxaldehyde; OPA: o-phthalaldehyde; PSE: pyronin succin-imidyl ester; DACCASE: 7-(diethylamino)coumarin-3-carboxylic acid succinimidyl es-ter. (A) Gassmann *et al.*, 1985; Gozel *et al.*, 1987; Gozel and Zare, 1988. (B) Nicker-son and Jorgenson, 1988b; Cheng and Dovichi, 1990; Hernandez *et al.*, 1991; Sweedler *et al.*, 1991; Higashijima *et al.*, 1992; Nouadje and Couderc, 1994. (C) Waldron *et al.*, 1990; Wu and Dovichi, 1992; Zhao *et al.*, 1992. (D) Lee and Yeung, 1992a. (E) Nick-erson and Jorgenson, 1988a, b; Ueda *et al.*, 1991. (F) Liu *et al.*, 1991; Bergquist *et al.*, 1994. (G) Nickerson and Jorgenson, 1988b; Kang and Buck, 1992. (H) Fuchigami *et al.*, 1993. (J) Jansson and Roeraade, 1993.

found in wine. The importance of aminoacids in wine is known. It is generally accepted that they contribute to the sensory properties and final characteristics, and arise from the maturation of grapes, sensitive as this is to climatic con-ditions, soil composition and agricultural practices (Chaves Das Neves, 1993). Only a few techniques are available to carry out a qualitative and quantitative analysis of such a complex sample, that contains not less than 500 differ-ent compounds. Moreover, we were able to find a published analysis of the presence of biogenic amines, like tyramine and histamine, in wine. These amines present a real danger to human health by modification of blood pres-sure (lethal dose of histamine: 50 mg/l of blood, tyramine: 25 mg/l of blood); their analysis would permit a fine selection of yeasts that do not provoke such biodegradation.

CE coupled with UV detection appeared to be a good technique to solve this problem because high-speed and high-resolution analysis are possible with easy dansyl derivatization of the sample. However, due to the complexity of the electropherograms obtained (Fig. 8A), it rapidly became apparent that the interpretation would be impossible. Thus we prefer to employ LIF detection because of its higher selectivity (Nouadje *et al.*, 1995b). Using our ball-lens collinear arrangement and without derivatization, it was possible to identify anthocyanins, that are naturally fluorescent at the 488 nm wavelength of the Ar$^+$ laser. In a second step the wines (diluted 1000-fold) were derivatized with

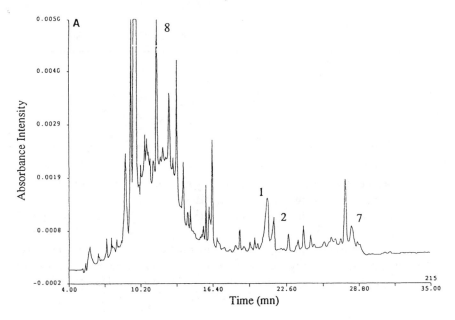

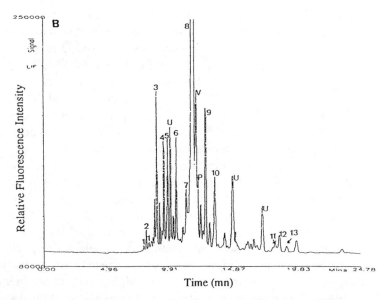

Figure 8. Amino acids and biogenic amines analysis. A: UV detection (215 nm) electrophero-gram of dansylated aminoacids and biogenic amines in red wine. Wine dilution factor: 2. B: Corresponding FITC-labeled sample with LIF detection electropherogram. Wine dilution factor: 10.000. In both cases: 82 cm long, 50 μm i.d. fused silica capillary, and 100 mM SDS, 100 mM borate (pH 9.2) buffer, were used. +24 kV Analysis: hydrodynamic injection 4 s. Peaks: 1 = lysine, 2 = arginine, 3 = histamine, 4 = histidine, 5 = tyramine, 6 = blank, 7 = tyrosine, 8 = proline, 9 = alanine, 10 = glycine, 11 = cysteine, 12 = glutamic acid, 13 = aspartic acid, V = valine, P = phenylalanine, S = serine, U = uknown.

Table 3.

Amount of histidine and tyrosine and their corresponding biogenic amines in wine

Wine	Histamine (mg/l)	Histidine (mg/l)	Tyramine (mg/l)	Tyrosine (mg/l)
Red	33.6	23.4	7.7	19.2
White	16.6	4.5	2.4	1.4
Mellow	13.3	4.5	1.4	4.3
Champagnised	32.5	24.6	6.2	6.7

FITC. A selective analysis of amino acids and biogenic amines is shown in Fig. 8B. The levels of histidine, tyrosine, histamine and tyramine were quantitated in different wines including red, white, sweet-white and champagnised wine (Table 3).

This study represents an improvement in quality control in the oenology and bromatology laboratory and will increase our knowledge of the use of fermentation yeast, and on the maturation of wine (Chaves Das Neves *et al.*, 1990).

Fluorescent dyes

To our knowledge few dyes have been studied using CE–LIF. Fluorescein, coumarin 102, and 4(dicyanomethylene)-2-methyl-6-(4-dimethylaminostyryl)-4H-pyran (DCM) have been used in sensitivity tests for arc-lamp-induced fluorescence detection with a CCD camera (Swaile and Sepaniak, 1989). More recently, a study concerning anionic and cationic dyes was published (Burkinshaw *et al.*, 1993; Croft and Hinks, 1993). Dovichi's group reported the analysis of sulforhodamine 101 (Cheng *et al.*, 1994). Using the sheath-flow cuvette detector and a 594 nm, He–Ne laser a LOCD of 3.8×10^{-13} M was obtained and the calculated performances were estimated to six molecules.

In order to determine the pollution pathway in rivers it is possible to employ biodegradable dyes. By dissolving a given amount of dye in a river upstream and sampling the water at different points downstream, the migration kinetics can be measured and the geographical repartition of the dye as a function of time can be inferred (Wanders *et al.*, 1993). This kind of study will be particularly useful to predict the scale and lifetime of a single pollution event.

In order to use the minimum amount of dye upstream and to be able to analyse the water samples as fast as possible downstream, a highly analytical technique is required. In a collaboration with Professor F. Everearts, we studied such dyes. The sensitivity obtained with rhodamine 123 using the 488 nm ray of an argon–ion laser is impressive. Figure 9A shows the electropherogram of a 3.5×10^{-13} M solution of rhodamine 123 using the ball-lens detector. The detector response is linear over five orders of magnitude. A separation of mixtures of rhodamines B and ωt and sulforhodamine B and G at a concentration of 10^{-11} M each is shown in Figs 9B and 9C. Detection at 10^{-12} M was not

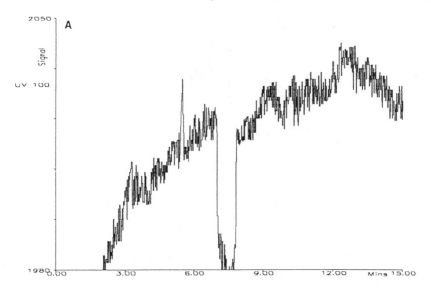

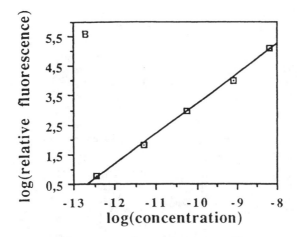

Figure 9. Rhodamine analysis. A: Electropherogram 3.5×10^{-13} M rhodamine 123. Buffer 40 mM sodium tetraborate (pH 9.2), +20 kV electrokinetic injection, 20 s analysis +15 kV. B: Calibration curve of rhodamine 123.

possible because the excitation wavelength used for these experiments did not match perfectly the maximum absorption wavelength of these rhodamines. The exitation wavelengths of each rhodamine dye are given in Table 4 (Green, 1990). A relatively expensive Ar–Kr ion laser at 569 nm, appears to be the best choise for this analysis.

We expect that rapid progress in diode lasers should lead to the development of much cheaper lasers, in this wavelength range, allowing us to reach the desired sensitivity at a reasonnable price.

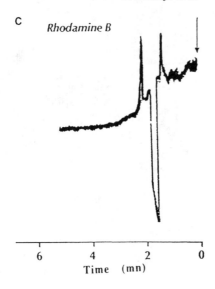

C   *Rhodamine B*

Time (mn)

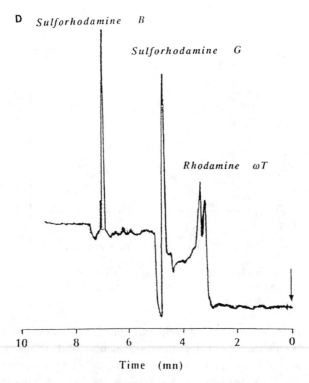

D   *Sulforhodamine   B*

Sulforhodamine   G

Rhodamine   ωT

Time   (mn)

Figure 9 (continued). Rhodamine analysis. C: 2.5×10^{-11} M rhodamines mixture analysis, borate buffer, +15 kV, +10 kV electrokinetic injection 10 s: rhodamine B eluted. D: 2.5×10^{-11} M rhodamines mixture analysis ε-amino caproic acid buffer analysis, −15 kV, −10 kV electrokinetic injection 10 s: rhodamine ωt, sulforhodamine B, sulforhodamine G eluted. C and D were realised with 70 cm long, 50 μm i.d. fused silica capillary.

Table 4.

Maximum excitation wavelength of rhodamine dyes

Dye (dilution solvent)	Maximum absorption wavelength (nm)
Rhodamine 123 (in water)	501
Rhodamine B (MeOH)	543
Sulforhodamine B (1.0 M NaOH)	554
Sulforhodamine (MeOH)	529

CONCLUSION

In this article we examine the different optical configurations for the CE–LIF detector. A collinear arrangement appeared to be the more appropriate design for a commercial instrument. We improved this design by using a ball lens coupled with a low NA objective instead of a high NA aperture. This improvement has led to the development of a detector available commercially at a reasonable price.

The applications of CE–LIF are numerous and cover the fields of molecular biology, analytical biochemistry, bromatology and ecology. The increase in the number of publications in this field shows that CE–LIF is leading to a new revolution in biotechnical science before the end of this century.

ACKNOWLEDGEMENT

We would like to acknowledge Leonora Poljak for comments on the text.

REFERENCES

Ansorge, W., Sproat, B. S., Stegemann, J. and Schwager, C. (1986). A non-radioactive automated method for DNA sequence determination. *J. Biochem. Biophys. Meth.* **13**, 315–326.

Bergquist, J., Gilman, S. D., Ewing, A. G. and Ekman, R. (1994). Analysis of human cerebrospinal fluid by capillary electrophoresis with laser-induced fluorescence detection. *Anal. Chem.* **66**, 3512–3518.

Bocek, P. and Chrambach, A. (1992). Capillary electrophoresis in agarose solutions: extension of size separations to DNA of 12 kb in length. *Electrophoresis* **13**, 31–34.

Burkinshaw, S. M., Hinks, D. and Lewis, D. M. (1993). Capillary zone electrophoresis in the analysis of dyes and other compounds employed in the dye-manufacturing and dye-using industries. *J. Chromatogr.* **640**, 413–417.

Butler, J. M., McCord, B. R., Jung, J. M., Wilson, M. R., Budowle, B. and Allen, R. O. (1994). Quantitation of polymerase chain reaction products by capillary electrophoresis using laser fluorescence. *J. Chromatogr.* **658**, 271–280.

Carson, S., Cohen, A. S., Belenkii, A., Ruiz-Martinez, M. C., Berka, J. and Karger, B. L. (1993). DNA sequencing by capillary electrophoresis: use of a two-laser-two-window intensified diode array detection system. *Anal. Chem.* **65**, 3219–3226.

Chan, K. C., Janini, G. M., Muschik, G. M. and Issaq, H. J. (1993). Pulsed UV laser-induced fluorescence detection of native peptides and proteins in capillary electrophoresis. *J. Liquid Chromatogr.* **16**, 1877–1890.

Chaves Das Neves, H. J. (1993). *Les acquisitions récentes en chromatographie du vin*, Doneche, B. (Ed.). Institut d'Oenologie Université de Bordeaux II, France, pp. 63–79.

Chaves Das Neves, H. J., Vasconcelos, A. M. P. and Costa, M. L. (1990). Racemisation of wine free amino acids as function of bottling age. In: *Chirality and Biological Activity*, Sandra, P. (Ed.). Alan R. Liss, New York, pp. 137–143.

Chen, D. A., Swerdlow, H. P., Harke, H. R., Zhang, J. Z. and Dovichi, N. J. (1991a). Single-color laser-induced fluorescence detection and capillary gel electrophoresis for DNA sequencing. *SPIE Optical Methods for Ultrasensitive Detection and Analysis: Techniques and Applications* **1435**, 161–167.

Chen, D. Y., Harke, H. R. and Dovichi, N. J. (1992). Two-label peak-height encoded DNA sequencing by capillary gel electrophoresis: three examples. *Nucleic Acids Res.* **18**, 4873–4880.

Chen, I. C. and Whang, C. W. (1993). Capillary electrophoresis with amperometric detection using a porous cellulose acetate joint. *J. Chromatogr.* **644**, 208–212.

Chen, J. W., Cohen, A. S. and Karger, B. L. (1991b). Identification of DNA molecules by pre-column hybridization using capillary electrophoresis. *J. Chromatogr.* **559**, 295–305.

Cheng, Y. F. and Dovichi, N. (1989). Subattomole amino acid analysis by capillary zone electrophoresis and laser-induced fluorescence. *Science* **242**, 562–564.

Cheng, Y. F. and Dovichi, N. J. (1990). Subattomole amino acid analysis by laser induced fluorescence: The sheath-flow cuvette meets capillary zone electrophoresis. *ASTM Spec. Tech. Publ.* **1066**, 151–159.

Cheng, Y. F., Piccard, R. D. and Vo Dinh, T. (1990). Charge-coupled device fluorescence detection for capillary zone electrophoresis (CCD-CZE). *Appl. Spectrosc.* **44**, 755–765.

Cheng, Y. F., Adelhelm, K., Cheng, X. L. and Dovichi, N. J. (1994). A simple laser-induced fluorescence detector for sulforhodamine 101 in a capillary electrophoresis system: Detection limits of 10 yoctomoles or six molecules. *Analyst* **119**, 349–352.

Chiari, M., Nesi, M. and Righetti, P. G. (1993). Movement of DNA fragments during capillary zone electrophoresis in liquid polyacrylamide. *J. Chromatogr.* **652**, 31–39.

Cohen, A. S., Najarian, D. R. and Karger, B. L. (1990). Separation and analysis of DNA sequence reaction products by capillary gel electrophoresis. *J. Chromatogr.* **516**, 49–60.

Croft, S. N. and Hinks, D. (1993). Analysis of dyes by capillary electrophoresis. *Text. Chem. Color* **25**, 47–51.

Drossman, H., Luckey, J. A., Kostichka, A. J., D'Cunha, J. and Smith, L. M. (1990). High speed separation of DNA sequencing reaction by capillary electrophoresis. *Anal. Chem.* **62**, 900–903.

Folestad, S., Johnson, L., Josefsson, B. and Galle, B. (1982). Laser induced fluorescence detection for conventional and microcolumn liquid chromatography. *Anal. Chem.* **54**, 925–929.

Fuchigami, T., Imasaka, T. and Shiga, M. (1993). Subattomole detection of amino acids by capillary electrophoresis based on semiconductor laser fluorescence detection. *Anal. Chim. Acta* **282**, 209–213.

Gassmann, E., Kuo, J. E. and Zare, R. N. (1985). Electrokinetic separation of chiral compounds. *Science* **230**, 813–814.

Gelfi, C., Righetti, P. G., Brancolini, V., Cremonesi, L. and Ferrari, M. (1994). Capillary electrophoresis in polymer networks for analysis of PCR products: Detection of ΔF508 mutation in cystic fibrosis. *Clin. Chem.* **40**, 1603–1604.

Gozel, P., Gassmann, E., Michelsen, H. and Zare, R. N. (1987). Electrokinetic resolution of amino acids enantiomers with copper(II)-aspartame support electrolyte. *Anal. Chem.* **59**, 44–49.

Gozel, P. and Zare, R. N. (1988). Electrokinetic resolution of amino acid enantiomers with copper(II)-aspartame support electrolyte. *ASTM Spec. Tech. Publ.* **1009**, 41–53.

Green, F. J. (1990). *The Sigma-Aldrich Handbook of Stains, Dyes and Indicators*. Aldrich Chemical Company Inc., Milwaukee, Wisc.

Green, J. S. and Jorgenson, I. C. (1986). Variable wavelength on-column fluorescence detector for open tubular zone electrophoresis. *J. Chromatogr.* **352**, 337–343.

Heiger, D. N., Cohen, A. S. and Karger, B. L. (1990). Separation of DNA restriction fragments by high performance capillary electrophoresis with low and zero crosslinked polyacrylamide using continuous and pulsed electric fields. *J. Chromatogr.* **516**, 33–48.

Hernandez, L. and Joshi, N. (1990). L'electrophorèse capillaire et la détection en fluorescence induite par laser. *Spectra 2000* **153**, 40–43.

Hernandez, L., Escalona, J. and Joshi, N. (1991). Laser induced fluorescence and fluorescence microscopy for capillary electrophoresis zone detection. *J. Chromatogr.* **559**, 183–196.

Higashijima, T., Fushigami, T., Imasaka, T. and Ishibashi, N. (1992). Determination of amino acids by capillary zone electrophoresis based on semi-conductor laser fluorescence detection. *Anal. Chem.* **64**, 711–714.

Hjerten, S. (1985). High-performance electrophoresis: Elimination of electroendosmosis. *J. Chromatogr.* **347**, 191–198.

Huang, X. C., Quesada, M. A. and Mathies, R. A. (1992a). Capillary array electrophoresis using laser-excited confocal fluorescence detection. *Anal. Chem.* **64**, 967–972.

Huang, X. C., Quesada, M. A. and Mathies, R. A. (1992b). DNA sequencing using capillary array electrophoresis. *Anal. Chem.* **64**, 2149–2154.

Hunkapiller, T., Kaiser, R. J., Koop, B. F. and Hood, L. (1991). Large-scale and automated DNA sequence determination. *Science* **254**, 59–67.

Jansson, M. and Roeraade, J. (1993). Laser-induced fluorescence detection in capillary electrophoresis with blue light from frequency doubled diode laser. *Anal. Chem.* **65**, 2766–2769.

Joosten, H. M. L. J. and Stadhouders, J. (1987). Conditions allowing the formation of biogenic amines in cheese. Decarboxylation properties of starter bacteria. *Neth. Milk Dairy J.* **41**, 247–258.

Jorgenson, I. C. and Lukacs, K. D. (1981a). Zone electrophoresis in open tubular glass capillaries. *Anal. Chem.* **53**, 1298–1302.

Jorgenson, I. C. and Lukacs, K. D. (1981b). High resolution separations based on electrophoresis and electrosmosis. *J. Chromatogr.* **218**, 209–216.

Kambara, H. and Takahashi, S. (1993). Multiple-sheathflow capillary array DNA analyser. *Nature* **361**, 565–566.

Kang, L. and Buck, R. H. (1992). Separation enantiomer determination of OPA-derivatised amino acids by using capillary zone electrophoresis. *Amino Acids* **2**, 103–109.

Karger, A. E., Harris, J. M. and Gesteland, R. F. (1991). Multiwavelength fluorescence detection for DNA sequencing reaction using capillary electrophoresis. *Nucleic Acids Res.* **18**, 4955–4962.

Kasper, T., Melera, M., Gozel, P. and Brownlee, R. (1988). Separation and detection of DNA by capillary electrophoresis. *J. Chromatogr.* **458**, 303–306.

Khrapko, K., Hanekamp, J. S., Thilly, W. G., Belenkii, A., Foret, F. and Karger, B. L. (1994). Constant denaturant capillary electrophoresis (CDCE): a high resolution approach to mutational ànalysis. *Nucleic Acids Res.* **22**, 364–369.

Kim, Y. and Morris, M. D. (1994). Separation of nucleic acids by capillary electrophoresis in cellulose solutions with mono- and bis-intercalating dye. *Anal. Chem.* **66**, 1168–1174.

Kuhn, R. and Hoffstetter-Kuhn, S. (1993). *Capillary Electrophoresis Principle and Practice*. Springer-Verlag, Berlin.

Kuhr, W. G. (1990). Capillary electrophoresis. *Anal. Chem.* **62**, 403R–414R.

Kuhr, W. G. and Monnig, C. A. (1992). Capillary electrophoresis. *Anal. Chem.* **64**, 389R–407R.

Lee, T. T. and Yeung, E. S. (1992a). Quantitative determination of native proteins in individual human erythrocytes by capillary zone electrophoresis with laser-induced fluorescence. *Anal. Chem.* **64**, 3045–3051.

Lee, T. T. and Yeung, E. S. (1992b). High sensitivity laser-induced fluorescence detection of native proteins in capillary electrophoresis. *J. Chromatogr.* **595**, 319–325.

Liu, J., Hsieh, Y. Z., Wiesler, D. and Novotny, M. (1991). Design of 3-(4-carboxybenzoyl)-2-quinolinecarboxaldehyde as a reagent for ultrasensitive determination of primary amines by capillary electrophoresis using laser fluorescence detection. *Anal. Chem.* **63**, 408–412.

Lu, W., Han, D. H. and Andrieu, J. M. (1994). Multi-target PCR analysis by capillary electrophoresis and laser-induced fluorescence. *Nature* **368**, 269–271.

Luckey, J. A. and Smith, L. M. (1993a). Optimization of electric field strength for DNA sequencing in capillary gel electrophoresis. *Anal. Chem.* **65**, 2841–2850.

Luckey, J. A. and Smith, L. M. (1993b). A model for the mobility of single-stranded DNA in capillary gel electrophoresis. *Electrophoresis* **14**, 492–501.

Luckey, J. A., Drossman, H., Kostichka, A. J., Mead, D. A., D'Cunha, J., Norris, T. B. and Smith, L. M. (1990). High speed DNA sequencing by capillary electrophoresis. *Nucl. Acids Res.* **18**, 4417–4421.

Luckey, J. A., Drossman, H., Kostichka, A. J. and Smith, L. M. (1993a). High-speed DNA sequencing by capillary gel electrophoresis. *Methods Enzymol.* **218**, 154–172.

Luckey, J. A., Norris, T. B. and Smith, L. M. (1993b). Analysis of resolution in DNA sequencing by capillary gel electrophoresis. *J. Phys. Chem.* **97**, 3067–3075.

Mathies, R. A. and Huang, X. C. (1992). Capillary array electrophoresis: an approach to high speed, high-throughput DNA sequencing. *Nature* **359**, 167–169.

McCord, B. R., McClure, D. L. and Jung, J. M. (1993). Capillary electrophoresis of polymerase chain reaction-amplified DNA using fluorescence detection with an intercalating dye. *J. Chromatogr.* **652**, 75–82.

McGregor, D. A. and Yeung, E. S. (1993). Optimization of capillary electrophoretic separation of DNA fragments based on polymer filled capillaries. *J. Chromatogr.* **652**, 67–73.

Monnig, C. A. and Kennedy, R. T. (1994). Capillary electrophoresis. *Anal. Chem.* **66**, 280R–314R.

Nickerson, B. and Jorgenson, J. (1988a). High speed capillary zone electrophoresis with laser-induced fluorescence detection. *J. High Resolut. Chromatogr., Chromatogr. Commun.* **11**, 533–534.

Nickerson, B. and Jorgenson, J. W. (1988b). High sensitivity laser-induced fluorescence detection in capillary zone electrophoresis. *J. High Resolut. Chromatogr., Chromatogr. Commun.* **11**, 878–881.

Nie, S., Dadoo, N. and Zare, R. N. (1993). Ultrasensitive fluorescence detection of polycyclic aromatic hydrocarbons in capillary electrophoresis. *Anal. Chem.* **65**, 3571–3575.

Nouadje, G. and Couderc, F. (1995a). Séparation chirale de fluoresceincarbamyle-acides aminés par electrophorèse capillaire couplée à la fluorescence induite par laser. *Spectra Analysis* **184**, 17–21.

Nouadje, G., Couderc, F., Puig, Ph. and Hernandez, L. (1995b). Combination of micellar electrokinetic chromatography and laser-induced fluorescence detection for the determination of pressor amines and some principal amino-acids in wine. *J. Cap. Elec.* **2**, 117–124.

Pariat, Y. F., Berka, J., Heiger, D. N., Schmitt, T., Vilenchik, M. and Cohen, A. S. (1993). Separation of DNA fragments by capillary electrophoresis using replaceable linear polyacrylamide matrices. *J. Chromatogr.* **652**, 57–66.

Pentoney, S. L., Jr., Konrad, K. D. and Kaye, W. (1992). A single-fluor approach to DNA sequence determination using high performance capillary electrophoresis. *Electrophoresis* **13**, 467–474.

Prober, J. M., Trainor, G. L., Dam, R. J., Hobbs, F. W., Robertson, C. W., Zagursky, R. J., Cocuzza, A. J., Jensen, M. A. and Baumeister, K. (1987). A system for rapid DNA sequencing with fluorescent chain-terminating dideoxynucleotides. *Science* **238**, 336–341.

Rocheleau, M. J., Grey, R. J., Chen, D. Y., Harke, H. R. and Dovichi, N. J. (1992). Formamide modified polyacrylamide gels for DNA sequencing by capillary electrophoresis. *Electrophoresis* **13**, 484–486.

Ruiz-Martinez, M. C., Berka, J., Belenkii, A., Foret, F., Miller, A. W. and Karger, B. L. (1993). DNA sequencing by capillary electrophoresis with replaceable linear polyacrylamide and laser-induced fluorescence detection. *Anal. Chem.* **65**, 2851–2856.

Schwartz, H. E. and Ulfelder, K. J. (1992). Capillary electrophoresis with laser-induced fluorescence detection of PCR fragments using thiazole orange. *Anal. Chem.* **64**, 1737–1740.

Schwartz, H. E., Ulfelder, K., Sunzeri, F. J., Busch, M. P. and Brownlee, R. G. (1991). Analysis of DNA restriction fragments and PCR products towards detection of the AIDS (HIV-1) virus in blood. *J. Chromatogr.* **559**, 267–274.

Seronie Vivien, S., Pradines, A., Couderc, B., Clamagirand, C., Berg, D., Soula, G. and Favre, G. (1995). Reversion of transformed phenotype of human adenocarcinoma A549 cells by expression of 3-hydroxy-3-methylglutaryl-CoA-reductase cDNA. *Cell Growth Differ.* **6**, 1415–1425.

Smith, L. M. (1991). High-speed DNA sequencing by capillary gel electrophoresis. *Nature* **349**, 812–813.

Smith, L. M., Sanders, J. Z., Kaiser, R. J., Hughes, P., Dodd, C., Connell, C. R., Heiner, C., Kent, S. B. H. and Hood, L. E. (1986). Fluorescence detection in automated DNA sequence analysis. *Nature* **321**, 674–678.

Srinivasan, K., Girard, J. E., Williams, P., Roby, R. K., Weedn, V. W. and Morris, S. C. (1993). Electrophoretic separation of polymerase chain reaction-amplified DNA fragments in DNA typing using a capillary electrophoresis–laser induced fluorescence system. *J. Chromatogr.* **652**, 83–91.

Starke, H. R., Yan, J. Y., Zhang, J. Z., Mühlegger, K., Effgen, K. and Dovichi, N. J. (1994). Internal fluorescence labeling with fluorescent deoxynucleotides in two-label peak-height encoded DNA sequencing by capillary electrophoresis. *Nucleic Acids Res.* **22**, 3997–4001.

Swaile, D. F. and Sepaniak, M. J. (1989). Fluorometric photodiode array detection in capillary electrophoresis. *J. Microcolumn Sep.* **1**, 155–158.

Sweedler, J. V., Shear, J. B., Fishman, H. A., Zare, R. N. and Scheller, R. H. (1991). Fluorescence detection in capillary zone electrophoresis using a charge coupled device with time delayed integration. *Anal. Chem.* **63**, 496–502.

Swerdlow, H. and Gesteland, R. (1990). Capillary gel electrophoresis for rapid, high resolution DNA sequencing. *Nucleic Acids Res.* **18**, 1415–1419.

Swerdlow, H., Wu, S., Harke, H. and Dovichi, N. J. (1990). Capillary gel electrophoresis for DNA sequencing. Laser-induced fluorescence detection with sheath-flow cuvette. *J. Chromatogr.* **516**, 61–67.

Swerdlow, H., Zhang, J. W., Chen, D. Y., Harke, H. R., Grey, R., Wu, S., Dovichi, N. J. and Fukker, C. (1991). Three DNA sequencing methods using capillary gel electrophoresis and laser-induced fluorescence. *Anal. Chem.* **63**, 2835–2841.

Swerdlow, H., Dew-Jager, K. E., Brady, K., Grey, R., Dovichi, N. J. and Gesteland, R. (1992). Stability of capillary gels for automated sequencing of DNA. *Electrophoresis* **13**, 475–483.

Takahashi, S., Murakami, K., Anazawa, T. and Kambara, H. (1994). Multiple sheath-flow gel capillary-array electrophoresis for multicolor fluorescent DNA detection. *Anal. Chem.* **66**, 1021–1026.

Taylor, J. A. and Yeung, E. S. (1992). Multiplexed fluorescence detector for capillary electrophoresis using axial optical fiber illumination. *Anal. Chem.* **64**, 1741–1744.

Taylor, J. A. and Yeung, E. A. (1993). Multiplexed fluorescence detector for in capillary electrophoresis using axial optical fiber illumination. *Anal. Chem.* **65**, 956–960.

Todoriki, H. and Hirakawa, A. Y. (1980). CW laser fluorometry using optical fiber: an application to HPLC. *Chem. Pharm. Bull.* (Japan) **4**, 1337–1339.

Ueda, T., Kitamura, F., Mittchell, R., Metcalf, T., Kuwana, T. and Nakamoto, A. (1991). Chiral separation of naphthalene-2,3-dicarboxaldehyde-labeled amino acid enantiomers by cyclodextrine-modified micellar electrokinetic chromatography with laser-induced fluorescence detection. *Anal. Chem.* **63**, 2979–2981.

Waldron, K. C., Wu, S., Earle, C. W., Harke, H. R. and Dovichi, N. J. (1990). Capillary zone electrophoresis separation and laser based detection of both fluorescein-thiohydantoin and dimethylaminoazobenzene-thiohydantoin derivatives of amino-acids. *Electrophoresis* **11**, 777–780.

Wanders, B. H., Rutten, T. P. A., Martins, H. H. P. A. and Everaerts, S. M. (1993). Ultrasensitive detection of tracers (rhodamines) in sea water using CE in combination with LIF detection. In: *Fourth International Symposium on HPCE*, Amsterdam, p. 164.

Wu, S. and Dovichi, N. (1989). High sensitivity fluorescence detector for fluorescein isothiocyanate derivatives of amino acids separated by capillary zone electrophoresis. *J. Chromatogr.* **480**, 141–156.

Wu, S. and Dovichi, N. J. (1992). Capillary zone electrophoresis separation and laser-induced fluorescence detection of zeptomole quantities of fluorescein thiohydantoin derivatives of amino acids. *Talanta* **39**, 173–178.

Xu, Y. (1993). Capillary electrophoresis. *Clin. Chem.* **65**, 425R–433R.

Yu, M. and Dovichi, N. (1988). Subfemtomole determination of dabsyl-amino acids with capillary zone electrophoresis separation and laser-induced thermooptical absorbance detection. *Mikrochim. Acta* **III**, 27–40.

Yu, M. and Dovichi, N. (1989). Attomole amino acids determination by capillary zone electrophoresis with thermooptical absorbance detection. *Anal. Chem.* **61**, 37–40.

Zhao, J. Y., Chen, D. Y. and Dovichi, N. J. (1992). Low-cost laser-induced fluorescence detection for micellar capillary zone electrophoresis detection at the zeptomol level of tetramethylrhodamine thiocarbamyl amino acids derivatives. *J. Chromatogr.* **608**, 117–120.

Zhu, H., Clark, S. M., Benson, S. C., Rye, H. S., Glazer, A. N. and Mathies, R. A. (1994). High sensitivity capillary electrophoresis of double-stranded DNA fragments using monomeric and dimeric fluorescent intercalating dyes. *Anal. Chem.* **66**, 1941–1948.

PROTEIN
AND PEPTIDE ANALYSIS

Progress in HPLC-HPCE, Vol. 5, pp. 75–96
H. Parvez *et al.* (Eds)
© VSP 1997.

Analysis of protein folding by capillary electrophoresis

MARK A. STREGE, CRAIG N. COLE and AVINASH L. LAGU

Lilly Research Laboratories – A Division of Eli Lilly and Co., Lilly Corporate Center, Indianapolis, IN 46285, USA

INTRODUCTION

Protein folding/unfolding studies have become increasingly important due to the impact of biotechnology in the research and development areas of many industries. Once synthesized, the protein must then go through conformational changes, directed by the sequence of amino acids, that will lead to the fully folded native structure. The predetermined folding pattern designated by the amino acid sequence is often referred to as the 'second half of the genetic code'. Academically, elucidation of this 'second half' could give researchers the ability to predict the three-dimensional structure and even the activity of the native protein from its amino acid sequence. From the biotechnology perspective, many of the recombinantly produced proteins are not isolated or purified in their native conformation. Therefore, detection and confirmation of properly folded (active) protein by *in vitro* studies are essential.

TYPES OF FOLDING STUDIES

The measurement of the activity of the folded protein is a direct approach toward the identification of a properly folded protein, since in many cases only the active protein is of interest. Activity assays can only demonstrate the difference between unfolded inactive enzyme and the completely folded active enzyme. Many other techniques such as circular dichroism (CD), UV and fluorescence spectroscopy, gel electrophoresis, and chromatography have been used to observe the folding process and in some cases the intermediate forms of the folding protein. It should be noted that only techniques such as chromatography, electrophoresis, or capillary electrophoresis (CE) can actually reveal the different intermediate forms of the folding/unfolding protein.

  There are several traditional modes to induce the unfolding or folding of a protein. In the study of chaotropic effects, a denaturant such as guanidine hydrochloride or urea is used to unfold the protein. Although the exact mechanism of these reagents to denature each specific protein is not known, Creighton (1984) has suggested that they increase the solubility of the hydrophobic regions of the molecule while sustaining the hydrogen bonding capacity of the aqueous solvent. Protein unfolding can also be induced by raising the temperature of the solution containing the protein. As the temperature of the solution rises, the protein begins to unfold, allowing newly charged sites to be accessible. A related technique to induce folding/unfolding is to change pH. As the pH of the solution containing a protein is altered from physiological conditions the protein can unfold, giving new charged residues access to the solvent.

These three techniques all involve non-covalent changes in the structure of the protein. Proteins varying in covalent structure may be generated through the use of the following procedure. Using a combination of a chaotropic reagent and a reducing reagent, proteins containing disulfide bonds can be denatured and reduced (i.e. disulfide bonds are broken). Upon making the solution less concentrated in the denaturant and creating a mild oxidizing environment the protein will simultaneously refold and undergo disulfide rearrangements. Acidification of the folding solution will stop the disulfide rearranging, locking the disulfides into position. By examining these samples over time, different folded states of the protein (different covalent forms generated by disulfide rearrangements) can be observed using various analytical techniques.

Another method for inducing folding/unfolding of a protein is the utilization of cofactor/ligand protein interaction. Many enzymes require the association of a cofactor/ligand for activity. If the cofactor/ligand is removed subtle structural changes may take place such that activity is lost. One can analyze both the active and inactive forms of the protein to determine the effects of the association of cofactor/ligand upon structural conformation.

ANALYTICAL TECHNIQUES USED TO OBSERVE PROTEIN FOLDING/UNFOLDING

Chaotropic interactions are commonly used for inducing protein unfolding and several methods have been utilized to follow urea or guanidine hydrochloride-induced denaturation. Gel electrophoresis is often used with chaotropic studies. For example, Katz and Denis (1969) used native gel electrophoresis employing polyacrylamide gels (PAGE) containing only the buffer and no denaturant to study bovine serum albumin dimers exposed to various concentrations of guanidine hydrochloride (GuHCl) or urea for a fixed time period. The results showed that increasing the concentration of the denaturant caused enough local unfolding such that the monomer transformed into several bands as visualized on the gel (see Figs 1 and 2). Another application of polyacrylamide gel electrophoresis is the detection of unfolding using urea gradient gels. Since urea is not ionic like GuHCl, it can be used in high concentrations within the gel matrix. Creighton (1979) utilized electrophoresis with a linearly increasing

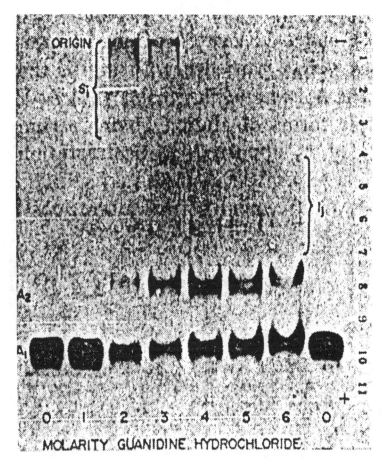

Figure 1. Electrophoretic patterns of 4% bovine serum albumin exposed to various concentrations of guanidine •HCl. The protein was exposed to the denaturant for 1 h at 25°C before electrophoresis. The symbol A_1 indicates the bovine serum albumin monomer; A_2, bovine serum albumin dimer; S_1, the forms of denatures albumin with electrophoretic mobility 5–30% of A_1; and I_j, the forms of denatured albumin with electrophoretic mobilities intermediate to S_1 and A_2. Electrophoresis was performed in 5% acrylamide, pH 8.3 buffer, for 2.5 h at 250 V. The scale is in cm. Reprinted from Katz and Denis (1969) with permission.

concentration of urea across the gel matrix to obtain a sigmoidal curve of protein visualized after staining the gel. This is due to the denatured form of the protein possessing a more unravelled conformation and therefore less mobility within the gel (see Fig. 3). Ion exchange HPLC can also be used to examine unfolding using a denaturing agent. Parente and Wetlaufer (1984a) employed a combination of chaotropic and ion exchange conditions to demonstrate denaturation of protein. They performed isocratic cation exchange HPLC to separate the unfolded protein from folded by using varying concentrations urea (0–8 M) in the mobile phase.

Protein denaturing profiles have been routinely generated by spectroscopic folding studies. The absorbance of the protein can be monitored by UV, flu-

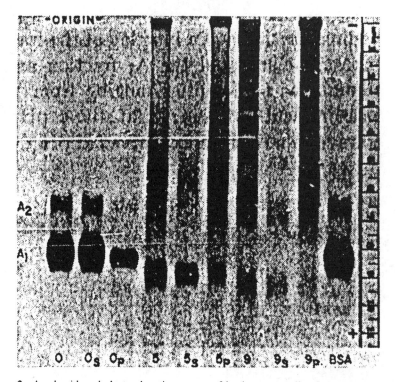

Figure 2. Acrylamide gel electrophoretic patterns of bovine serum albumin exposed to 0–9 M urea for 7 days at 25 °C. The procedure is similar to the previous figure. The symbol O represents bovine serum albumin in water after 7 days; O_s, the pattern exhibited by the retentate after thin thin-film dialysis; O_p, the pattern of the precipitate found after dialysis. Similar protocols were adopted for systems exposed to 5 and 9 M urea with the numerals indicating the urea molarity. The precipitates, indicates by the subscript, p, were dissolved in 0.03 M HCl prior to electrophoresis. BSA indicated native bovine serum albumin used as control. Conditions for electrophoresis were as described in Fig. 1. Reprinted from Katz and Denis (1969) with permission.

orescence, or circular dichroism spectroscopy in various concentrations of a denaturant, such as guanidine hydrochloride or urea. An 'S'-shaped curve where the absorbance drops quickly as the protein transforms for the folded to the unfolded state is seen, demonstrating that the protein rarely exists in a transition phase and spends most of the time in the completely unfolded or entirely folded state.

Another technique to monitor the folded state of a protein is the biological or enzymatic activity assay. Mendoza *et al.* (1991) employed an enzyme activity assay for rhodanese to determine if proper folding had occurred. It should be noted that activity checks are not exclusive to chaotropic studies. Activity methods can be utilized to monitor folding in pH, disulfide rearrangement, and temperature studies. Finally, size exclusion chromatography can be use to observe the difference in folded and unfolded states. Al-Obeidi and Light (1988) showed that by incubating trypsinogen in various concentrations of urea for 24 h the various protein folded states will have different retention times

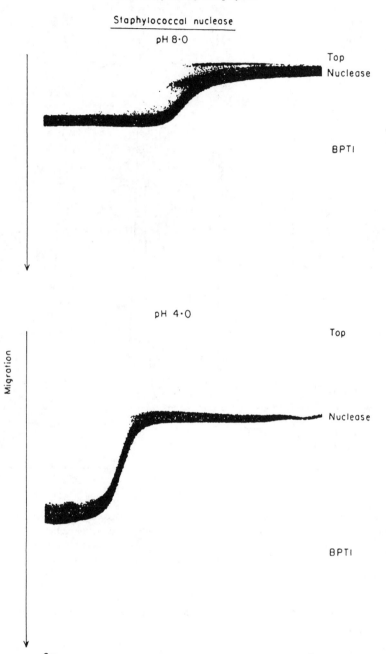

Figure 3. Urea-gradient electrophoresis of staphylococcal nuclease. The buffer at the top was 0.05 M Tris–acetate (pH 8.0), that at the bottom 0.05 M Tris–acetate (pH 4.0). BPTI was added to both protein samples as an internal standard. The linear gradients of 0 M to 8 M were superimposed on inverse linear gradients of 15% to 11% (top) or to 10% (bottom) acrylamide. Electrophoresis was towards the cathode at 15% C. Reprinted from Creighton (1979) with permission.

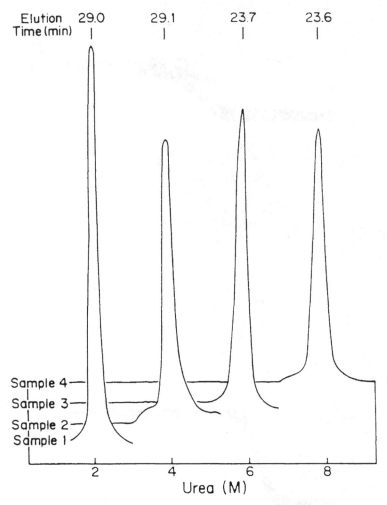

Elution Time (min): 29.0, 29.1, 23.7, 23.6

Urea (M): 2, 4, 6, 8

Sample 4, Sample 3, Sample 2, Sample 1

Figure 4. SE–HPLC of trypsinogen-(S-S) at various urea concentrations. The four samples were injected onto two TSK G-2000SW columns in series. The elution buffer was 0.2 M Tris, 0.2 M NaCl, pH 8.6, containing the appropriate concentration of urea. The flow rate was 1 ml/min. The elution times for the peaks are shown at the top. Reprinted from Al-Obeidi and Light (1988) with permission.

when chromatographed in the presence of appropriate concentrations of urea using a TSK G-2000 column (see Fig. 4).

Similarly, protein folding/unfolding reactions achieved using temperature as the denaturant can also be monitored using conventional analytical techniques. Spectroscopic measurements of protein solutions can be taken at different temperatures to observe the differences in absorption maxima or absorbance using a temperature controlled cell. Ion exchange HPLC can also be used to observe folding and unfolding of proteins using temperature as the denaturant. Parente and Wetlaufer (1984b) have shown that by raising the temperature of the ion

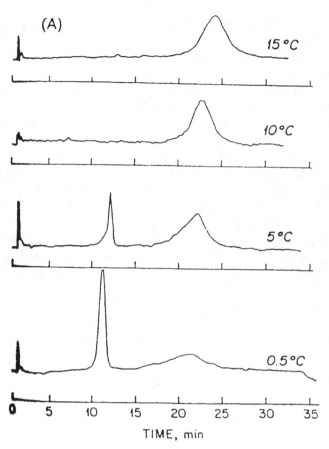

Figure 5. Effect of temperature on chromatographic (HIC) behavior of α-lactalbumin. Reprinted from Wu *et al.* (1986) with permission.

exchange column a shift in the retention time of alpha-chymotrypsinogen was observed, indicating unfolding of the protein.

Hydrophobic interaction chromatography performed over a range of temperatures has been also used to separate native from unfolded protein. Wu *et al.* (1986) injected alpha-lactalbumin onto an HIC column and raised the temperature from 0.5 to 15°C (see Fig. 5). They observed that the unfolded form of alpha-lactalbumin had a stronger binding to the stationary phase.

Changes in the covalent disulfide bond structure of a protein are known to result in significant conformational differences. Representative of this relationship is the fact that many enzymes with disulfide bond structure are rendered inactive by reducing agents such as mercaptoethanol or dithiothreitol. As discussed earlier, variations in covalent protein structure result in significant changes in the physical dimensions, solvent-accessible net charge, and surface charge distribution of the biomolecule. Differences in these characteristics have facilitated the separation of dissimilar disulfide conformations by a variety of analytical techniques. One of the most popular has been native–PAGE, as has been mentioned previously. This type of gel electrophoresis uses no denaturing reagents and no reducing agents in the gel matrix or in the running buffers. If experiments have been conducted to generate various disulfide forms of the folded or incorrectly 'folded' protein and the fully reduced to fully oxidized protein, these forms can be visualized as different bands on the stained gel as reported by Creighton (1974) and Strege and Lagu (1993a). An activity assay again can be used to check if the protein has the correct disulfide arrangements. Finally Bean and Carr (1992) have reported the use of mass spectrometry to probe the positions of cysteine bridges in proteins. Mass spectrometry can give rapid analysis of the disulfide bond rearrangement during folding. This may be especially useful for industrial processes where a fast, definitive analysis is essential for the assurance of correct folding.

Chromatography can also be used to separate different folded forms of protein. Light and Higaki (1987) utilized SEC and Strege and Lagu (1993a) employed SEC and cation exchange chromatography to study the folding of trypsinogen over time. To observe samples from 10 min to 22 h of folding time, both SEC and cation exchange were used to separate the native from the unfolded forms of the protein (see Fig. 6). The distribution of sample components eluting between the native and the unfolded forms seen in Fig. 6 probably represent intermediate folding forms of the protein.

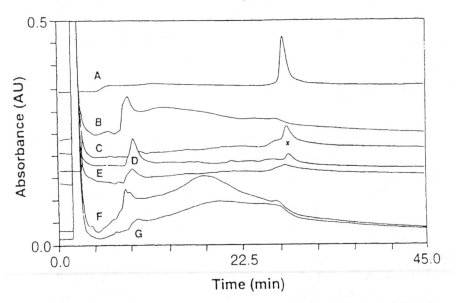

Figure 6. Cation-exchange HPLC profiles of 50 µg/ml samples (100-µl injections) of (A) native trypsinogen, (B) reduced trypsinogen, (C) 22-h, (D) 7-h, (E) 1-h, (F) 30-min and (G) 10-min refolded trypsinogen obtained via elution through a Progel-TSK SP-5PW packing (7.5 × 0.75 cm) in the presence of 25 mM sodium acetate pH 4.0, 2 M urea, 20% acetonitrile and a 45 min 0.00–0.35 M sodium chloride gradient at a flow-rate of 1.0 ml/min monitored at 280 nm. Refolded trypsinogen is marked by an 'x'. Reprinted from Strege and Lagu (1993a) with permission.

It has been well established that the biological activity of enzymes is highly dependent upon pH. Indirect analytical methods which provide an average system property, such as mass spectrometry and various spectroscopic techniques, have been utilized successfully for the determination of conformational changes undergone by proteins in response to changes in pH. For example, Choudhary *et al.* (1990) observed differences in mass spectra by changing the pH of a solution of a protein, thus resulting in change in the availability of ionizable basic sites. Separation techniques such as slab gel electrophoresis and liquid chromatography, however, have not traditionally been utilized for the investigation of pH-induced protein folding/unfolding because the separation mechanisms of these methods are based upon the ionic and/or hydrophobic characteristics of the analytes. In the case of molecules such as proteins which possess weakly ionizable moieties within their structure, these characteristics change in response to pH variation, even in the absence of changes in structural conformation, thereby limiting the sensitivity of these techniques for the detection of folding/unfolding.

Finally, methods have been developed to observe proteins which require a cofactor for activity. Activity assays have been commonly used to check for proper conformation in these cases. Electrospray mass spectrometry has also been used to detect conformational changes in proteins after the removal of a cofactor. Katta and Chait (1991) reported differences in mass spectra when the pH was lowered to cause the removal of the heme moiety from myoglobin. Strege *et al.* (1994) employed mass spectrometry for the analysis of serine hydroxymethyltransferase as a pyridoxal phosphate-associated holo enzyme and as an apoenzyme which did not have the cofactor.

APPLICATIONS OF CE TO PROTEIN FOLDING ANALYSIS

Capillary electrophoresis (CE) has recently emerged as a powerful analytical technique which may be successfully utilized for the analysis of proteins as well as other biomolecules. The potential for the analysis of biomolecules by CE has been reviewed by Novotny *et al.* (1990) and by Landers *et al.* (1993a). A significant asset of CE is that virtually all of the protein folding/unfolding investigations described in the previous two sections can be conducted in some format via this technique. Because the solution environment inside a silica capillary is intrinsically non-denaturing, CE analyses will not interfere with the structure or conformation of the protein during high voltage separations. The issue of protein structure denaturation is a factor which has limited the use of popular traditional protein analytical methods such as sodium dodecyl sulfate – polyacrylamide gel electrophoresis (SDS–PAGE) or reversed phase chromatography for the analysis of protein folding/unfolding. Accurate quantitation, which is not easily possible using slab gel electrophoresis, can be achieved through the use of CE. Most CE separations are relatively rapid, often requiring time intervals of only minutes compared to lengthy chromatography or slab gel analyses. CE electropherogram peak efficiencies are known to routinely attain levels at least one order of magnitude higher than that possible via HPLC. Although CE is considered a micro-analytical technique, sample fractions can be collected for further analyses. These and other attributes have made CE an attractive method to utilize for the analysis of protein folding/unfolding.

CE METHODS

Thermally induced protein folding/unfolding

The feasibility of employing capillary electrophoresis for the characterization of thermally induced protein folding/unfolding processes has been demonstrated by Hilser *et al.* (1993). In this investigation, chicken egg white lysozyme was analyzed in comparison to a heptapeptide mobility marker (assumed to display minimal folded structure) by capillary electrophoresis in 20 mM sodium citrate pH 2.07 buffer at 19 different temperatures between 37 and 67°C (see Fig. 7). In addition to the peak corresponding to native lysozyme, a second component, presumed to be a denatured form of protein, appeared at *ca.* 52°C and achieved a maximal area at temperatures of about 57–60°C. Plots of the electrophoretic mobilities of lysozyme and the peptide reference marker obtained over the range of temperatures were constructed and are displayed in Fig. 8. The mobility of the peptide standard was a linear function of temperature, while lysozyme mobility was a linear function of temperature both before and after a transition temperature region existing between 50–60°C. Within this transition temperature range, lysozyme mobility changed in a sigmoidal fashion, similar to the transition curves obtained for protein denaturation using other techniques.

The intrinsic lysozyme peak in the separations described above also appeared to broaden as the system temperature was increased, and then became narrow again at higher temperatures. This phenomenon was explained by considering that at low and high temperatures, the only significantly populated states were those of the completely folded and unfolded conformations, respectively. At intermediate temperatures, the two-state transition was near equilibrium and a distribution of conformations possessing different electrophoretic mobilities was present, resulting in the observed peak broadening.

Quantitative thermodynamic characterization of the lysozyme folding/unfolding transitions achieved via CE compared well with similar analyses obtained by direct calorimetric measurements. CE offered significant advantages for generating thermodynamic information of this nature, since it required only very small amounts of protein. The entire temperature folding/unfolding study of lysozyme was determined by CE with only 40 ng of protein, compared to the approximately 100 µg typically used in spectroscopic assays.

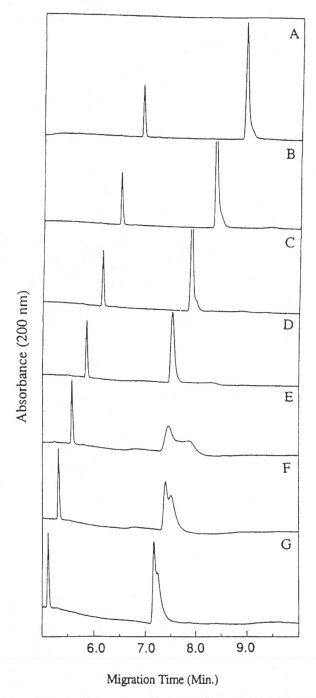

Figure 7. Capillary electropherograms of lysozyme at (A) 37°C, (B) 42°C, (C) 47°C, (D) 52°C, (E) 57°C, (F) 62°C, and (G) 67°C. The first peak corresponds to the peptide mobility standard (RKRSRKE) and the second peak is lysozyme. Reprinted from Hilser *et al.* (1993) with permission.

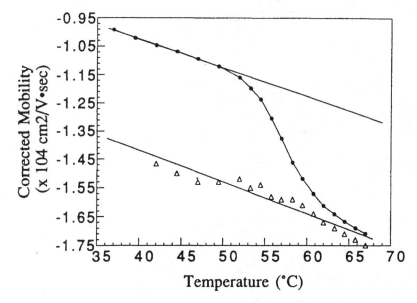

Figure 8. Corrected mobility values for lysozyme (closed circles) at various temperatures. Corrected mobility values for the unfolded state (open triangles) are also shown at different temperatures. The line drawn through these points results from regression analysis of the data. Reprinted from Hilser *et al.* (1993) with permission.

In a less extensive investigation, Rush *et al.* (1991) observed the thermodynamic folding/unfolding of alpha-lactalbumin using CE. In this study, the unfolded state displayed a greater negative charge than did the native conformation, and it was speculated that the enhanced electrophoretic mobility of the protein may have been caused by the exposure of acidic residues following denaturation. A plot of the viscosity corrected electrophoretic mobility of the protein *vs* temperature yielded a sigmoidal curve indicating a transition (between two conformations) temperature of 32–33 °C. These results correlated well with intrinsic fluorescence measurements obtained under the same buffer conditions.

Chaotrope-induced protein folding/unfolding

Chaotropic agents such as guanidine hydrochloride and urea disrupt the ionic or hydrophobic forces which hold a protein in its native conformation, and their use in protein folding/unfolding studies has been extensive. Although the use of guanidine hydrochloride in investigations utilizing electrophoretic methods has been limited by the high conductivity of concentrated solutions of this salt, urea gradient gel electrophoresis has been demonstrated to be convenient method for the visualization of the conformational transition states of proteins. Recently, the urea denaturation of proteins has been studied by CE as well.

Kilar and Hjerten (1993) used an alkaline Tris-borate buffer and capillaries coated to eliminate electroosmotic flow to obtain CE analyses of human serum transferrin isoforms in the presence of 0–8 M urea (see Fig. 9). Since the pro-

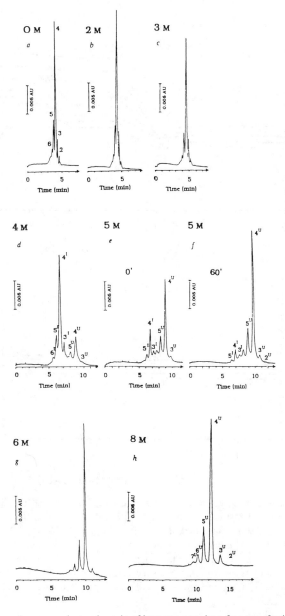

Figure 9. High-performance electrophoresis of human serum iron-free transferrin in free solution in the absence (a) and presence (b–h) of urea. Experimental conditions: electrophoresis buffer, 18 mM Tris – 18 mM boric acid – 0.3 mM EDTA, pH 8.4; tube dimensions, 0.1 (i.d.) × 0.3 (o.d.) × 200 mm; voltage, 8000 V; on-tube detection at 280 nM. The current decreased with the increase of the urea concentration from 10 μA to 4.5 μA. Transferrin isoforms (2-sialo, 3-sialo, 4-sialo, 5-sialo, and 6-sialo marked by 2, 3, 4, 5, 6, respectively) are better resolved at higher urea concentrations. The electrophoresis shows two conformations of the isoforms (I, intermediate; U, unfolded) in 4 M (d) and 5 M (e and f) urea. The ratio of the amounts of the conformational states changed significantly after 1 h in 5 M urea. Reprinted from Kilar and Hjerten (1993) with permission.

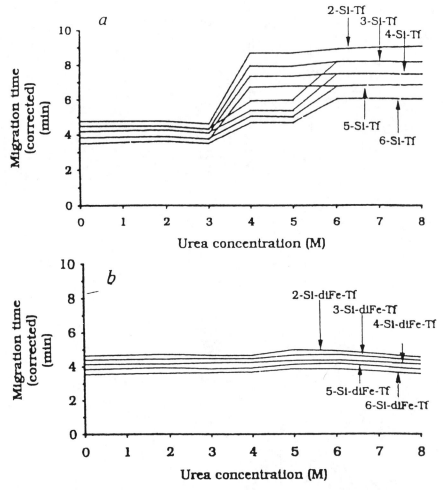

Figure 10. Denaturation curves of the human serum (a) iron-free and (b) diferric transferrin isoforms between 0 and 8 M urea. Corrected migration times for the 2-sialo, 3-sialo-, 4-sialo-, 5-sialo-, 6-sialo-transferrin forms (marked as 2-Si-Tf, 3-Si-Tf, 4-Si-Tf, 5-Si-Tf, 6-Si-Tf or 2-Si-diFe-Tf, 3-Si-diFe-Tf, 4-Si-diFe-Tf, 5-Si-diFe-Tf, 6-Si-diFe-Tf, respectively) *vs* urea concentration are plotted as shown. Intermediate transition states of the iron-free transferrin (a) isoforms exist between 3 and 6 M urea together with the unfolded states. The iron-saturated (diferric) isoforms of transferrin (b) are resistant to urea denaturation. Reprinted from Kilar and Hjerten (1993) with permission.

tein migration rates decreased in the viscous urea solutions, migration times were corrected by dividing them by the relative viscosity of the urea buffer solutions. The corrected transferrin migration times were plotted *vs* urea concentration to generate denaturation curves (see Fig. 10). An intermediate folded conformation of iron-free transferrin was observed in solutions of 2–6 M urea, while no unfolded or intermediate states of diferric transferrin were observed. It was noted that the rapidly interconverting intermediate folded state present

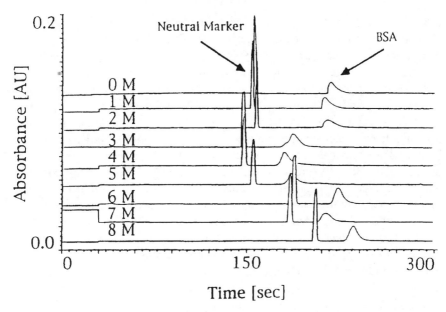

Figure 11. Capillary electrophoretic separations of bovine serum albumin and 0.5% mesityl oxide obtained in 100 mM tricine pH 7.5, in the presence of 0–8 M urea in a 57-cm, 50-μm i.d. bare silica capillary at 30-kV applied potential, monitored at 214 nm. Reprinted from Strege and Lagu (1994) with permission.

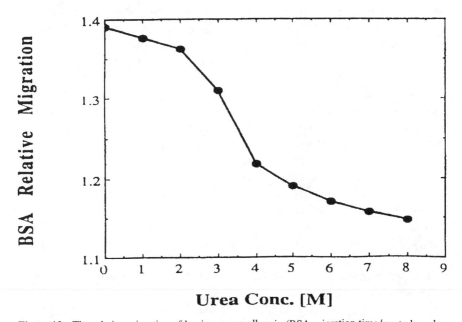

Figure 12. The relative migration of bovine serum albumin (BSA migration time/neutral marker migration time) plotted versus buffer urea concentration. Reprinted from Strege and Lagu (1994) with permission.

in solutions of 4–6 M urea was observed during the 15 min CE analysis but not on a slab gel which required a 15 h run time. Thus CE enabled the acquisition of information which was not available through the use of the more traditional technique of slab gel urea gradient electrophoresis.

Strege and Lagu (1994) demonstrated that protein urea unfolding curves could be generated through the use of tricine pH 7.5 buffers containing urea at concentrations ranging from 0–8 M (see Fig. 11). In the presence of electroosmotic flow in a bare silica capillary, the electrophoretic mobility of bovine serum albumin (BSA) was calculated relative to a mesityl oxide neutral marker. The unfolding of BSA by urea resulted in a decrease in the solvent accessible negative charge of the protein, causing a decrease in protein electrophoretic mobility. Two components (folded and unfolded BSA) could be observed in the separation in 3 M urea. A plot of BSA relative migration *vs* urea concentration demonstrated an unfolding transition curve similar to those previously reported for the protein but which were obtained through the use of more labor intensive traditional techniques (see Fig. 12).

Covalent disulfide bond folding/unfolding

Strege and Lagu (1993a) studied the disulfide bond refolding of reduced bovine trypsinogen (MW = 22 500, 14 cysteine residues) by CE, and compared these analyses to those obtained through the use of liquid chromatography, native slab gel electrophoresis, isoelectric focusing, and a bioactivity assay. Reduced trypsinogen was slowly refolded in a mild oxidizing environment, and aliquots of the folding solution, removed over time, were quenched to generate samples of intermediate folded disulfide bond conformations. By employing sodium phosphate buffers at pH 2.0 in bare silica capillaries, native, reduced, and intermediate refolded trypsinogen were resolved (see Fig. 13). Under these conditions in free solution, the reduced protein was observed to demonstrate decreased net solvent-accessible positive charge relative to the native conformation. The incorporation into the separation buffer of polyethylene glycol (PEG), a neutral hydrophilic polymer, was found to enhance resolution (see Fig. 14). Intermediate refolded species were isolated via size exclusion chromatography, and subsequent CE analyses of these samples demonstrated that the presence of PEG had enabled the CE buffer to act as a size-based sieving medium. Quantitative CE analyses of trypsinogen refolding demonstrated excellent correlation with determinations by biological activity and size exclusion and cation-exchange chromatography assays (see Fig. 15), and the CE separations also compared well qualitatively to the results of native gel electrophoresis and isoelectric focusing.

CE separations of a much simpler covalent protein folding system, the monomer and disulfide-linked homo-dimer of two different peptides (one with an amino-terminal cysteine residue, the other with a cysteine at the carboxy-terminal), were reported by Landers *et al.* (1993b). The time-course for homo-dimer formation with both peptides was easily monitored, and the co-oxidation

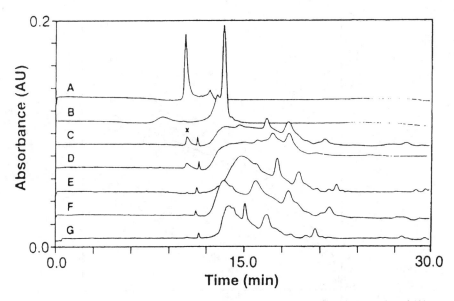

Figure 13. Capillary electrophoretic separations of 5-s injections of 2 mg/ml samples of (A) native trypsinogen, (B) reduced trypsinogen, and (C) 22-h, (D) 7-h, (E) 1-h, (F) 30-min and (G) 10-min refolded trypsinogen obtained using 25 mM sodium phosphate pH 2.0 in a 37 cm, 50 μm bare silica capillary, 10 kV potential, and 200 nm detection. Refolded trypsinogen is marked by an 'x'. Reprinted from Strege and Lagu (1993a) with permission.

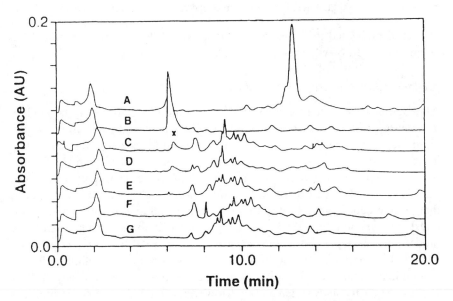

Figure 14. Capillary electrophoretic separations of 15-s injections of 2 mg/ml samples of (A) reduced trypsinogen, (B) native trypsinogen, and (C) 22-h, (D) 7-h, (E) 1-h, (F) 30-min and (G) 10-min refolded trypsinogen obtained using 25 mM sodium phosphate pH 2.0, 5% PEG in a 37 cm, 50 μm bare silica capillary, 20 kV potential, and 200 nm detection. Refolded trypsinogen is marked by an 'x'. Reprinted from Strege and Lagu (1993a) with permission.

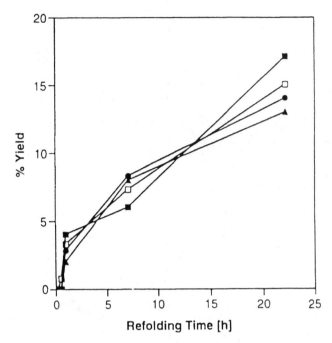

Figure 15. Rate of appearance of refolded trypsinogen from analyses via (□) enzymatic activity following activation with trypsin, (●) SEC, (▲) cation-exchange HPLC, (■) CE in 25 mM sodium phosphate pH 2.0, 5% PEG. Reprinted from Strege and Lagu (1993a) with permission.

of both peptides led to the expected production of both homo-dimers and the hetero-dimer, all of which were resolved.

pH-Induced protein conformational changes

Although slab gel electrophoresis and liquid chromatography may be applied to monitor gross protein conformational changes induced by changing the pH, there are limitations to the use of these separation techniques since significant changes in separation characteristics can occur independent of folding when observed over a range of pH. Such limitations also apply to the use of CE. For a pH study to be successfully monitored using CE, it is crucial that protein-capillary wall interactions be minimized over the pH range of interest. Strege and Lagu (1993b) have reported the analysis of a mixture of five model proteins separated within the pH range of pH 2 to pH 7. Since cetyl trimethyl ammonium chloride (CTAC) was present in the separation buffers as a solublizing and anti-adsorption agent, the environment within the capillary in these experiments was denaturing. However, even under these conditions, a plot of lysozyme migration relative to a neutral marker *vs* pH revealed marked differences in electrophoretic mobilities suggestive of the presence of two conformations of the protein, one existing at pH 2, 6, and 7 and the other within the range of pH 3–5 (see Fig. 16). These results agreed with those previously reported by Benedek *et al.* (1984) and by Nimura *et al.* (1991) who utilized

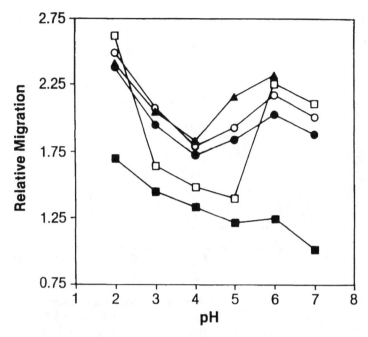

Figure 16. The dependence of relative protein migration (protein migration time/neutral marker migration time) upon buffer pH, for separations of 2-s injections of 0.5% mesityl oxide and 5-s injections of the protein mixture (1 mg/ml each protein, (■) ribonuclease; (□) lysozyme; (●) BSA; (○) β-lactoglobulin; (▲) myoglobin) obtained in 50 mM sodium phosphate or sodium acetate, 0.1% CTAC inside a 120-cm C18-derivatized capillary at 30 kV (outlet = anode). Reprinted from Strege and Lagu (1993b) with permission.

HPLC and with separations obtained by CE described by Strege and Lagu (1993c) which demonstrated that lysozyme can exist as two conformations in solution.

Cofactor-induced protein folding/unfolding

Many proteins require the association of a cofactor for the attainment of the conformation required for biological activity. The use of CE for the analysis of recombinant serine hydroxymethyltransferase (SHMT), an *E. coli*-derived enzyme which exists as a dimer with a pyridoxil phosphate (PLP) cofactor, has been reported by Strege *et al.* (1994). In this study, CE analyses of the enzyme were compared to those obtained via slab gel isoelectric focusing and mass spectrometry. Resolution of the holo- and apoenzyme forms of SHMT were achieved using a polyacrylamide-coated capillary and sodium borate pH 8.0 buffer (see Fig. 17). At least 50 mM sodium borate was required to achieve the separation of the two SHMT complexes, and it was proposed that a sufficient salt concentration may have stabilized a conformation of the SHMT apoenzyme which possessed a significantly reduced solvent-accessible negative charge relative to that of the holoenzyme conformation within the same environment.

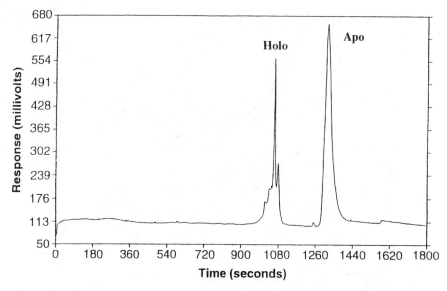

Figure 17. Electropherogram of a 5 s injection of a mixture of holoenzyme and apoenzyme (1.0 mg/ml concentration of each component) obtained in 50 mM sodium borate pH 8.0 in a 50 μ i.d. 67 cm polyacrylamide-derivatized capillary at 30 kV (outlet = anode) with 200 nm UV detection. Reprinted from Strege *et al.* (1994) with permission.

CONCLUSIONS

The reports of the use of CE for the analysis of protein folding/unfolding have demonstrated that the technique can provide advantages relative to the use of more traditional analytical techniques. Unlike indirect methods such as spectroscopy which provide an average system property, CE can provide an average property in addition to the distribution of values around the average. The ability of CE to provide population distributions may prove to be valuable for the analysis of partially folded protein intermediates for non-two-state transitions, as observed in the study of the disulfide refolding of trypsinogen reported by Strege and Lagu (1993a). Methods such as liquid chromatography and slab gel electrophoresis may also provide resolution of folding intermediates, but CE is unique in that it exists as a hybrid technique, offering the advantages of rapid, high resolution, non-denaturing, quantitative separations. Limitations of CE using UV absorbance detection include its relatively low mass sensitivity due to the short detector path length, an issue important for the analysis of disulfide-bond folding solutions where the protein concentration is often low (< 100 μg/ml) to prevent protein aggregation. However, improvements in sensitivity achieved through the use of detection methods such as laser-induced fluorescence may eliminate this limitation for future studies. Another disadvantage of CE, and electrophoretic techniques in general, is the intolerance of the method to samples of high conductivity. This characteristic has prevented the utilization of capillary electrophoresis for the analysis of proteins undergoing unfolding in the presence of high concentrations of guanidine hydrochloride, for example. However, it is expected that because the advantages of CE far outweigh its limitations, the use of the technique for the analysis of proteins varying in structural conformation will result in significantly better understanding of the complex processes of folding/unfolding.

REFERENCES

Al-Obeidi, A. M. and Light, A. (1988). Size exclusion high performance liquid chromatography of native trypsinogen, the denatured protein, and partially refolded molecules. *J. Biol. Chem.* **263**, 8642–8645.

Bean, M. F. and Carr, S. A. (1992). Characterization of disulfide bond position in proteins and sequence analysis of cystine-bridget peptides by tandem mass spectrometry. *Anal. Biochem.* **201**, 216–226.

Benedek, K., Dong, S. and Karger, B. L. (1984). Kinetics of unfolding of proteins on hydrophobic surfaces in reversed-phase liquid chromatography. *J. Chromatogr.* **317**, 227–243.

Choudhary, S. K., Katta, V. and Chait, B. T. (1990). Probing conformational changes in proteins by mass spectrometry. *J. Am. Chem. Soc.* **112**, 9012–9013.

Creighton, T. E. (1974). Intermediates in the refolding of reduced pancreatic trypsin inhibitor. *J. Mol. Biol.* **87**, 579–602.

Creighton, T. E. (1979). Electrophoretic analysis of the unfolding of proteins by urea. *J. Mol. Biol.* **129**, 235–264.

Creighton, T. E. (1984). *Proteins–Structure and Molecular Principles*. W. H. Freeman, New York, USA, pp. 149–152 and 291–292.

Hilser, V. J., Worosila, G. D. and Freire, E. (1993). Analysis of thermally induced protein folding/unfolding transitions using free solution capillary electrophoresis. *Anal. Biochem.* **208**, 125–131.

Katta, V. and Chait, B. T. (1991). Observation of the heme-globin complex in native myoglobin by electrospray-ionization mass spectrometry. *J. Am. Chem. Soc.* **113**, 8534–8535.

Katz, S. and Denis, J. (1969). Mechanism of denaturation of bovine serum albumin by urea and urea-type reagents. *Biochim. Biophys. Acta* **188**, 247–254.

Kilar, F. and Hjerten, S. (1993). Unfolding of human serum transferrin in urea studied by high-performance capillary electrophoresis. *J. Chromatogr.* **638**, 269–276.

Landers, J. P., Oda, R. P., Spelsberg, T. C., Nolan, J. A. and Ulfelder, K. J. (1993a). Capillary electrophoresis: A powerful microanalytical technique for biologically active molecules. *Biotechniques* **14**(1), 98–111.

Landers, J. P., Oda, R. P., Liebnow, J. A. and Spelsberg, T. C. (1993b). Utility of high resolution capillary electrophoresis for monitoring peptide homo- and hetero-dimer formation. *J. Chromatogr.* **A652**, 109–117.

Light, A. and Higaki, J. N. (1987). Detection of intermediate species in the refolding of bovine trypsinogen. *Biochemistry* **26**, 5556–5564.

Mendoza, J., Rogers, E., Lorimer, G. H. and Horowitz, P. M. (1991). Unassisted refolding of urea unfolded rhodanese. *J. Biol. Chem.* **266**, 13587–13591.

Nimura, N., Itoh, H., Kinoshita, T., Nagae, N. and Nomura, M. (1991). Fast protein separation by reversed phase high performance liquid chromatography on octadecylsilyl-bonded non-porous silica gel – Effect of particle size of column packing on column efficiency. *J. Chromatogr.* **585**, 207–211.

Novotny, M. V., Cobb, K. A. and Liu, J. (1990). Recent advances in capillary electrophoresis of proteins, peptides and amino acids. *Electrophoresis* **11**, 735–749.

Parente, E. S. and Wetlaufer, D. B. (1984a). Influence of urea on the high-performance cation-exchange chromatography of hen egg white lysozyme. *J. Chromatogr.* **288**, 389–398.

Parente, E. S. and Wetlaufer, D. B. (1984b). Effects of urea-thermal denaturation on the high-performance cation-exchange chromatography of α-chymotrypsinogen-A. *J. Chromatogr.* **314**, 337–347.

Rush, R. S., Cohen, A. S. and Karger, B. L. (1991). Influence of column temperature on the electrophoretic behavior of myoglobin and α-lactalbumin in high-performance capillary electrophoresis. *Anal. Chem.* **63**, 1346–1350.

Strege, M. A. and Lagu, A. L. (1993a). Capillary electrophoresis as a tool for the analysis of protein folding. *J. Chromatogr.* **A652**, 179–188.

Strege, M. A. and Lagu, A. L. (1993b). Micellar electrokinetic capillary chromatography of proteins. *Anal. Biochem.* **210**, 402–410.

Strege, M. A. and Lagu, A. L. (1993c). Capillary electrophoretic protein separations in polyacrylamide-coated silica capillaries and buffers containing ionic surfactants. *J. Chromatogr.* **630**, 337–344.

Strege, M. A. and Lagu, A. L. (1994). Capillary electrophoretic analysis of the urea unfolding of bovine serum albumin. *Amer. Lab.* Feb., 48C–49C.

Strege, M. A., Schmidt, D. F., Kruezman, A., Yeh, W. K., Kaiser, R. E. and Lagu, A. L. (1994). Capillary electrophoretic analysis of serine hydroxymethyl transferase in *E. coli* fermentation broth. *Anal. Biochem.* **223**, 198–204.

Wu, S., Figueroa, A. and Karger, B. L. (1986). Protein conformational effects in hydrophobic interaction chromatography – Retention characterization and the role of mobile phase additives and stationary phase hydrophobicity. *J. Chromatogr.* **371**, 3–27.

Progress in HPLC-HPCE, Vol. 5, pp. 97–113
H. Parvez *et al.* (Eds)
© VSP 1997.

HPCE monitoring of phosphorylated histone H1 variants

HERBERT LINDNER* and WILFRIED HELLIGER

Institute of Medical Chemistry and Biochemistry, Fritz-Preglstrasse 3, A-6020 Innsbruck, Austria

INTRODUCTION

HPCE continues to become more widely used as a technique for the separation of peptides and proteins. The objectives of the work described in this paper were to investigate the effects of different buffer concentrations and compositions, respectively, on the elution order and separation of H1 histone subtypes and their phosphorylated modifications isolated from two different mammalian species. Excellent resolution was achieved by using: 1) an untreated capillary, 2) a low pH (pH = 2.0), 3) a high-ionic strength phosphate buffer (500 mM) or a perchloric acid/sodium perchlorate system and 4) the dynamic coating agent hydroxypropylmethyl cellulose (0.02%–0.03%) to prevent inconvenient wall-interactions with the highly basic histone proteins. The HPCE technique described looks attractive because it is a simple system which enables high-efficiency separations with theoretical plate counts up to 1.7 million/m.

Histones are basic nuclear proteins and are primarily responsible for organizing the nucleosomal structure of eukaryotic chromatin (Igo-Kemenes *et al.*, 1982). The nucleosomal core region contains the histone octamer (H2A,H2B, H3,H4)$_2$, while the nucleosomal linker region is associated with histone H1 (Felsenfeld, 1978). These H1 histones can be resolved into a number of closely related subtypes which vary in amino acid sequence and molecular size (Rall and Cole, 1971; Lindner *et al.*, 1990). A synopsis of the H1 histone family is shown below.

---

*To whom correspondence should be addressed.

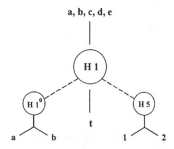

Histone H1 subtypes undergo covalent modification reactions (i.e. phosphorylation and ADP-ribosylation). These modifications alter the local charge and
structure of histones substantially. The functional significance of these multiple
H1s and their modifications, however, are not yet fully understood (Von Holt *et
al.*, 1979; Cole, 1984; Lennox, 1984). Therefore, a precise analysis of these
very lysine-rich proteins is an essential prerequisite for studying the biological
role of the histone family.

The above mentioned complexity of H1 histones and their phosphorylated
forms makes high demands on the analytical methods applied. In the past,
the most widely utilized methods for the study of histones and histone modifications have been polyacrylamide gel electrophoreses (Panyim and Chalkley, 1969; Laemmli, 1970; Zweidler, 1978; Lennox *et al.*, 1982). All these
methods, however, are labour intensive and time consuming and, moreover,
unsatisfactory for precise quantification.

High performance capillary electrophoresis (HPCE), combining the advantages of conventional gel electrophoresis and high performance liquid chromatography, may offer substantial improvements over other analytical techniques.

In HPCE, compounds are separated according to their ability to migrate in an
electric field inside a fused-silica capillary (Tsuda *et al.*, 1982, 1983; Jorgenson
and Lukacs, 1983; Gassmann *et al.*, 1985; Karger *et al.*, 1989; Novotny *et al.*,
1990; Chen, 1991; Deyl and Struzinski, 1991; Landers, 1993). Highly efficient
and rapid separations can be achieved for sample quantities in the nanolitre
range. Furthermore, the technique is well suited for accurate quantification
and automation.

However, the separation of protein mixtures by this method can be disturbed
by wall adsorption of the protein components. This troublesome effect is primarily caused by the attraction between positively charged proteins and the
negative silanol groups of the capillary surface, resulting in peak broadening
and tailing. A variety of techniques have been developed to reduce the adsorption effects, among them being the use of buffer pH values higher than
the pI values of the proteins (Lauer and McManigill, 1986), the use of low pH
buffer systems (McCormick, 1988; Grossmann *et al.*, 1989), the application of
'dynamic coating' agents (Hjerten *et al.*, 1989; Gordon *et al.*, 1991), the use of

high-ionic strength buffer systems (Rasmussen and McNair, 1990; Chen *et al.*, 1992), and the chemical modification of the silanol groups (Hjerten, 1985; Swedberg, 1990; Schomburg, 1991; Hjerten and Kubo, 1993).

To separate histone variants and their modified forms Lindner *et al.* (Lindner *et al.*, 1992a, b, 1993) have developed strategies which minimize coulombic interactions with the capillary surface, using low pH phosphate buffers in the presence of the dynamic coating agent hydroxypropylmethyl cellulose (HPMC).

This chapter summarizes the effects of buffer concentration and buffer composition on analysis of phosphorylated histone H1 variants by HPCE using the dynamic coating agent HPMC.

EXPERIMENTAL

Chemicals

Hydroxypropylmethyl cellulose (HPMC; 4000 cP), and trifluoroacetic acid (TFA) were obtained from Sigma (Munich, Germany), ethylene glycol mono-methyl ether (EGME) from Aldrich (Steinheim, Germany), Triton X-100, Tris and phenylmethanesulfonyl fluoride (PMSF) from Serva (Heidelberg, Germany), and colcemid from Boehringer (Mannheim, Germany). All other chemicals were purchased from Merck (Darmstadt, Germany).

Sample preparations

To isolate rat testis H1 histones, the tissue was homogenized in 2 vol. of buffer A (50 mM Tris, pH 7.5, 25 mM KCl, 10 mM $CaCl_2$, 10 mM $MgCl_2$, 250 mM sucrose, 10 mM 2-mercaptoethanol, 1 mM phenylmethanesulfonyl fluoride) in a potter homogenizer. This and all following procedures were carried out at $0–4°C$. After centrifugation at 800 *g* for 5 min, the cells were washed twice with buffer A and then homogenized in buffer A containing 0.1% Triton X-100 in a Dounce homogenizer. Nuclei thus obtained were collected by centrifugation at 900 *g* for 10 min. The nuclear pellet was washed once with buffer A without Triton X-100, centrifuged down and thereupon extracted with 1 vol. of 10% $HClO_4$ and 2 vol. of 5% $HClO_4$ for 1 h with occasional vortex-mixing. $HClO_4$-insoluble material was removed by centrifugation at 12 000 *g* for 10 min, and the soluble H1 histones were precipitated for one hour by the addition of trichloroacetic acid (20% final concentration). The H1 histones were centrifuged down at 20 000 *g* for 15 min, washed twice with pure acetone, dissolved in water containing 10 mM 2-mercaptoethanol and lyophilized.

Mouse NIH 3T3 and rat fibroblast cell histones were isolated as described by Talasz *et al.* (1993). Mouse erythroleukemic cell histones (line B8) were isolated according to Helliger *et al.* (1992).

Cell lines and culture conditions

Mouse and rat fibroblasts were grown in monolayer cultures and cultivated in DMEM (Biochrom; Berlin, Germany) supplemented with 10% fetal calf serum (FCS), penicillin (60 µg/ml) and streptomycin (100 µg/ml) in the presence of 5% CO_2. To obtain mitotic enriched cells, colcemid (0.06 µg/ml) was added to the cultures for 8 h, thereafter the cells were harvested.

Mouse erythroleukemic cells (MEL cells), line B8, were a generous gift from Dr. T. F. Sarre (Institut für molekulare Zellbiologie, Freiburg, Germany). Cells were grown in liquid suspension culture in Dulbecco's MEM containing 2 × non-essential amino acids, 1 × penicillin/streptomycin and 10% fetal calf serum at 37°C and 5% CO_2. Cells were diluted every two days to 10^5 cells per ml in fresh medium, and harvested in the log-phase.

Incubation of histones with alkaline phosphatase

Approximately 100 µg of histones, in 0.25 ml of 10 mM Tris-HCl (pH 8.0), and 1 mM phenylmethanesulphonyl fluoride were mixed with 210 µg of E. coli alkaline phosphatase (60 units/mg, Sigma) for 12 h at 37°C according to Sherod et al. (1970). The digested samples were separated by reverse-phase chromatography and the corresponding histone fractions subjected to capillary electrophoresis.

High performance liquid chromatography

Chromatography was performed with a Beckman HPLC gradient system (two Model 114M pumps, a Model 421A system controller and a Model 165 variable-wavelength UV-VIS detector). The effluent was monitored at 210 nm and the peaks were recorded using System Gold software (Beckman Instruments, Palo Alto, CA, USA). The H1 histones from rat testis were separated using a Nucleosil 300-5 C_4 column, 4.6 mm × 100 mm (Macherey Nagel, Düren, Germany). The lyophilized proteins were dissolved in water containing 0.1% TFA, and samples of 20 µg were injected onto the column. At a constant flow-rate of 0.7 ml/min the H1 histones were eluted within 35 min using a two-step gradient starting at 40% B (solvent A is water containing 10% EGME and 0.1% TFA; solvent B is 10% EGME/70% acetonitrile (90%) and 0.1% TFA). The concentration of solvent B was increased to 54% (22 min) and finally to 68% (13 min).

The mouse H1 histone separations were performed on a Nucleosil 300-5 C_4 (8 mm × 125 mm) column and samples of 100 µg histones were injected. The flow-rate was 1 ml/min. A linear acetonitrile gradient was used with an increase in solvent B (composition of solvents A and B the same as described above) from 43% to 63% over 45 min.

High performance capillary electrophoresis

HPCE was performed on a Beckman system P/ACE 2100 controlled by an AT486 computer. Data collection and post-run data analysis were carried out using P/ACE and System Gold software (Beckman Instruments). The capillaries varied in length from 50 cm to 60 cm (75 μm i.d.). Protein samples were injected by pressure and on-column detection was performed by measuring UV absorption at 200 nm. Sample amounts and voltage applied varied and are given in the corresponding figure legends. An untreated capillary was used in all experiments, but after every five to ten injections the capillary was rinsed with water, 0.1 M NaOH, water, 0.5 M H_2SO_4, water and finally with the running buffer. Washing was done for 2 min with each solvent. Phosphate buffers of different concentrations with the appropriate amounts of HPMC were prepared with phosphoric acid adjusting pH with sodium hydroxide to pH 2.0. To analyze MEL cell histones a 10 mM perchloric acid/90 mM sodium perchlorate system (pH 2.0) containing 0.02% HPMC was used.

SEPARATION PROCEDURES

Separation of phosphorylated rat testis H1 variants

The use of a 100 mM *phosphate buffer.* The aim of the first studies has been to determine whether HPCE can be used for the simultaneous separation of both rat testis histone H1 subtypes and their different phosphorylated forms. To reduce inconvenient adsorption effects between positively charged histones having a pI of 10–11 and the capillary wall, a low-pH buffer system was tested. Using an untreated capillary and a 100 mM phosphate buffer (pH = 2.0) we attempted to separate H1 histones isolated from rat testis nuclei. In Fig. 1 the electropherogram obtained with this system is depicted. As can be seen,

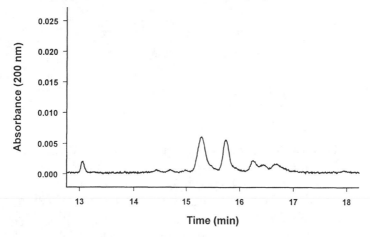

Figure 1. Electropherogram of rat testis H1 histones in 100 mM phosphate buffer (pH = 2.0) without HPMC. Other conditions: sample concentration, 0.5 mg/ml; injection time, 2 s; voltage applied, 12 kV; temperature, 25 °C; detection at 200 nm; untreated capillary (50 cm × 75 μm i.d.).

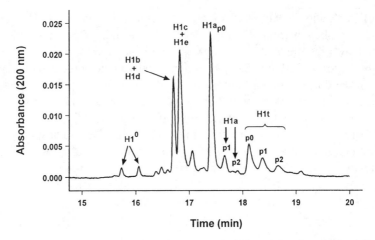

Figure 2. Electropherogram of rat testis H1 histones in 100 mM phosphate buffer (pH = 2.0) containing 0.03% HPMC. Other conditions were the same as in Fig. 1. The designations p0, p1, and p2 indicate the non-, mono- and diphosphorylated forms of the corresponding H1 histone.

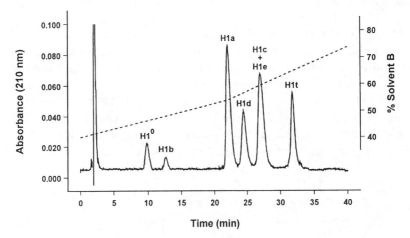

Figure 3. Reverse-phase HPLC chromatogram of H1 histones obtained from rat testis nuclei. Column: Nucleosil 300-5 C$_4$ (4.6 mm × 100 mm). Amount injected, 20 μg; flow-rate, 0.7 ml/min; detection at 210 nm. A multi-step acetonitrile gradient system was used (for details, see EXPERIMENTAL).

only two major and several minor peaks could be achieved. It should be taken into account, however, that this H1 sample consisted of eight sequence variants and, in addition, some phosphorylated forms. We assume therefore, that even at that low pH value, essential electrostatic histone protein-wall interactions occur, which are responsible for the modest resolution. To enhance resolution, the influence of an addition of various dynamic coating agents such as ethylene glycol, polyethylene glycol, and HPMC, respectively, to the separation buffer was studied. Figure 2 reveals the separation of H1 histones with analysis conditions as described in Fig. 1 except that 0.03% HPMC was added

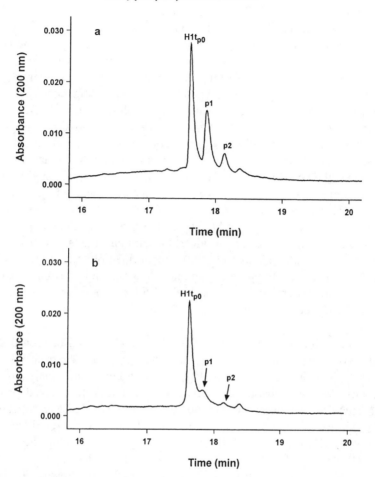

Figure 4. Identification of phosphorylated histone H1t. a) Electropherogram of the H1t subfraction obtained by reverse-phase HPLC (Fig. 3). b) Electropherogram of the H1t subfraction obtained by reverse-phase HPLC after treatment with alkaline phosphatase. Conditions for a) and b) as described in Fig. 2.

to the phosphate buffer. As a result a series of clearly resolved and sharp peaks could be obtained within 20 min. To identify the individual peaks of the electropherogram of Fig. 2 pure histone H1 fractions were prepared by separating rat testis H1 histones by reverse-phase HPLC (as shown in Fig. 3), which has been demonstrated to be an excellent technique for the separation of unmodified histones (Lindner *et al.*, 1986a, b, 1988, 1990; Helliger *et al.*, 1988; Lindner and Helliger, 1990). Subsequently the assignment was performed in two ways: on the one hand the histone fractions were analyzed individually by HPCE comparing their mobilities (demonstrated for H1t in Fig. 4a). On the other the same histone fractions were added to the whole rat testis H1 histone sample and separating these mixtures, a characteristic change of one or more peaks in the electropherogram was found (data not shown). It should be noted, however, that histones H1a and H1t, represented by a single peak in the chromatogram (Fig. 3) were further resolved into three peaks by HPCE (shown in Fig. 2 and Fig. 4a, respectively). We assumed, therefore, that the multiple peaks were due to a mixture of nonphosphorylated and different phosphorylated forms. To confirm this hypothesis, the H1t subfraction obtained by reverse-phase HPLC was digested with alkaline phosphatase. The treatment with the enzyme resulted in a dramatic decrease of the two slower migrating peaks (shown in Fig. 4b). Therefore, it seems reasonable to suspect that the prominent peak (designated as $H1t_{p0}$) consists of nonphosphorylated histone H1t, whereas the peaks $H1t_{p1}$ and $H1t_{p2}$ are different phosphorylated forms of $H1t_{p0}$. This outcome is not unexpected, when one considers that the phosphorylation of H1 histones results in a decrease of the overall positive charge of the molecule caused by the covalent binding of the negatively charged phosphate groups to serine and/or threonine residues, thus diminishing the electrophoretic mobility. These identifications were further confirmed by traditional acid urea polyacrylamide gel electrophoresis (data not shown). Similar extensive procedures of digestion were performed with all HPLC fractions and led finally to the assignments shown in Fig. 2.

The use of a 500 mM *phosphate buffer.* From Fig. 2 it can be seen that the separation of histone H1b from H1d and that of histone H1c from H1e is a problematic part of total rat H1 separation. In earlier studies analyzing core histones we observed remarkable differences in separation efficiency, depending on phosphate buffer concentration (unpublished results). We were interested, therefore, in examining the influence of different buffer concentrations in the separation of H1 histones as well. In fact, changing the buffer concentration from 100 mM to 200 mM and 350 mM, respectively, had a marked effect on elution order of certain histone H1 subtypes (data not shown). The most

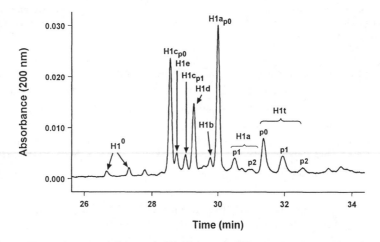

Figure 5. Electropherogram of rat testis H1 histones in 500 mM phosphate buffer (pH = 2.0) containing 0.03% HPMC. Other conditions as in Fig. 2.

successful results were obtained using a 500 mM phosphate buffer. Under these conditions all eight H1 histones and their phosphorylated forms (as far as they are present in detectable amounts) were clearly separated. For example, using the high-salt buffer an unambiguous assignment of both nonphosphorylated H1c/H1e and monophosphorylated H1c was possible. Figure 5 shows the electropherogram obtained using the optimum conditions. We should underline, however, that an increase of buffer concentration to 500 mM does not in itself account completely for the gain in resolution. Moreover, separation efficiency is being enhanced in this high degree only in the presence of 0.03% HPMC.

Separation of hyperphosphorylated histone H1 variants isolated from mitotic enriched rat fibroblasts

The use of a 100 mM phosphate buffer. The rat testis H1 proteins used for HPCE separations in the preceding section were histones revealing a low level of phosphorylation. To examine the analytical potential of the HPCE method developed we subsequently attempted to separate hyperphosphorylated histone H1 variants into their different modified forms. As it is known that degree of phosphorylation is dependent on cell cycle and that interphase and mitotic cells contain highly phosphorylated H1 histones, we isolated histones from mitotic enriched rat fibroblasts. Due to the complexity of the hyperphosphorylated H1 histone sample, individual H1 subfractions were pre-separated using the chromatographic procedure as described for the separation of rat testis H1 histones in Fig. 3 (data not shown). Figure 6 shows a CE separation of such an

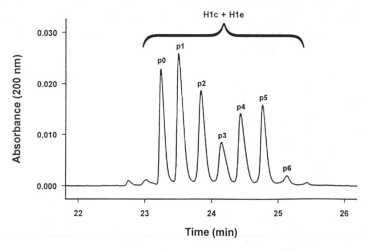

Figure 6. Electropherogram of multi-phosphorylated histone subtypes H1c/H1e in 100 mM phosphate buffer (pH = 2.0) containing 0.02% HPMC. Histone subfraction H1c/H1e obtained from mitotic enriched rat fibroblasts was isolated under the same chromatographic conditions as described for Fig. 3 (data not shown). Other conditions: sample concentration, 0.6 mg/ml; injection time, 4 s; voltage applied, 12 kV; temperature, 25 °C; detection at 200 nm; untreated capillary (60 cm × 75 μm i.d.).

HPLC fraction containing a complex mixture of H1c plus H1e as well as their phosphorylated forms. We have found that a 100 mM sodium phosphate buffer (pH = 2.0) with 0.02% HPMC using an untreated capillary (60 cm × 75 μm i.d.) permits a resolution into seven peaks. It should be noted, however, that in RP–HPLC the same histone fraction appears as a single peak.

To further characterize the seven peaks obtained, the protein sample was treated with alkaline phosphatase. In the subsequent CE analysis we have found only one prominent peak revealing a migration time similar to the first peak in Fig. 6. As a result of this experiment, and in accordance with the analyses data obtained from rat testis H1 histones in a 100 mM phosphate buffer, we must conclude that the first peak in Fig. 6 consists of a mixture of nonphosphorylated H1c plus H1e. The other slower migrating peaks, however, designated as p1 to p6 represent different phosphorylated forms of the parent proteins H1c/H1e. As clearly can be seen, the 100 mM phosphate buffer system is not suitable to resolve the histone subtypes H1c and H1e from each other; however, the modified forms having up to six phosphate groups were fully separated.

Applying this CE method the other HPLC subfractions containing the subtypes $H1^0$, H1b, H1a and H1d can easily be resolved into their non- and different phosphorylated forms as well (data not shown).

The use of a 500 mM *phosphate buffer.* The next approach was to take advantage of the resolving power of the high-ionic strength buffer for the separation of the histone variants H1c and H1e as well. In fact, using the 500 mM phosphate buffer containing 0.02% HPMC we succeeded in separating not only nonphosphorylated histone subtypes H1c and H1e (designated as $H1c_{p0}$ and $H1e_{p0}$) but also the mono-($H1c_{p1}$ and $H1e_{p1}$) and diphosphorylated ones

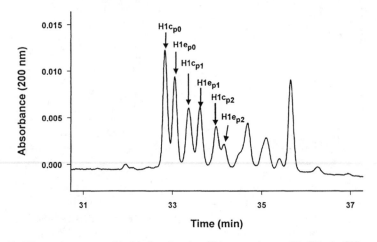

Figure 7. Electropherogram of multi-phosphorylated histone subtypes H1c/H1e in 500 mM phosphate buffer (pH = 2.0) containing 0.02% HPMC. Other conditions and isolation of the histone subfraction H1c/H1e as in Fig. 6.

($H1c_{p2}$ and $H1e_{p2}$) from each other as demonstrated in Fig. 7. An unambiguous assignment of the residual peaks, however, was not possible, because the extent of modification especially of the higher phosphorylated forms of the two proteins H1c and H1e is still unknown.

In contrast to the separation of whole rat testis H1 histones (Fig. 5) the electrophoreses of the hyperphosphorylated histone fractions H1c/H1e (Figs 6 and 7) were performed using a 60 cm × 75 μm capillary. Increasing the length of capillary from 50 cm to 60 cm resulted in a significant enhancement of resolution, whereas the migration times were increased only negligibly. It should be noted that the H1c/H1e fractions used for the studies shown in Fig. 6 and Fig. 7 originate from two different biological experiments. That accounts for the different ratio mainly of the nonphosphorylated and the monophosphorylated forms of the proteins.

Separation of hyperphosphorylated histone H1 variants isolated from mitotic enriched mouse NIH 3T3 fibroblasts

The use of a 100 mM phosphate buffer. In view of the fact that the histone H1 family exhibits a remarkable species specificity (Lennox, 1984) we were interested in investigating the electrophoretic behaviour of highly phosphorylated *mouse* H1 subtypes isolated from mitotic enriched NIH 3T3 fibroblasts. In this context it should be pointed out that differences in the amino acid sequence of certain rat and mouse histone H1 variants are responsible for a changed elution order in RP–HPLC as well. For example, in the HPLC run of rat testis H1 histones (shown in Fig. 3) H1e coelutes with H1c, whereas in the case of mouse H1 histones H1e and H1d remain unresolved, as demonstrated in Fig. 8.

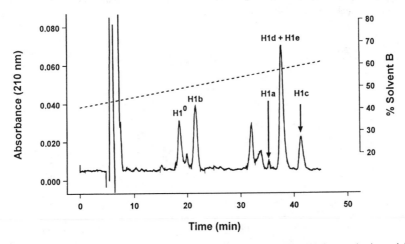

Figure 8. Reverse-phase HPLC chromatogram of H1 histones obtained from mitotic enriched mouse NIH 3T3 fibroblasts. Column: Nucleosil 300-5 C_4 (8 mm × 125 mm). Amount injected, 100 μg; flow-rate, 1 ml/min; detection at 210 nm. A linear acetonitrile gradient was used with an increase in solvent B (10% EGME/70% acetonitrile (90%)/0.1% TFA) from 43% to 63% over 45 min.

108 *H. Lindner and W. Helliger*

Figures 9 and 10 display CE runs of the histone fractions H1b and H1c, respectively, obtained by RP–HPLC (see Fig. 8). These separations were performed in 50 cm × 75 μm untreated capillaries with a 100 mM phosphate buffer (pH = 2.0) containing 0.03% HPMC within about 15 min. In Fig. 9 six nearly baseline separated peaks are discernible. The fastest migrating component represents the nonphosphorylated histone H1b (designated as p_0), followed by five different phosphorylated forms p_1 to p_5. In the case of histone H1c a nonphos-

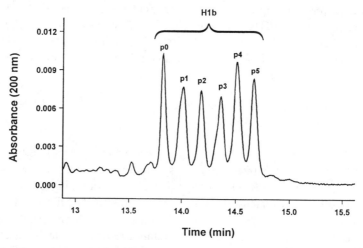

Figure 9. Electropherogram of multi-phosphorylated histone H1b subtype in 100 mM phosphate buffer (pH = 2.0) containing 0.03% HPMC. Histone H1b obtained from mitotic enriched mouse NIH 3T3 fibroblasts was isolated by RP–HPLC (see Fig. 8). Other conditions: sample concentration, 0.5 mg/ml; injection time, 5 s; voltage applied, 16 kV; temperature, 20°C; detection at 200 nm; untreated capillary (50 cm × 75 μm i.d.).

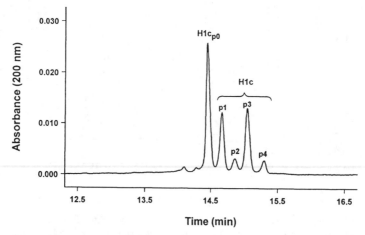

Figure 10. Electropherogram of multi-phosphorylated histone H1c subtype in 100 mM phosphate buffer (pH = 2.0) containing 0.03% HPMC. Other conditions and isolation of histone H1b as in Fig. 9.

phorylated (p_0) and four different phosphorylated H1 histones (p_1–p_4) could be separated, as illustrated in Fig. 10. In order to prove the presence of phosphorylated H1 proteins, the histones were digested with alkaline phosphatase as already outlined in Section EXPERIMENTAL. In addition, we isolated both histone subfractions H1b and H1c from quiescent NIH 3T3 fibroblasts by RP–HPLC, because it is known that slowly or nongrowing cells contain H1 histones primarily in their nonphosphorylated forms. Applying the same HPCE separation described above a single peak each was obtained (data not shown).

Finally, it should be noted that in contrast to the longer migration times of phosphorylated rat H1 subtypes (shown in Figs 6 and 7) the HPCE separation of phosphorylated mouse histones H1b and H1c took less than 16 min (Figs 9 and 10, respectively). This result is due to both the use of a shorter capillary (50 cm × 75 μm) and a higher running voltage (16 kV).

Separation of phosphorylated mouse erythroleukemic cell histone H1 variants

The use of phosphate buffers. As shown in Fig. 8, the most troublesome part of mouse histone H1 analysis by RP–HPLC is the separation of H1d from H1e. To resolve these two variants at present it is necessary to subject the isolated HPLC fraction to acetic acid urea gel electrophoresis. Applying this laborious and time consuming procedure it was possible to separate unphosphorylated histones H1d and H1e. This method, however, is unable to resolve phosphorylated histone samples, because the phosphorylated forms of histone H1e overlap with both unphosphorylated and phosphorylated forms of H1d thus preventing a calculation of the ratio of the two subtypes.

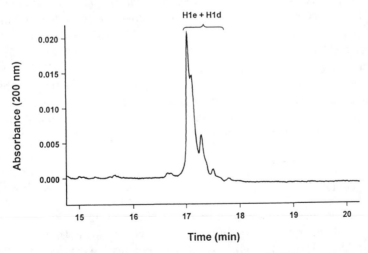

Figure 11. Electropherogram of phosphorylated histone subtypes H1d/H1e in 100 mM phosphate buffer (pH = 2.0) containing 0.02% HPMC. Histone H1d/H1e subfraction obtained from mouse erythroleukemic cells was isolated under the same chromatographic conditions as described for Fig. 8 (data not shown). Voltage applied, 12 kV; temperature, 25 °C; other conditions as in Fig. 9.

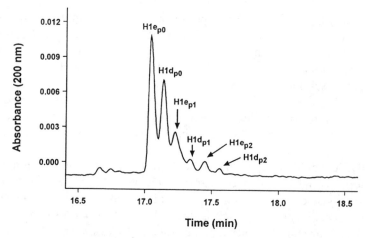

Figure 12. Electropherogram of phosphorylated histone subtypes H1d/H1e in 10 mM perchloric acid/90 mM sodium perchlorate (pH = 2.0) containing 0.02% HPMC. Other conditions and isolation of the histone H1d/H1e subfraction as in Fig. 11.

Consequently, we examined the potential of the CE technique described above for the separation of the HPLC subfraction consisting of phosphorylated and nonphosphorylated histone subtypes H1d and H1e derived from log-phase erythroleukemic cells. Unfortunately, no satisfying separation of the H1d/H1e fraction was achieved using the 100 mM phosphate buffer system (Fig. 11). Even the application of high-ionic strength phosphate buffers up to 700 mM did not substantially increase resolution (data not shown).

The use of a 10 mM *perchloric acid/90* mM *sodium perchlorate system.* In an attempt to overcome the problem of H1d/H1e separation we tested the influence of various running buffer systems. As shown in Fig. 12, a remarkable enhancement of resolution could be achieved using a 10 mM perchloric acid/90 mM sodium perchlorate system (pH = 2.0) in the presence of HPMC. This considerable effect of perchlorate on selectivity led to good separation not only of nonphosphorylated, but also of mono- and diphosphorylated histones H1e from H1d. In this context it should be stressed that up to now no other analytical method has achieved such a fine resolution.

CONCLUSION

In conclusion, the results demonstrate that HPCE using the dynamic coating agent HPMC in combination with certain buffer systems is an attractive technique for the analysis of highly basic histone proteins and their posttranslational modifications. The simplicity of the methods using untreated fused-silica capillaries, the quick separation, the very small sample volumes, the plate counts up to 1.7 million/m and the highly reproducible results are the most important attributes of this HPCE technique. Hence, the described new CE methods appear to be an excellent alternative to the traditional gel electrophoresis of H1 histones and their modified forms.

ACKNOWLEDGEMENTS

The authors gratefully acknowledge the help of Astrid Devich, Mag. Christoph Meraner and Dr Heribert Talasz.

REFERENCES

Chen, F. T. A. (1991). Rapid protein analysis by capillary electrophoresis. *J. Chromatogr.* **559**, 445–453.

Chen, F. A., Kelly, L., Palmieri, R., Biehler, R. and Schwartz, H. (1992). Use of high ionic strength buffers for the separation of proteins and peptides with capillary electrophoresis. *J. Liq. Chromatogr.* **15**, 1143–1161.

Cole, R. D. (1984). A minireview of microheterogeneity in H1 histone and its possible significance. *Anal. Biochem.* **136**, 24–30.

Deyl, Z. and Struzinski, R. (1991). Capillary zone electrophoresis: its applicability and potential in biochemical analysis. *J. Chromatogr.* **569**, 63.

Felsenfeld, G. (1978). Chromatin. *Nature (London)* **271**, 115–122.

Gassmann, E., Kus, J. E. and Zare, R. N. (1985). Electrokinetic separation of chiral compounds. *Science* **230**, 813–814.

Gordon, M. J., Lee, K.-J., Arias, A. A. and Zare, R. N. (1991). Protocol for resolving protein mixtures in capillary zone electrophoresis. *Anal. Chem.* **63**, 69–72.

Grossmann, P. D., Colburn, J. C., Lauer, H. H., Nielsen, R. G., Riggin, R. M., Sittampalam, G. S. and Rickard, E. C. (1989). Application of free-solution capillary electrophoresis to the analytical scale separation of proteins and peptides. *Anal. Chem.* **61**, 1186–1194.

Helliger, W., Lindner, H., Hauptlorenz, S. and Puschendorf, B. (1988). A new h.p.l.c. isolation procedure for chicken and goose erythrocyte histones. *Biochem. J.* **255**, 23–27.

Helliger, W., Lindner, H., Grübl-Knosp, O. and Puschendorf, B. (1992). Alteration in proportions of histone H1 variants during the differentiation of murine erythroleukaemic cells. *Biochem. J.* **288**, 747–751.

Hjerten, S. (1985). High-performance electrophoresis. Elimination of electroendosmosis and solute adsorption. *J. Chromatogr.* **347**, 191–198.

Hjerten, S. and Kubo, K. (1993). A new type of pH- and detergent-stable coating for elimination of electroendosmosis and adsorption in (capillary) electrophoresis. *Electrophoresis* **14**, 390–395.

Hjerten, S., Valtcheva, L., Elenbring, K. and Eaker, D. (1989). High-performance electrophoresis of acidic and basic low-molecular-weight compounds and of proteins in the presence of polymers and neutral surfactants. *J. Liq. Chromatogr.* **12**, 2471–2499.

Igo-Kemenes, T., Hörz, W. and Zachau, H. G. (1982). Chromatin. *Ann. Rev. Biochem.* **51**, 89–121.

Jorgenson, J. W. and Lukacs, K. D. (1983). Capillary zone electrophoresis. *Science* **222**, 266–272.

Karger, B. L., Cohen, A. S. and Guttman, A. (1989). High performance capillary electrophoresis in the biological sciences. *J. Chromatogr.* **492**, 585.

Laemmli, V. K. (1970). Cleavage of structural proteins during the assembly of the head of bacteriophage T4. *Nature (London)* **227**, 680–685.

Landers, J. P. (1993). Capillary electrophoresis pioneering new approaches for biomolecular analysis. *Trends Biochem. Sci.* **18**, 409–414.

Lauer, H. H. and McManigill, D. (1986). Capillary zone electrophoresis of proteins in untreated fused silica tubing. *Anal. Chem.* **58**, 166–170.

Lennox, R. W. (1984). Differences in evolutionary stability among mammalian H1 subtypes. *J. Biol. Chem.* **259**, 669–672.

Lennox, R. W., Oshima, R. G. and Cohen, L. H. (1982). The H1 histones and their interface phosphorylated state in differentiated and undifferentiated cell lines derived from murine teratocarcinomas. *J. Biol. Chem.* **257**, 5183–5189.

Lindner, H. and Helliger, W. (1990). Effects of eluent composition, ion-pair reagent and temperature on the separation of histones by high performance liquid chromatography. *Chromatographia* **30**, 518–522.

Lindner, H., Helliger, W. and Puschendorf, B. (1986a). Rapid separation of histones by high-performance liquid chromatography on C4 reversed-phase columns. *J. Chromatogr.* **357**, 301–310.

Lindner, H., Helliger, W. and Puschendorf, B. (1986b). Histone separation by high-performance liquid chromatography on C4 reverse-phase columns. *Anal. Biochem.* **158**, 424–430.

Lindner, H., Helliger, W. and Puschendorf, B. (1988). Separation of Friend erythroleukaemic cell histones and high-mobility-group proteins by reversed-phase high-performance liquid chromatography. *J. Chromatogr.* **450**, 309–316.

Lindner, H., Helliger, W. and Puschendorf, B. (1990). Separation of rat tissue histone H1 subtypes by reverse-phase h.p.l.c. Identification and assignment to a standard H1 nomenclature. *Biochem. J.* **269**, 359–363.

Lindner, H., Helliger, W., Dirschlmayer, A., Jaquemar, M. and Puschendorf, B. (1992a). High-performance capillary electrophoresis of core histones and their acetylated modified derivatives. *Biochem. J.* **283**, 467–471.

Lindner, H., Helliger, W., Dirschlmayer, A., Talasz, H., Wurm, M., Sarg, B., Jaquemar, M. and Puschendorf, B. (1992b). Separation of phosphorylated histone H1 variants by high-performance capillary electrophoresis. *J. Chromatogr.* **608**, 211–216.

Lindner, H., Wurm, M., Dirschlmayer, A., Sarg, B. and Helliger, W. (1993). Application of high-performance capillary electrophoresis to the analysis of H1 histones. *Electrophoresis* **14**, 480–485.

McCormick, R. M. (1988). Capillary zone electrophoretic separation of peptides and proteins using low pH buffers in modified silica capillaries. *Anal. Chem.* **60**, 2322–2328.

Novotny, M. V., Cobb, K. A. and Lui, J. (1990). Recent advances in capillary electrophoresis of proteins, peptides and amino acids. *Electrophoresis* **11**, 735–749.

Panyim, S. and Chalkley, R. (1969). High resolution acrylamide gel electrophoresis of histones. *Arch. Biochem. Biophys.* **130**, 337–346.

Rall, S. C. and Cole, R. D. (1971). Amino acid sequence and sequence variability of the amino terminal regions of lysine-rich histones. *J. Biol. Chem.* **246**, 7175–7190.

Rasmussen, H. T. and McNair, H. M. (1990). Influence of buffer concentration, capillary internal diameter and forced convection on resolution in capillary zone electrophoresis. *J. Chromatogr.* **516**, 223–231.

Schomburg, G. (1991). Polymer coating of surfaces in column liquid-chromatography and capillary electrophoresis. *Trends Anal. Chem.* **10**, 163–169.

Sherod, D., Johnson, G. and Chalkley, R. (1970). Phosphorylation of mouse ascites tumor cell lysine-rich histone. *Biochemistry* **9**, 4611–4615.

Swedberg, S. A. (1990). Characterization of protein behavior in high-performance capillary electrophoresis using a novel capillary system. *Anal. Biochem.* **185**, 51–56.

Talasz, H., Helliger, W., Puschendorf, B. and Lindner, H. (1993). G1- and S-Phase synthesis of histone H1 subtypes from mouse NIH fibroblasts and rat C6 glioma cells. *Biochemistry* **32**, 1188–1193.

Tsuda, T., Nomura, K. and Nakagawa, G. (1982). Open-tubular microcapillary liquid chromatography with electroosmosis flow using a UV detector. *J. Chromatogr.* **248**, 241–247.

Tsuda, T., Nomura, K. and Nakagawa, G. (1983). Separation of organic and metal ions by high voltage capillary electrophoresis. *J. Chromatogr.* **246**, 385–392.

Von Holt, C., Strickland, W. N., Brandt, W. F. and Strickland, M. S. (1979). More histone structures. *FEBS Lett.* **100**, 201–218.

Zweidler, A. (1978). Resolution of histones by polyacrylamide gel electrophoresis in presence of nonionic detergents. In: *Methods in Cell Biology*, Stein, G., Stein, J. and Kleinsmith, L. J. (Eds). Academic Press, New York, Vol. 17, pp. 223–233.

Progress in HPLC-HPCE, Vol. 5, pp. 115–154
H. Parvez *et al.* (Eds)
© VSP 1997.

Application of HPCE-analysis to process monitoring in biotechnology

RUTH FREITAG

Eidgenössische Technische Hochschule, EPFL-Ecublens, IGC IV CH-1015 Lausanne, Switzerland

INTRODUCTION

A vast number and variety of analytical procedures is involved in the production of the so-called 'high value' products of modern biotechnology. There is, for example, process monitoring with regard to process control in fermentation proper as well as for the down-stream process. Besides the more conventional challenge of supervising the pH and the oxygen content, the necessity of quantifying certain nutrients and metabolites, including the product itself, must be met. Especially when mammalian cells and their inherantly complex environments and low product concentrations are concerned, the latter may be obstructed by the presence of similar or even closely related substances such as malexpressed or degenerated product molecules and other proteins and peptides. Once successfully isolated, the purity of the product needs to be ascertained as well as its molecular structure characterized and/or verified. Process validation as well as pharmacological studies may also have to be considered.

The more complex of these analytical problems have been met for the last three decades by ever improving liquid chromatography (HPLC). The progress in stationary phase development, instrumental design, detection methodology, data handling and general theoretical understanding together with the arrival of the various hyphenated techniques have provided analytical tools yet unsurpassed in their versatility and adaptability. Fully automated systems exist, which may be interfaced directly via a sampling module to the bioprocess. In spite of the recent progress in, for example, perfusion and membrane chromatography, the fast (on line) analysis of biomacromolecules by HPLC is still severely handicapped by band broadening due to mass transport resistances caused by the low diffusivity of such molecules. The elaborate sample preparation necessary to protect the expensive HPLC-columns, the difficulties encountered in column regeneration and the usually prerequisite gradient elution render such techniques rather unwieldy for fast process monitoring of biopolymers.

Another common means of biopolymer separation is electrophoresis performed in a rod or slab gels, which serve primarily to stabilized the flow pattern. Sometimes the separation is aided by the provision of a sieving effect by the gel. In electrophoresis the molecules are separated due to differences in their mobility in an electrical field. Low diffusivity constitutes a bonus. Several samples are run simultaneously, usually in the presence of a standard. Gel electrophoresis may be run in two dimensions (2D electrophoresis). In the first dimension the sample components are often separated according to their isoelectric points, followed by a separation of each band thus obtained according to size. The resolution obtained is still unsurpassed. However, other than HPLC, conventional slab gel electrophoresis is difficult to apply for routine analysis in biotechnology, whenever a fast, robust, fully automated technique, highly reproducible and reliable at line quantification is desired. In slab gel electrophoresis sample components may be quantified only after separation and by an elaborate staining procedure. The required density scanning is time consuming and comparatively unreliable. The method is inheritantly unsuited to the consecutive processing of samples. Also, automation is difficult. The advantage of the 2D gels cannot be exploited in the case of the analysis of a complex, constantly changing fermenter supernatant, especially when a technical or serum supplemented — i.e. not fully characterized — culture medium is used.

During the 1980s some of the above mentioned analytical reservations were solved by the arrival of biosensors. However, long term stability, cost, and process interfacing constituted problems in this area. So when it was realized, largely after the landmark paper by Jorgenson and Lukacs in 1981, how the performance of electrophoresis in capillaries with inner diameters of less than 200 μm would yield a technique which combines the separation principle of electrophoresis with the instrumental advantages until then only known from HPLC, research was immediately started on the development of CE methods for the separation and analysis of biomacromolecules (nucleic acids, proteins/peptides, polysaccharides). In rapid succession, capillary zone electrophoresis (CZE), capillary isoelectric focussing (CIEF), capillary gel electrophoresis (CGE) and micellar electrokinetic chromatography joined the already existing capillary isotachophoresis (CITP). The suitability for automation and on capillary detection, the unsurpassed separation efficiency (in terms of plate number per unit column), the small sample volumes, the simple sample preparation as well as the high versatility and the speed were among the more attractive features of the various new techniques. Questions of reducing the CE dimension to chip size and of the development of auxiliary techniques to interface the CE to some other unit were soon addressed (see, e.g. Li, 1992).

While the various challenges of CE continue to attract an ever increasing number of analytical scientists, there continues to be a comparative lack of application of CE to the analysis of real life samples in the life and biosciences including biotechnology, where such applications are limited at present to less than a dozen papers in the pertinent journals (e.g. Frenz *et al.*, 1989; Banke *et al.*, 1991; Hurni and Miller, 1991; Guzmann *et al.*, 1991, 1992; Yim, 1991; Vinther *et al.*, 1992; Tsuij, 1993). The application of the CE-principle to enzyme assays, immunoanalysis or the determination of binding constants is just getting started (e.g. Chen *et al.*, 1988; Nielsen *et al.*, 1991; Bao and Regnier, 1992; Arentoft *et al.*, 1993; Foret *et al.*, 1993; Schultz and Kennedy, 1993; Regnier *et al.*, 1994).

For some time now our group has been developing CE methods for biotechnological process analysis. Below, methods are discussed for product monitoring in mammalian cell cultures and the ensuing isolation procedure, for the quantification of amino acids in fermenter supernatants of *Cephalosporium acremonium*, for the supervision of the plasmid stability of recombinant *E. coli*, as well as for the characterization of proteineous products in terms of molecular mass, isoelectric point, biological activity and peptide map. Conditions for quantifying an antigen (IgG) in a complex matrix (human serum) by FACE (Fluorescence Affinity Capillary Electrophoresis) are also included.

PROCESS ANALYSIS

PRODUCT MONITORING

Product concentration and quality are among the most important parameters in a bioproduction process. The final product concentration in the fermenter supernatant, for example, has been shown to determine the cost for the down-stream process and thus the total production cost to a large extent (Dwyer, 1984). The monitoring of the production rate is mandatory during process development, but might later still be an important parameter for process validation or process control. The problem is especially pressing for proteineous products, which are not easily isolated and analyzed and in the case of mammalian cell cultures, where complex culture media prevail.

The standard method for product quantification in mammalian cell cultures is the ELISA, a heterogeneous sandwich immuno assay with enzyme reaction enhanced photometric detection. Usually culture samples are collected during the fermentation, stored below $0°C$ and analyzed batchwise afterwards. Thus, no process control and only limited process monitoring is possible. An alternative method is to use immunosensors or immuno-FIAs, both of which have been successfully used in the past for product monitoring (Karube, 1988; Bradley *et al.*, 1991; Mattiasson and Hakanson, 1992; Middendorf *et al.*, 1993). However, all immuno assays are limited to the sole detection of the product and may give a biased signal due to the presence of other molecules, such as product fragments still presenting the correct antigenic determinant to the

detection antibody, which resemble the product to some degree. In the case
of a biologically active product, i.e. an enzyme, an activity assay may also be
used for product quantification.

Antithrombin III (AT III) is a factor of the blood clotting cascade which
counteracts thrombin and various other clotting factors (Damus *et al.*, 1973;
Conard *et al.*, 1983; Lane and Caso, 1989). The interaction is enhanced by the
presence of heparin; hence the label heparin cofactor (Rosenberg and Damus,
1973; Blackburn *et al.*, 1984; Hoylaerts *et al.*, 1984). The substance is of
pharmacological interest and given in the case of a natural or acquired chronic
AT III deficiency as well as during surgery to prevent the untimely formation
of blood clots. Human AT III is currently isolated from fractions collected
early on in the Cohn (Cohn *et al.*, 1946) or Kistler and Nitchmann (1962)
plasma fractionation scheme (Burnouf, 1991). The danger inherent to blood
derived substances makes AT III production by recombinant (mammalian) cells
a desirable biotechnological option. A number of AT III producing cells have
been described in the literature (Stephens *et al.*, 1987; Wasley *et al.*, 1987;
Wirth *et al.*, 1987; Zettelmeissl *et al.*, 1987, 1988); the investigation of the
process parameters on product quantity and quality is a rather active area of
mammalian cell culture technology.

For two of these cell lines atline product monitoring by CE was attempted.
One cell line was an adherent BHK (Baby Hamster Kidney) line grown in
microcarrier culture, which required some 10% FCS (fetal calf serum) to be
added to their media during growth and at least 5% during production, else the
product be of inferior quality. Bovine serum albumin (BSA, Mw 69 kg/mol,
IEP 4.9), bovine transferrin (Mw 75 kg/mol, IEP 5.8) and bovine immunoglob-
uline G (b-IgG, Mw 150 kg/mol, IEP *ca.* 7) thus constitute up to 70% of the
proteins extant in the culture supernatant. Other proteins, peptides, amino acids,
carbohydrates, salts, etc. are also present. In all cultures considered, the r-AT
III constitutes less than 1% of the protein content, i.e. 0.5 µg/ml to 15 µg/ml.
The other cell line was a CHO (Chinese Hamster Ovary) cell line, which could
be grown in suspension using a standardized serum free culture medium. This
culture medium also contained BSA (*ca.* 50 µg/ml), transferrin (*ca.* 50 mg/ml)
and several unspecified peptides, amino acids and sugars, albeit at a significantly lower concentration. A considerably higher product concentration of up
to 80 µg/ml r-AT III was usually found in the CHO cell culture supernatants.

Experimental

Bioprocess. The BHK 21-c13 cell line was grown in repeated batch and
continuous perfusion mode in a 2 l Biostat MC bioreactor (B. Braun Melsungen,
Germany) suitable for microcarrier culture. The medium exchange rate was
determined by the lactate concentration in the medium, as determined by a
selfassembled online FIA analyzer (Ludwig *et al.*, 1992).

The CHO cells were grown in a 2 l batch culture at 37 °C. In this case 50 µl samples were removed, clarified by centrifugation (2 min, 10 000 rpm) and injected directly into the CE.

Downstream process. 250 ml of BHK cell culture supernatant were clarified by centrifugation and ultrafiltration (membrane cut off: 1000 Da), dialyzed against the sample buffer for 10 min (membrane cut off: 12 000 Da) and further purified using ion exchanger and heparin affinity membrane adsorber (MA) as described previously (Reif and Freitag, 1994a). After each step, 500 µl of the AT III containing fraction were removed and analyzed by CZE using conditions identical to those given below for process monitoring, save for the application of a linear voltage ramp (5 to 15 kV within 10 min).

Sample pre-concentration. If necessary, the sample was preconcentrated by IEXMA (ion exchanger membrane adsorber chromatography). Buffer A was a 0.05 M borate buffer, pH 9.5 and buffer B a 0.05 M borate buffer, pH 9.5 with 1 M NaCl added. The MA (stack of 10, each with an effective filtration area of 3.4 cm^2) were integrated into a conventional Pharmacia FPLC. 1 ml of centrifuged culture supernatant were applied and a gradient run from 0% to 50% buffer B. 100 µl fractions were collected and the AT III containing fraction analyzed by CE.

Capillary electrophoresis. The instrument used was a Beckmann P/ACE 2210. The electrophoresis buffer was a 0.05 M borate buffer, pH 9.5, which contained 0.1% modified cellulose (HPMC). Capillary dimensions were 50 µm i.d. and 37 cm length (30 cm from inlet to detector). Sample injection was by pressure (50 mbar for 2 s; approximately 5 nl). The detection wave length was 200 nm, the applied voltage 15 kV for the CHO and 10 kV for the BHK cell culture supernatant and the temperature 22 °C.

Results

Capillary zone electrophoresis in untreated fused silica capillaries is the most simple and most common variety of CE analysis. The sample components are separated in free solution due to differences in their migration velocities, which in turn depend on the sign of their net-charge, their mass-to-charge ratio and the strength of the electroendosmotic flow (EOF), a plug-like flow of the aqueous electrophoresis buffer to the cathode, prevalent in such systems.

Monitoring of CHO cell cultures. Preliminary experiments on the separation of AT III, BSA and transferrin had yielded the above CZE-conditions (Reif, 1994). The HPMC is added to reduce unspecific protein adsorption to the capillary walls as well as to increase the buffer's viscosity and thus to improve the separation. By using a detection wavelength of 200 nm, a minimum of

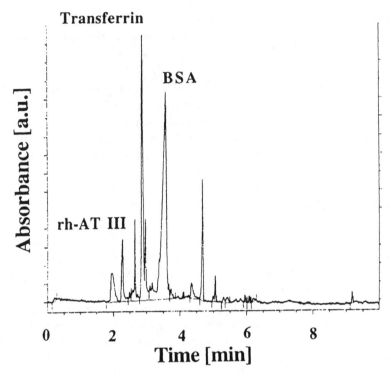

Figure 1. CZE-analysis of a CHO cell culture sample containing 16 µg/ml r-AT III (Reif, 1994).

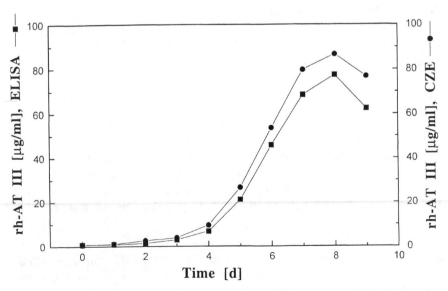

Figure 2. r-AT III concentrations determined by CZE and ELISA during a CHO-cell batch fermentation (Reif, 1994).

5 μg/ml AT III could be detected. This enabled the direct quantification of AT III in the CHO cell culture supernatants (Fig. 1).

The AT III concentration is established after 2.26 min. The concentration can be reliably determined by either the peak height or the peak area. A relative standard deviation of 1.33% was established for the retention time by repeated analysis of a sample containing approximately 30 μg/ml r-AT III. The deviation of the peak height and area from the average was less than 1%. BSA (*ca.* 3.7 min) and transferrin (*ca.* 2.9 min) are quantified concomitantly to the AT III. Several other signals are well resolved within the electropherogram; however, we were not yet able to ascertain the identity of these substances.

Good correlation was found between the results obtained by CZE and those obtained by the standard AT III-ELISA (Fig. 2). Some discrepancies are observed in the low concentration region, where the CZE is operated close to the detection limit. However, that region is passed after a couple of hours in an average fermentation and should not constitute a general obstacle to the application of CZE to product monitoring. On the other hand, the necessary sample dilution of 1:10 000 renders the ELISA impractical for the major part of the CHO cell fermentations.

Monitoring of BHK cell cultures. In comparison to the CHO-cell cultures, product monitoring is more of a challenge in the case of AT III producing BHK cells (Freitag *et al.*, 1996). The complex culture medium prevents, for example, the use of the sensitive but rather unspecific detection wavelength of 200 nm. The r-AT III concentration would altogether be below even the detection limit determined for 200 nm during the larger part of any fermentation. A simple one-dimensional CZE analysis does not suffice here. It was attempted to adapt CITP and CIEF to the problem, both of which allow the injection of larger sample volumes and thus higher absolute amounts of sample (Reif, 1994; Reif and Freitag, 1994b). The analyte zones are concomitantly concentrated. However, neither method yielded satisfactory results. In CITP the product gave three distinct signals, of which at least two appeared to be contaminated by other proteins; in CIEF no reliable identification of the product peaks was possible.

Better results were obtained in a two-dimensional system, when the BHK cell culture samples were preconcentrated by a cheap and fast IEXMA step. Membrane adsorbers (MA) constitute a new type of stationary phase design. Thin filter membranes are functionalized by covalent linkage of the desired interactive groups, i.e. strong anion exchanger groups (quarternary ammonium) in this case, and used as analog to an LC column. The low back pressure of these systems together with their extremely favorable mass transfer properties allow flow rates of up to 50 ml/min cm^2 to be used (Reif and Freitag, 1993). By setting the mobile phase pH to 9.5 and using a flow rate of 1 ml/min the AT III concentration of a given sample was increased ten-fold within minutes. When the AT III containing fraction of the eluate was analyzed by CZE, the results shown in Fig. 3 were obtained. A relative standard deviation of 0.91%

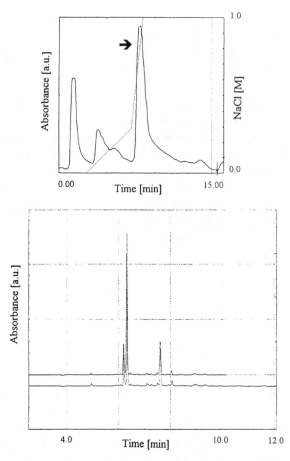

Figure 3. Top: Chromatogram of the cell culture supernatant on the IEXMA. The rh-AT III containing fraction is indicated by arrow. Bottom: Analysis of the rh-AT III containing fraction by CZE (Freitag *et al.*, 1996).

was calculated from the repeated analysis of a sample containing approximately 6 µg/ml r-AT III.

When the results obtained for the CZE were correlated to the AT III-ELISA, (Fig. 4), a correlation coefficient of 0.965 was calculated. In Fig. 5 examples are given for the CZE-monitoring of a repeated batch and a continuous cultivation of AT III producing BHK cells. Table 1 compiles a comparison of the suitability of the developed CZE for product monitoring in biotechnology with that of the respective ELISA. In CZE the results are available within minutes. The assay-to-assay reproducibility is high, even when sample preconcentration is required. The CZE has excellent long term stability, is well suited to automation and may be directly interfaced to the bioprocess for true online monitoring. The somewhat higher detection limit should constitute no problem in actual product monitoring, where high product concentrations are usually aimed for. The good correlation to the ELISA and the low assay costs should also be noted.

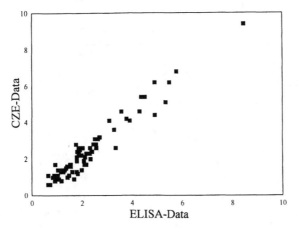

Figure 4. Correlation plot of the AT III data collected by CZE and ELISA during two repeated batch and one continuous culture of BHK cells (Freitag *et al.*, 1996).

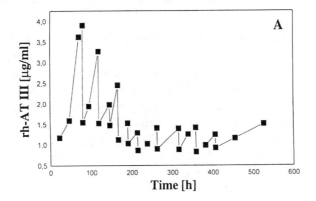

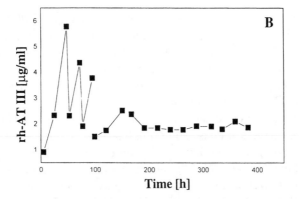

Figure 5. Determination of the AT III concentration in a repeated batch (A) and a continuous perfusion (B) culture of BHK cells (Freitag *et al.*, 1996).

Table 1.

A comparison of CZE and ELISA for product monitoring

	CZE	ELISA
Analyte concentration	> 0.5 µg/ml with IEXMA > 5 µg/ml without IEXMA	10–100 ng/ml
Time	3 min/assay	4 h/plate (4–15 samples)
Cost	low	high
Relative standard deviation	0.91%	on plate: 2–3% between plates: 10–20%
Long term stability	excellent	one-time-use only
Online/atline suited	yes	no
Correlation to ELISA	0.965	

Downstream process monitoring. The same CZE method, save for the application of a voltage ramp rather than a constant voltage, was used to monitor product concentration and product purity during the down-stream-processing of r-AT III produced in BHK cell cultures (Reif and Freitag, 1994b). The voltage ramping was necessary to improve the separation. The isolation procedure incorporated preparative IEXMA and heparin affinity chromatography (Reif and Freitag, 1994a). Figure 6 compiles the electropherograms taken after each step. Whereas no AT III peak can be found in the raw supernatant, the AT III (*ca.* 4.4 min) is already enriched to approximately 0.3% by the IEXMA step. After the heparin affinity step an electrophoretically pure AT III is obtained.

AMINO ACID MONITORING

Antibiotics are among the most important products in biotechnology. At present well over 8000 different microbial antibiotics have been described; however, only a few, including the β-lactam derivatives penicillin and cephalosporin, have ever been produced at large scale. Antibiotics are a product of secondary microbial metabolism. In Fig. 7 the steps involved in the biogenesis of cephalosporin C (CPC) are compiled.

The product, the intermediates and the amino acids are routinely monitored by RP–HPLC (Reversed Phase HPLC) in our institute during fermentations of CPC-producing *Cephalosporium acremonium* (Holzhauer-Rieger, 1990). While the CPC and the later intermediates are accessible by the online HPLC, the di- and tripeptide as well as the amino acids require elaborate sample preparation including a derivatization step with OPA (*o*-phthalaldehyde) before they can be quantified in a 60 min gradient RP–HPLC with fluorescence detection.

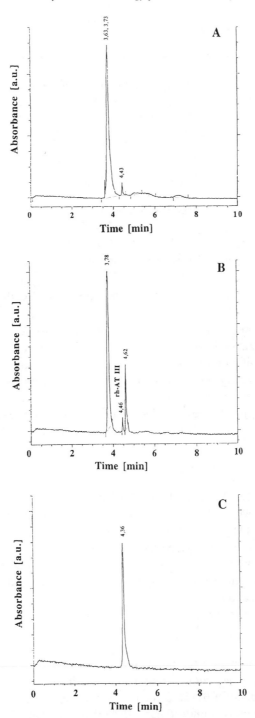

Figure 6. Monitoring of the AT III purification by CZE. A: culture supernatant, B: after the IEXMA step, C: after the heparin affinity step (Reif and Freitag, 1994b).

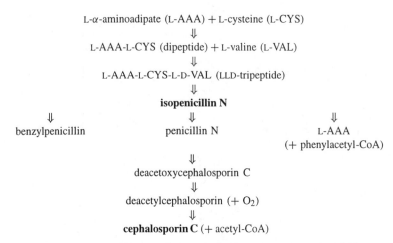

L-α-aminoadipate (L-AAA) + L-cysteine (L-CYS)
⇓
L-AAA-L-CYS (dipeptide) + L-valine (L-VAL)
⇓
L-AAA-L-CYS-L-D-VAL (LLD-tripeptide)
⇓
isopenicillin N

⇓   ⇓   ⇓
benzylpenicillin   penicillin N   L-AAA
(+ phenylacetyl-CoA)

⇓
deacetoxycephalosporin C
⇓
deacetylcephalosporin (+ O₂)
⇓
cephalosporin C (+ acetyl-CoA)

Figure 7. Biogenesis of the antibiotic cephalosporin C (According to Demain, 1983).

Detection limits of approximately 100 nM are found for the various amino acids.

Amino acids and small peptides can, however, also be quantified by CE. Without derivatization lower detection limits of as little as 200 fmol have been reported, depending on the structure of the molecule (Bergman *et al.*, 1991). Exceptionally low detection limits have been observed for the aromatic amino acids. Even lower detection limits have been described for amino acids after derivatization with a fluorescent dye (Fuchigami *et al.*, 1993). However, the application of such methods to fermentation monitoring is still missing.

Experimental

The Beckmann P/ACE 2210 CE was used throughout the experiments. Capillary dimensions were 50 μm i.d. and 27 cm length, injection was by pressure (3 s, *ca.* 10 nl), the detection wave length was 214 nm, the temperature was 25°C and the applied voltage 20 kV. A 50 mM phosphate buffer, pH 2.5 was used as electrophoresis buffer. Samples were diluted 1:1 with cold methanol to enable protein removal by precipitation/centrifugation and injected directly into the CE afterwards.

Results

CZE was used to monitor the production of the antibiotic cephalosporin C by *Cephalosporium acremonium*. Preliminary experiments had yielded the detection limits compiled in Table 2 for the various amino acids under the electrophoresis conditions employed here (Brüggemann, 1994). Valin and threonin were not detectable by CZE, whereas the cystein and the tryptophan, which in turn were not accessible in the culture samples by the RP–HPLC could be quan-

Table 2.
Detection limits of amino acids in CZE (Brüggemann, 1994)

Amino acid	AAA	ARG	ASN	ASP	CYS	GLN	HIS	ILE
Detection limit, pmol	8	10	5	50	3	20	0.1	15

Amino acid	LEU	LYS	MET	PHE	SER	THR	TRP	CPC
Detection limit, pmol	10	100	1	0.1	20	30	0.05	5

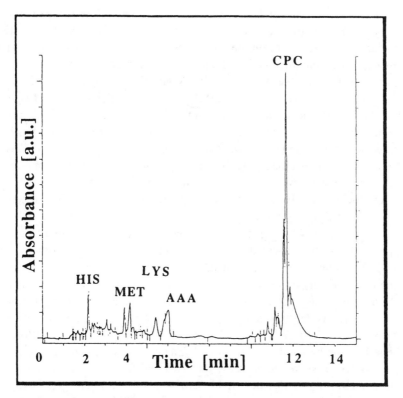

Figure 8. Electropherogram of a cultivation sample taken after 132 h from a culture of *Cephalosporium acremonium* (Brüggemann, 1994).

tified without difficulty. The CE detection limits are lower than in RP–HPLC in most cases including the important amino acid L-α-aminoadipate (AAA).

Samples of the fermenter supernatant were taken every 20 min during the first 122 h of a 132 h fermentation of *Cephalosporium acremonium* and every 10 min during the remaining 10 h. Selected samples were analyzed by RP–HPLC and CZE. An exemplary electropherogram is given in Fig. 8. The amino acids, certain peptides and the product CPC itself could be positively identified in the electropherogram. Other signals were reproducible recorded, but could not be related to any substance. These signals are presumably caused by the di- and tripeptide; however, due to the lack of the corresponding standards, no positive identification was possible. The agreement between RP–HPLC and CZE is adequate in most cases. The sample volume is much smaller and the required reagents are less expensive. The product and various intermediates can be quantified simultaneously. An analysis takes 20 min. This is a considerable improvement over the conventional analysis by RP–HPLC.

MONITORING OF THE PLASMID STABILITY OF RECOMBINANT E. COLI

In recent years, recombinant DNA technology has extended the scope of biotechnology considerably by empowering, for example, bacteria to produce proteins such as interferon α, antithrombin III, etc., which were previously synthesized exclusively by mammalian cells. The necessary additional genetic information is introduced into the bacteria – mostly *Escherichia coli* (*E. coli*) are used – through plasmids, i.e. short circular DNA sequences, which should not become part of the bacterial genome. Usually, organisms carrying the plasmid are at a metabolic disadvantage and a general tendency to lose the plasmids has been observed in many strains of genetically modified bacteria. The surveillance of the plasmid stability becomes thus an important aspect in biotechnological process control whenever genetically modified bacteria are used. Previously, DNA fragments have been analyzed using agarose slab gels. The separation itself is followed by a staining and density scanning procedure.

The analysis of DNA-fragment by CGE was, among the first applications of CE, partly due to the expected beneficial effect on the human genome project. DNA-CGE continues to be a most active area of CE-research. While many CE methods can rely on differences in the mass-to-charge ratio of the analytes for differences in the electrophoretic velocities and thus separation, this is not possible in the case of nucleic acids, where – save for small oligonucleotides – only insignificant differences in this parameter are found. Instead, differences in the size of the molecules are exploited. The capillaries are filled with a sieving matrix which serves to increase the frictional forces working opposed to the acceleration in the electrical field. Cross-linked and linear gels have been successfully used for that purpose (Cohen *et al.*, 1988; Kasper *et al.*, 1988; Drossmann *et al.*, 1990; Heiger *et al.*, 1990; Bocek and Chrambach, 1991; Grossmann and Soane, 1991; Hebenbrock *et al.*, 1991, 1993; Motsch *et al.*, 1991; Schwartz *et al.*, 1991; Garner and Chrambach, 1992; Nathakarnkitkool *et al.*, 1992; Righetti *et al.*, 1992). The former are synthesized within the capillary and cannot be removed once their limited lifetime is spent. In the case of the linear polymers, the physical entanglement and association occurring in aqueous polymer solution is responsible for the formation of a more or less impervious capillary filling. Linear polyacrylamide (PAA), molten agarose and cellulose derivatives have been used. While highly viscous, such polymer solutions may be replaced after a separation by pumping a fresh charge into a given capillary.

Surveillance of the plasmid stability was desired for two strains of *E. coli*, a one-plasmid and a three-plasmid system (Hebenbrock *et al.*, 1993). The molecular weights of these plasmids varied between 3000 and 20 000 base pairs (bp). Good reproducibility, an analysis time of no more than 60 min and the option to exchange the sieving matrix if necessary were additional requirements.

Experimental

Instrumentation. The Beckmann P/ACE 2210 system was used. A constant current between 25 μA (*ca.* 4.7 kV) was applied during the separation. The DNA samples were injected electrokinetically. Capillaries with 100 μm i.d. and 27 cm (20 cm to the detector) length were used. A 90 mM Tris-borate-EDTA (TBE) buffer, pH 8.3, containing 0.5 μg/ml ethidium bromide, served as electrophoresis and sample buffer (Schwartz *et al.*, 1991; Nathakarnkitkool *et al.*, 1992).

Preparation of the linear PAA-gel. A 10% (w/w) solution of acrylamide in bidist. water was prepared. The polymerization was started by the addition of 5 μl/ml of N,N,N′,N′-tetramethylethylenediamine (TEMED) in water followed by 5 μl/ml of a freshly prepared 10% solution of ammonium persulfate in water. To remove any remaining small ions, the PAA-gel solution was electrodialyzed (cut off 12 000 Da). Prior to use, the gels were degassed under vacuum.

To prevent the linear PAA-gel from moving out of the capillary the EOF was suppressed by coating of the capillary walls. 1 ml of a 5% acrylamide solution was mixed with 0.05 ml of the TEMED and the ammonium peroxide solutions respectively. The mixture was sucked into a capillary, which had previously been activated with methacryloxypropyltrimethoxysilane and left overnight (Chen *et al.*, 1991; Hebenbrock *et al.*, 1991). For storage, the capillaries were rinsed and filled with water.

Biological strains. Two strains were purchased from the Deutsche Sammlung für Mikroorganismen (DSM). System I was *E. coli* K12W3110 containing the plasmid pCHv1 (13 000 bp, including the genes for luciferase and tetracycline resistance); system II was *E. coli* JM103 containing three different plasmids (pMTC48 carrying the gene encoding for a fusion protein of staphylococcus Protein A and EcoR I; 4860 bp, pEcoR 4 protecting the host against the EcoR I, 5929 bp; pRK248cI controlling expression with the temperature sensitive cI857 repressor, 10 040 bp).

Sample preparation. For the molecular weight standard, 1.26 μg of lambda DNA was digested with EcoR I using the restriction enzyme buffer kit from

Boehringer-Mannheim (Germany). After incubation, the DNA was precipitated from isopropanol, washed with 70% ice-cold ethanol, dried and redissolved in 50 µl of distilled water containing 0.5 µl/ml ethidium bromide. The ethidium bromide served to stiffen the DNA molecules.

The plasmid DNA was isolated with the Quiagen > plasmid < mini kit (Quiagen, Chatsworth, CA), mixed with the internal standard (pBr322, 4363 bp), restricted to give the linearized fragments and further processed as given for the molecular weight standard. For linearization, the restriction enzymes PSTI and EcoR V were used in the case of system I and system II respectively.

Results

The size of the plasmids to be analyzed was 13 000 bp in the case of system I and 4860 bp, 5929 bp as well as 10 040 bp in the case of system II. Treatment of lambda-DNA with the restriction endonuclease EcoR I yields six DNA fragments with sizes of 3530 bp, 4878 bp, 5643 bp, 5804 bp and 21 226 bp and lambda-DNA was thus chosen as an example system. Preliminary experiments on the separation of the lambda-DNA digest using capillaries filled with crosslinked PAA and agarose gels were soon abandoned (Hebenbrock, 1993). The former gels were found to give a poor signal-to-noise ratio and tended to become unstable after a few runs, at which point, the entire capillary had to be abandoned. Agarose had the advantage of being replacable in the molten stage. However, here too, poor reproducibility was observed for the separation.

Better results were obtained with linear PAA gels. The gels were polymerized externally and sucked into the capillary for the measurements. Between measurements the gels were removed and the capillaries refilled with fresh gel. The time required for such an replacement strongly depended on the viscosity and thus on the concentration of the gel. While a 4% PAA gel could be replaced within minutes using a hand-held syringe, the exchange of a 6% gel required already several hours of time and concomitantly a capillary filling station. The retention time of a given analyte also increased with the gel concentration.

Using the lambda-DNA digest as a standard, a minimum gel concentration of 4% PAA was established necessary to achieve separation (Fig. 9). Even then, the 5643 bp fragment and the 5804 bp fragment are not resolved at all, while there is no base line separation for the 4878 bp fragment. A yet better resolution was achieved with a 6% gel; however, the time necessary for an analysis was almost an hour. Together with the time requirement for gel exchange this was judged inadequate for process monitoring, while the resolution of the 4% gel should suffice for the intended analytical purpose. Using the shortest possible capillary length of 27 cm and a current of 25 µA, i.e. a voltage of approximately 4.7 kV, an entire separation including a change of the capillary filling is possible within 30 min.

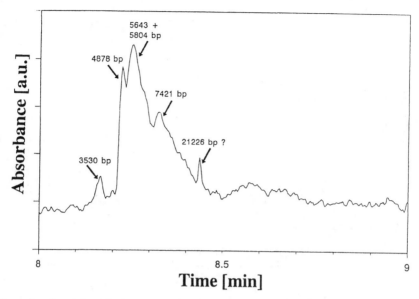

Figure 9a. Separation of a lambda-DNA digest using a 2% linear PAA-gel (Hebenbrock *et al.*, 1993).

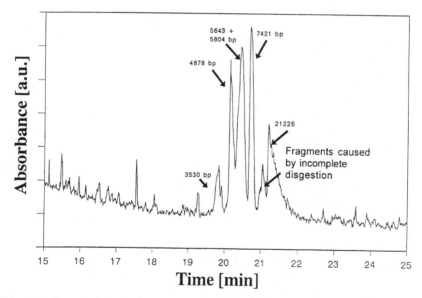

Figure 9b. Separation of a lambda-DNA digest using a 4% linear PAA-gel (Hebenbrock *et al.*, 1993).

Monitoring of E. coli *fermentations.* The preparation of the plasmids from the *E. coli* bacteria leads to three different subspecies: supercoiled, open circular and linear DNA, in each case. Since each species would have a different mobility, a linearization was necessary. To reduce the complexity of the fi-

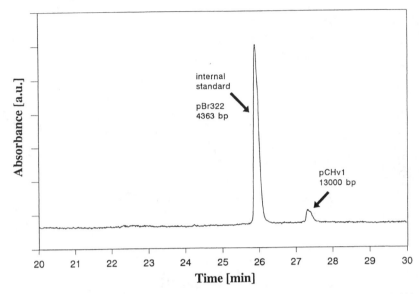

Figure 10. Analysis of the plasmid content of *E. coli* system I in the presence of pBr322 as internal standard (Hebenbrock *et al.*, 1993).

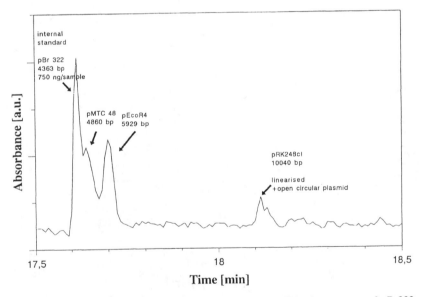

Figure 11. Analysis of the plasmid content of *E. coli* system II in the presence of pBr322 as internal standard (Hebenbrock *et al.*, 1993).

nal electropherogram, especially in the case of system II, only one fragment per plasmid was desired. From the restriction charts available for each of the plasmids except pRK248cI, the restriction enzymes PSTI and EcoR V were chosen for system I and system II, respectively. To prevent the various sample preparation steps from biasing the results, a known quantity of plasmid pBr322 (4363 bp) was added to each sample as internal standard immediately before the linearization.

Examples for the electropherograms obtained for both of the *E. coli* systems are given in Figs 10 and 11. Clearly, the analysis of the plasmid content of system I presents no problem. For system II all plasmids are isolated. However, the internal standard and fragment pMTC 48 are not well separated. It is also doubtful whether the largest plasmid was successfully linearized. The alignment of the signals and the plasmids was verified by agarose slab gel electrophoresis.

Pressure injection of the sample is difficult and unreliable in the case of CGE. Electrokinetical injection, where the analyte molecules are moved into the gel-filled capillary through the application of the electrical field, is the norm. This type in injection is, however, biased towards the smaller and thus faster molecules. Differences in the size of an internal standard and the analyte may be corrected for by running experiments using several concentrations of the internal standard an identical analyte concentration and followed by extrapolation to zero standard concentration.

Taking these precautions, the plasmid copy number of *E. coli* system I was determined during a batch cultivation. The results are given in Fig. 12. The increase in the copy number observed after 36 h is the result of a change in the feeding protocol. The copy number determined in the preculture was 2.

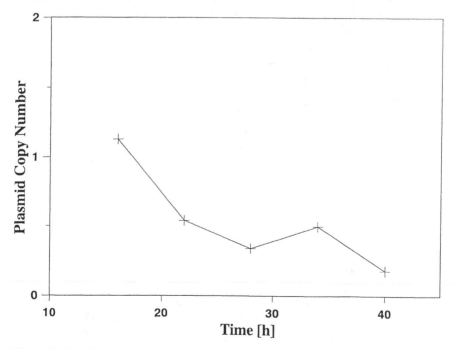

Figure 12. Development of the plasmid copy number during a batch fermentation of recombinant *E. coli* (Hebenbrock, 1993).

In spite of the addition of the antibiotic chloramplenicol for selection of the genetically modified bacteria, a steady reduction of the plasmid copy number is observed during the fermentation.

CGE constitutes a comparatively fast and reliable means for the determination of the plasmid copy number of genetically modified microorganisms. Such a continuous monitoring is necessary, since as demonstrated above, the addition of an antibiotic resistance for the genetically modified organism does not always suffice to ensure plasmid stability.

SAMPLE INJECTORS FOR CAPILLARY ELECTROPHORESIS

Ideally process monitoring should be a fully automated procedure. The analytical device is coupled directly via a sampling module to the bioprocess; the required sample volumes are injected by a suitable injection module; finally, the data are evaluated by a data handling system and made available for process documentation or process control. The very smallness of the sample volume required in CE renders most of the established injection modules, e.g. the rotary type injectors commonly used in HPLC, unsuited for that purpose in CE analysis. The electric sample splitter introduced by Deml et al. (1985), the feeder sampling device introduced by Verheggen et al. (1988), the split flow syringe injector by Tehrani et al. (1991), and the various microinjectors by Ewing and coworkers (Wallingford and Ewing, 1987, 1988, 1990; Ewing et al., 1989) are among the isolated attempts to realize an automated CE-injector, e.g. by adapting the sample injection principles known from FIA.

Two injection systems are used at our institute. For the direct interfacing of CE and HPLC the rotary vial system shown in Fig. 13 is used. The direct coupling of IEXMA and CZE involving such an injector has, for example, been used in the case involving the quantification of r-AT III in culture supernatants of BHK cells. A sufficient amount of sample was preconcentrated on a strong anion exchanger MA and eluted in a salt gradient with a flow rate of 1 ml/min. The eluate was passed through the 6-port injection module fitted with the capillary and into the waste. When the AT III containing fraction passed through,

Electrophoresis   **Injection from LC to CE**

Figure 13. HPCL–CE interface (Reif, 1994).

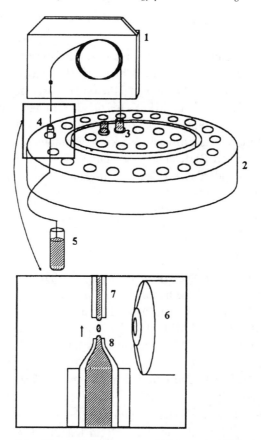

Figure 14. Diagram of the CE-microinjector system (Sziele *et al.*, 1994). 1 – Capillary holder, 2 – Autosampler, 3 – Buffer storage, 4 – Injection system, 5 – Sample, 6 – Microscope, 7 – Capillary, 8 – Microdrop ejector.

the valve was switched to the injection mode and the eluate injected into the capillary for two seconds. Afterwards the electric field was activated and the AT III content determined. A standard deviation of 1.7% was calculated for the AT III concentration, which is only slightly above the deviation calculated in case of the manual sample transfer between HPLC and CE.

The second module resulted from the adaption to CE of a commercially available microinjector (Microdrop GmbH, Norderstedt, Germany) originally developed for ink jet printers, Fig. 14 (Sziele *et al.*, 1994).

Singular droplets (size 113 pl, RST: 1%) are injected directly into the capillary over a distance. No actual contact takes place between the nozzle of the module and the capillary during injection; thus no contamination is possible. Repeated CZE measurements of a 10 mM arginine solution in a 50 mM phosphate buffer, pH 2.5, gave a relative standard deviation of 3% for the peak area.

PRODUCT CHARACTERIZATION

The verification or determination of some characteristic feature of an isolated but yet unidentified substance or even of a well known protein are common analytical problems in biotechnology and the other biosciences. Below, the potential of CZE, CIEF, CGE and MECC for the determination of the molecular weight, the isoelectric point, the biological activity, and the peptide map of a (recombinant) protein are discussed, taking recombinant h-AT III produced by genetically modified BHK cells as example. The results are compared to those obtained for standard slab gels and with data already published for AT III isolated from human plasma.

Experimental

Determination of the isoelectric point by CIEF. The sample was desalted by precipitation from cold methanol and the protein precipitate recovered by centrifugation. The pellet was resuspended in 50 μl of a buffer consisting of: 0.01% HPMC, 0.1% TEMED, 0.001% Triton X-100, 2% Ampholine 4/6, 0.5% Pharmalyte 3/10, 0.5% Pharmalyte 2.5/5 and 0.5% Pharmalyte 4/6.5 in deionized water. For calibration, a mixture of trypsin inhibitor (IEP 4.6), β-lactoglobulin A (IEP 5.1) and carbonic anhydrase II (IEP 5.4 and 5.9) was treated correspondingly.

10 mM phosphoric acid served as anolyte, 20 mM NaOH as catholyte. A dextrane coated capillary (50 μm i.d., 27 cm length, 20 cm to the detector) was used. For focussing, a voltage of 12 kV was applied for 2 min. Hydrodynamic mobilization was done by applying pressure in the presence of an applied voltage of 8 kV. The detection wave length was 254 nm. For comparison, isoelectric focussing (according to Bailey, 1984) was done in precast 7% slab gels for a pH-range from 4 to 6 (Serva). Silver staining was used to visualize the protein zones.

Determination of the molecular weight by CGE. Samples were diluted 1:1 in a 50 mM Tris/Ches buffer, pH 8.6, containing 1% SDS and heated to 90°C for 5 min. For calibration, four standard proteins (pepsin, Mw 34.7 kg/mol, ovalbumin, Mw 45 kg/mol, BSA, Mw 69 kg/mol and the phosphorylase subunit B, Mw 97.4 kg/mol) were dissolved in the same buffer containing an additional 5% of mercaptoethanol. Electrokinetic sample injection was done by applying a voltage of 10 kV for 10 s.

The electrophoresis buffer was 100 mM Tris/Ches, pH 8.6, containing 0.1% SDS. Capillaries (100 μm i.d., 27 cm length, i.e. 7 cm from injection to the detector) were coated with dextrane and filled with a dextrane gel (0.1 g/ml, Mw dextrane 2000 kg/mol). A fresh portion of the degassed gel was injected into the capillary by applying pressure before each measurement and removed afterwards. The detection wave length was 200 nm.

Biological activity. 10 μl of r-AT III (1 mg/ml in a 50 mM Tris/HCl buffer, pH 7.5) were mixed with 10 μl of a thrombin solution (250 μg/ml in a 50 mM Tris/HCl buffer, pH 7.5). After 10 s, 5 min and 10 min incubation the samples were mixed with 10 μl of a 50 mM Tris/Ches buffer, pH 8.6, 1% SDS and analyzed by CGE. The influence of heparin addition was investigated by adding 150 μg of solid heparin to the otherwise identical AT III/thrombin mixtures.

Peptide mapping by CZE and MECC. AT III was lysed by exposure to trypsin (Judd, 1990). To prevent the inhibition of the protease by the protease inhibitor AT III, the AT III is beforehand heated to 100°C for 10 min in the presence of 1% mercaptoethanol or 3 M urea.

CZE: Capillary: 50 μm i.d. × 67 cm, buffer: 50 mM phosphate, pH 2.5, containing 0.01% HPMC, voltage: 15 kV, detection: 200 nm, injection: pressure for 2 s.

MECC: Capillary: 50 μm i.d. × 67 cm buffer: 50 mM borate, pH 9.5, containing 0.1% SDS, voltage: 15 kV, detection: 280 nm, injection: pressure for 2 s.

All CE-experimets were performed with the Beckmann P/ACE 2210.

Results

Isoelectric points. The isoelectric points of the r-AT III isolated from the supernatant of a BHK cell culture were determined by capillary isoelectric focussing (CIEF). In CIEF the analytes are focussed into sharp zones according to their isoelectric points in the stable pH-gradient that is formed across the capillary by a mixture of amphoteric molecules (e.g. amino acids) with different IEP values. The commercially available ampholytes all have a high buffer capacity at their isoelectric point. A strongly acidic buffer, the anolyte, is placed at the anodic side of the capillary, while a strongly basic buffer, the catholyte, is placed at the cathodic side. The EOF needs to be suppressed in CIEF to prevent the disturbance of the pH gradient during focussing. Afterwards, the analyte zones need to be moved past the detector. Mobilization may be achieved by a disturbance of the pH-gradient through salt addition or pH-change in the anolyte or catholyte. Mobilization by pressure is also possible. The reproducibility and the resolution possible in CIEF depend strongly on the mobilization procedure.

Unstable capillary coatings due to the extreme pH-values of the electrode buffers are a problem in CIEF. Here the cellulose coating suggested by Hjerten and Kubo (1993) was used to advantage. While being rather laborious, this method yielded capillaries that could be used well over a hundred times without a significant decrease in resolution or reproducibility.

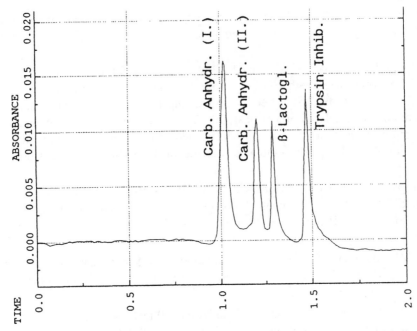

Figure 15. CIEF of the four standard proteins: trypsin inhibitor, IEP 4.6, β-lactoglobulin A, IEP 5.1 and carbonic anhydrase II, IEP 5.4 and 5.9 (Reif and Freitag, 1994b).

Rather than opting for pH or salt mobilization the usually less reproducible pressure mobilization was used. Excessive zone broadening as a result of the parabolic flow profile was prevented by the application of a voltage of 8 kV throughout the mobilization. The proteins are thus still being focussed during mobilization. Triton X was added to the sample buffer to prevent protein precipitation at the respective substance IEP, and TEMED to prevent sample components from focussing behind the detector. The applied spacer mixture was established to be a workable compromise between resolving power and stability of the pH-gradient. Four standard proteins were repeatedly separated by CIEF to provide a calibration curve (Fig. 15). A relative standard deviation of 2.1% was determined for the retention times. The focussing was achieved within 2 min, the mobilization required another 2 min.

When the CIEF method was used to determine the isoelectric point of the r-AT III (Fig. 16), three major peaks corresponding to IEPs of 4.7, 4.75, and 4.85 were found as well as three minor peaks at 5.0, 5.1 and 5.3. The results agree well with the bands found in the corresponding slab gel. AT III from human blood is reported to have three major and three minor subfractions in a similar region (IEP of: 4.75, 4.8, 4.85 and 5.0, 5.05, 5.2) (Daly and Hallinan, 1985). In addition, a doublet is usually found at 4.7; however, this subfraction is not seen here either in the CIEF or in the slab gel. The fraction might, of course, be lost due to poor resolution in the major peak at 4.7. It is also possible, that this particular subfraction was indeed underrepresented in the batch of recombinant h-AT III considered here.

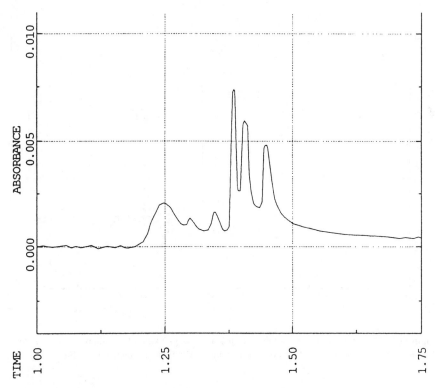

Figure 16. Determination of the isoelectric points of recombinant h-AT III by CIEF (Reif and Freitag, 1994b).

Molecular weight. The separation of molecules according to their size by CGE has already been discussed for DNA fragments. Compared to the separation of nucleotides, CGE is less commonly used for the separation of proteins, where several other methods are available. Resolution and the adaptation of the gel matrix to a wide variety of molecular weights are still problematic. However, the existing CGE techniques may be used for the determination of the molecular weight of a protein such as AT III.

Using a dextrane (0.1 g/ml) gel, four standard proteins (pepsin, Mw 34.7 kg/mol; ovalbumin, Mw 45 kg/mol; BSA, Mw 69 kg/mol; and the phosphorylase subunit B, Mw 97.4 kg/mol) with molecular weights ranging from 30 kg/mol to 100 kg/mol were separated in less than 3 min (Fig. 17, Lausch *et al.*, 1993; Reif and Freitag, 1994b). A linear correlation with a correlation coefficient of 0.9989 was established between the molecular weight and the retention time in the gel-filled capillary. A relative standard deviation of less than 0.8% was calculated for the retention times. Taking the calibration into account, a molecular weight of 59 kg/mol was calculated for the r-AT III from the CGE (Fig. 18). This is close to the value published for human AT III (58 kg/mol), which is, in turn, supported by protein and cDNA sequencing (Mourey *et al.*, 1990).

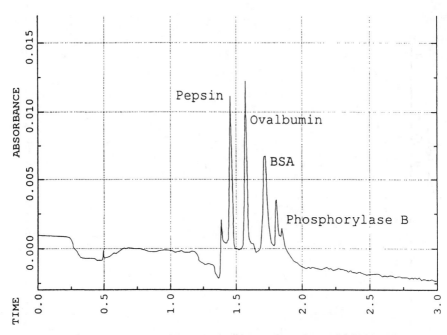

Figure 17. CGE of the four standard proteins pepsin, Mw 34.7 kg/mol, ovalbumin, Mw 45 kg/mol, BSA, Mw 69 kg/mol and the phosphorylase subunit B, Mw 97.4 kg/mol (Reif and Freitag, 1994b).

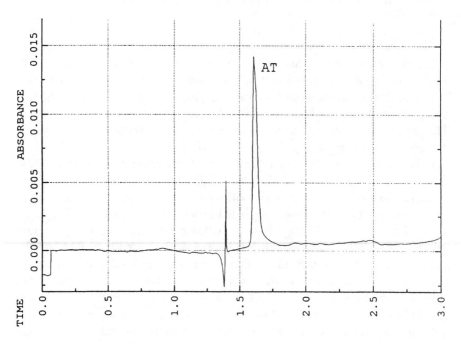

Figure 18. Determination of the molecular weight of r-AT III by CGE (Reif and Freitag, 1994b).

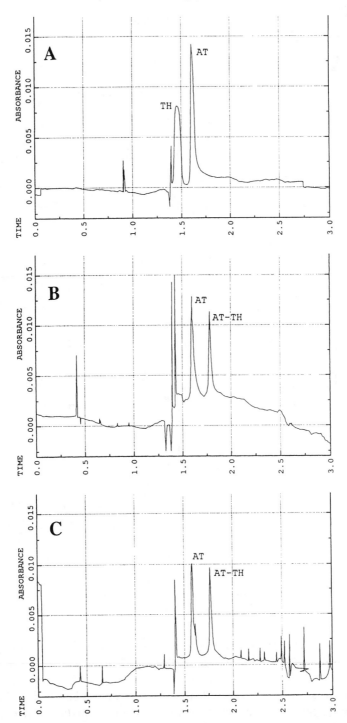

Figure 19. Complex (AT-TH) formation between r-AT III (AT) and thrombin (TH). A: 10 s incubation, B: 5 min incubation, C: 10 min incubation (Reif and Freitag, 1994b).

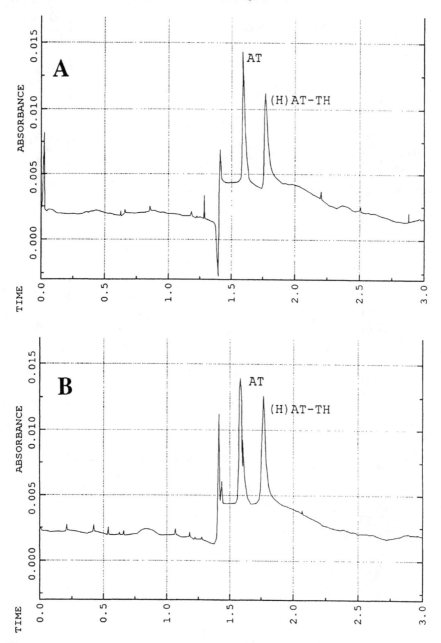

Figure 20. Complex (AT-TH) formation between r-AT III (AT) and thrombin (TH) in the presence of heparin A: 10 s incubation, B: 5 min incubation (Reif and Freitag, 1994b).

The speed is a major advantage of the methods described so far. Taking a single coated capillary filled with either the ampholyte mixture or the dextrane gel, both the isoelectric point and the molecular weight of a protein may be determined within 30 min.

Biological activity. Besides accessing the AT III's molecular weight proper, CGE may also be used to investigate complex formation between AT III (59 kg/mol) and thrombin (32 kg/mol) and thus allow a statement regarding the biological activity. This is possible because the affinity complex is stable under CGE conditions. The complex formation was investigated *per se* (Fig. 19) and in the presence of heparin, a substance known to expedite the affinity reaction (Fig. 20).

In the absence of heparin, only the two protein peaks are seen after 10 s of extracapillary incubation. After an incubation of 5 min, a third signal appears, which corresponds to a molecular weight of 92 kg/mol, i.e. the AT III-thrombin complex. When heparin is added to the mixture, however, significant complex formation has already taken place after 10 s of incubation.

Peptide mapping. Peptide maps are well established means for characterizing the 'structure' of a protein. The protein is cut into peptide fragments by a protease with high cleavage site specificity. Here trypsin was used, which is known to lyse the peptide bond at the N-terminus of arginine and lysine. The fragments are usually analyzed by RP–HPLC. The analysis of the peptide mix by CZE or MECC constitutes an interesting alternative. Both CE-methods are usually faster, less expensive and require much smaller sample volumes — a distinct advantage in protein characterization.

In CZE the peptide fragments are resolved due to the differences in electrophoretic mobility. In MECC (micellar electrokinetic chromatography), similar to RP–HPLC, the hydrophobicity of the analytes is the decisive factor. A micelle forming substance, usually SDS, is added to the electrophoresis buffer. Depending on the charge of their surface and the strength of the EOF, the micelles will (slowly) move in the electric field. Depending on their hydrophobicity, analyte molecules are distributed between the bulk electrophoresis buffer and the micellar phase. Thus the velocity of the more hydrophobic molecules will be determined by the speed of the micelles, while the less hydrophobic molecules will move with a velocity largely determined by their own mass-to-charge ratio and the strength of the EOF.

The tryptic maps obtained by CZE and MECC for r-AT III produced by BHK cells in the presence of 10% FCS are presented in Fig. 21 and Fig. 22 respectively. Both CE-methods yield approximately 50 peaks, i.e. roughly the number of peptide fragments expected. The time required for an analysis is 10 min less in MECC. The pattern generated by CZE in the case of genuine h-AT III derived from human serum (Sigma) is quite similar to that of the digest obtained for r-AT III produced with 10% FCS. Significant discrepancies are seen for the AT III produced with 3% FCS (Fig. 23). Apparently, the reduction of the FCS content, while yielding a more easily purified product and a cheaper means of production, does not lead to a high product quality.

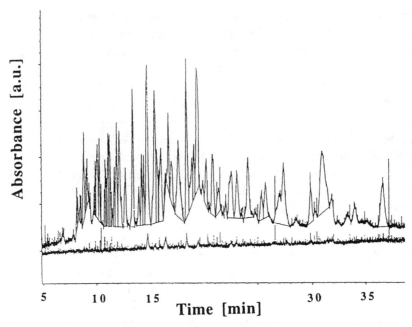

Figure 21. Analysis by CZE of the tryptic digest of r-AT III produced in the presence of 10% FCS (Reif, 1994).

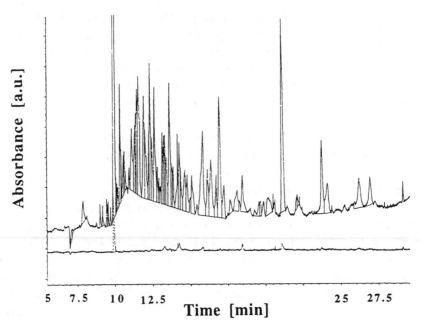

Figure 22. Analysis by MECC of the tryptic digest of r-AT III produced in the presence of 10% FCS (Reif, 1994).

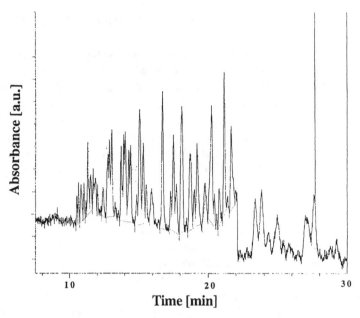

Figure 23. Analysis by CZE of the tryptic digest of r-AT III produced in the presence of 3% FCS (Reif, 1994).

FLUORESCENCE AFFINITY CAPILLARY ELECTROPHORESIS (FACE)

The high specificity of immuno and affinity reactions had led in the past to a number of sensitive assays which allow the detection and quantification of certain analytes even at low concentrations in complex sample matrices. Often the antibody or affinity reactant is labelled with an enzyme or a fluorescent dye to improve the detection. The union of the principle of immuno/affinity based analysis with the methodology of CE yields a powerful analytical technique. FITC-labeled Protein G may, for example, be used to quantify IgG directly in serum samples, since Protein G is known to bind to all subgroups of h-IgG, but not to other human immunoglobulins such as IgA or IgM.

Experimental. For the experiments described here, the Beckmann P/ACE 2210 CE was equipped with an argon laser detector (Beckmann; excitation 488 nm, emission 520 nm). The electrophoresis buffer was a 0.05 M borate buffer, pH 10.5. Capillaries were uncoated and of 50 μm i.d. and 27 cm length (20 cm to the detector). Sample was injected by pressure (50 mbar for 2 s, *ca.* 5 nl). The applied voltage was 10 kV or 15 kV as indicated, the temperature was maintained at 22°C.

When pre-incubation of the sample was called for, 250 μl of a solution containing 300 μg/ml reagent (FITC labeled Protein G) in a 100 μM phosphate buffer, pH 5.8 and 250 μl of a solution containing the analyte (between 2 and 2000 μg/ml) in the same buffer were mixed and incubated for the indicated amount of time, usually 4 min, at the indicated temperature, usually 22 °C. The mixture was injected into the capillary and analyzed.

For the quantification of h-IgG in human serum (kindly donated by the Red Cross, Springe, Germany) the samples were diluted 1:10 (v:v) with a 50 mM phosphate buffer, pH 5.8. FITC-Protein G was dissolved at a concentration of 1600 µg/ml in the same buffer. Both solutions were injected consecutively into the capillaries by applying pressure for 2 s in each case. Immediately afterwards the electrical field was activated and the mixture analyzed. For comparison, a SRID (Single Radial Immunodiffusion) IgG quantification was carried out according to Hobbs (1970).

Results

The transfer of immuno/affinity analysis to CE faces certain difficulties. The complex and the reaction partners differ in electrophoretic velocity: hence the separation. However, since the complex is not covalently linked, but the result of a dynamic equilibrium, complex dissociation and reformation occurs throughout the assay. As soon as a complex dissociates during the CZE-analysis, the reactants are rapidly moved apart into a region where the concentration of the complex partner is low (Schultz and Kennedy, 1993). Thus, complex reformation becomes unlikely. Moreover, conditions which are favorable to the formation of stable complexes, e.g. a pH value of 7 and a fairly high salt concentration, are unfit for efficient CE. At near neutral pH the EOF is low during the ensuing long separation so many chances for complex dissociation will occur. In addition, the tendency of proteins to adhere to the capillary walls is high. A high salt concentration, which is also helpful for reducing the protein-wall-interaction, leads to increased Joule heating and thus to irreproducible separations and diminished complex stability.

A somewhat different approach was taken by Reif *et al.* (1994). It was found that whenever the concentration of a pre-incubated Protein G-IgG affinity complex was evaluated by CZE as a function of a variable such as the capillary's length, which influences mainly the duration of the separation, an asymptotic approach to a maximum value was observed with decreasing analysis time (Reif, 1994). If the separation was completed within 3 min, a maximum signal was detected. This value was assumed to correspond to a situation where the complex dissociation enforced by the electrophoretic conditions was at minimum, i.e. where the complex was influenced by the electrophoretic conditions for too short a time for the slow reaction kinetics to allow more than an insignificant amount of complex dissociation to take place.

The detection range of an affinity reaction in solution depends on the concentration of the labeled reaction partner. Using a 10^{-6} M solution of FITC labeled Protein G, h-IgG in concentrations between 10^{-8} M and 10^{-6} M could be quantified (Fig. 24). A voltage of 15 kV was applied in cases A through C and one of 10 kV in cases D through G. The complex is observed after *ca.* 1 min in the former and after *ca.* 2 min in the latter cases. The peak following

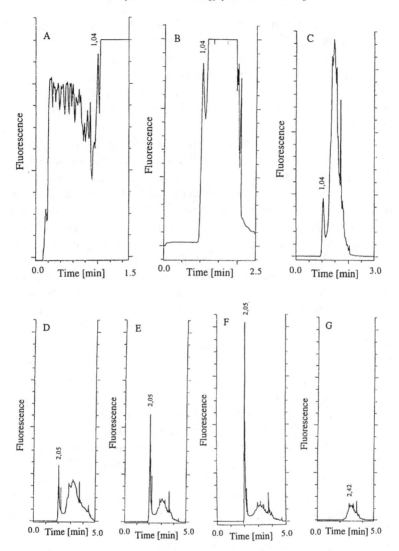

Figure 24. Quantification of h-IgG standards using FITC labeled Protein G. A: 8×10^{-9} M h-IgG, B: 8×10^{-8} M h-IgG, C: 8×10^{-7} M h-IgG, D: 2×10^{-6} M h-IgG, E: 4×10^{-6} M h-IgG, F: 8×10^{-6} M h-IgG, G: blank.

the complex signal is caused by unreacted FITC-Protein G. Due to varieties in the mass-to-charge ratio of the FITC-Protein G, the result of differences in the number of affixed FITC molecules, a comparatively broad signal is observed for the reagent, whereas, due to the predominance of the mass-to-charge ratio of the antibody, a sharp signal is observed for the affinity complex.

The application of voltages of more than 15 kV led to increased Joule heating and influenced the complex stability adversely. An IgG concentration of 8×10^{-9} M h-IgG appears to constitute the detection limit, as the signal from the complex is already hard to distinguish from the base line noise in Fig. 24A. When a total of 28 (10 kV) or 40 runs (15 kV) respectively were evaluated, relative standard deviations of 1.93% and 1.64% were calculated for the retention times, while the relative standard deviation for the peak height was 1.72% for a IgG concentration of 4×10^{-6} M. A linear relationship with a correlation factor of $r = 0.9988$ could be established between the concentration and the peak height as taken from the baseline.

The formation of the Protein G/antibody complex is known to be a comparatively slow reaction. When the influence of the incubation time on the complex formation as evaluated from the height of the signal detected in the electropherogram was investigated, it became clear that even after 30 min the reaction had not yet reached its equilibrium. However, more than 90% of the final obtainable peak height were already observed after 4 min incubation. Reaction temperature and pH had a strong influence on the complex formation. Incubation at 35 °C or even 45 °C yielded a significantly smaller signal and a non-linear relationship between the peak height and the analyte concentration. Increased Joule heating caused either by the application of too high a voltage or an electrophoresis buffer too rich in salt, is thus to be avoided at all costs. Incubation at low pH value, e.g. 2.5, as expected, prevented the formation of a detectable amount of affinity complex.

Analyte determination in human sera. Protein G is known to bind to all subgroups of h-IgG but not to the various other immunoglobulins and proteins present in a serum sample. The FITC labeled Protein G should thus be suitable for the direct quantification of h-IgG in human serum. A calibration curve obtained through the analysis of eight h-IgG standards was used as basis for the calculations of the IgG-concentration in a given sample. The results were compared to those obtained with SRID. To reduce sample handling as much as possible, both the sample (standard) and the reagent were consecutively injected by pressure (2 s) into the capillary without pre-incubation. Due to the dispersive effect and the laminar flow profile present during the pressure injection, the zones are sufficiently mixed. The injection order was established to be of no consequence for the final electropherogram. Immediately after injection the field was switched on and the affinity complexes separated from the reagent. Further mixing of the two reaction partners may have taken place due to the different electrophoretic velocities.

As shown in Fig. 25 the resolution is here inferior to that obtained for the premixed samples. Nevertheless, a correlation factor of 0.9986 was found for the height of the affinity complex peak and the analyte concentration. When the data obtained for various sera by the FACE assay were compared to those obtained for the SRID, Fig. 26, good agreement was found. In regard to the analysis time, the sample handling, the sample volume requirement, and the potential for automation the FACE awaits a number of applications in biotechnology.

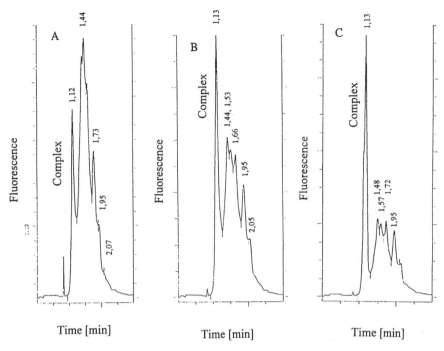

Figure 25. Analysis of the IgG content of human serum by FACE. A: IgG-deficient serum, B: Normal serum, C: Serum with abnormally high IgG-content (Reif, 1994).

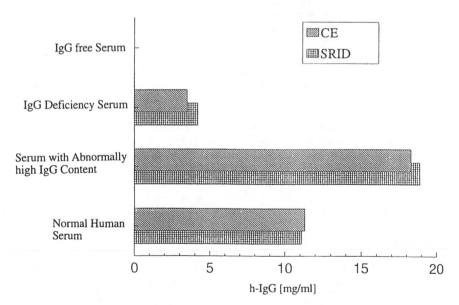

Figure 26. Comparison of the IgG concentrations measured in human serum samples by SRID and FACE, using FITC-labeled Protein G as reagent (Reif, 1994).

REFERENCES

Arentoft, A. M., Frokar, H., Michaelsen, S., Sorensen, H. and Sorensen, S. (1993). HPCE for the determination of trypsin and chymotrypsin inhibitors and their association with trypsin, chymotrypsin and monoclonal antibodies. *J. Chromatogr. A* **652**, 189.

Bailey, G. S. (1984). Immunodiffusion in gels. In: *Methods in Molecular Biology*, Vol. 1 (Proteins), Walker, J. M. (Ed.). Humana Press, Clifton, NJ.

Banke, N., Hansen, K. and Diers, I. (1991). Detection of enzyme activity in fractions collected from free solution capillary electrophoresis of complex samples. *J. Chromatogr.* **559**, 325.

Bao, J. and Regnier, F. E. (1992). Ultramicro enzyme assays in a capillary electrophoretic system. *J. Chromatogr.* **608**, 217.

Bergman, T., Agerberth, B. and Jornvall, H. (1991). Direct analysis of peptides and amino acids from capillary electrophoresis. *FEBS Lett.* **283**, 100.

Blackburn, M. N., Smith, R. L., Carson, J. and Sibley, C. C. (1984). The heparin-binding site of AT III: Identification of a critical tryptophane in the amino acid sequence. *J. Biol. Chem.* **259**, 939.

Bocek, P. and Chrambach, A. (1991). Capillary electrophoresis of DNA in agarose solutions at 40°C. *Electrophoresis* **12**, 1059.

Bradley, J., Stöcklein, W. and Schmid, R. D. (1991). Biochemistry based analysis systems for bioprocess control. *Process Control Qual.* **1**, 157.

Brüggemann, O. (1994). *Einsatz kapillarelektrophoretischer Verfahren in der Prozeßanalytik.* Diploma-Thesis, Institut für Technische Chemie, University of Hannover, Hannover, Germany.

Burnouf, T. (1991). Integration of chromatography with traditional plasma protein fractionation methods. *Bioseparation* **1**, 383.

Chen, S. M., Lee, T. D. and Shively, J. E. (1988). Structure-function relationship of NAD(P)H quinone reductase; characterization of NH2-terminal blocking group and essential tyrosine and lysine residues. *Biochemistry* **27**, 6877.

Chen, J. W., Cohen, A. S. and Karger, B. L. (1991). Identification of DNA-molecules by precolumn hybridization using capillary electrophoresis. *J. Chromatogr.* **559**, 295.

Cohen, A. S., Najarian, P. R., Paulus, A., Guttmann, A., Smith, J. A. and Karger, B. L. (1988). Rapid separation of DNA restriction fragments using capillary electrophoresis. *J. Chromatogr.* **458**, 323.

Cohn, E. J., Strong, L. E., Hughes, W. L., Mulford, D. J., Ashworth, J. N., Strong, L. E. and Taylor, H. L. (1946). Preparation and properties of serum and plasma proteins. IV: A system for the separation into fractions of the protein and lipoprotein components of biological tissues and fluids. *J. Am. Chem. Soc.* **8**, 159.

Conard, J., Bosstad, F., Larsen, M. L., Samama, M. and Abildgaard, U. (1983). Molar antithrombin concentration in normal human plasma. *Haemostasis* **13**, 363.

Daly, M. and Hallinan, F. (1985). Analysis of antithrombin III microheterogeneity by isoelectric focusing in polyacrylamide gels and immunoblotting. *Thromb. Res.* **40**, 207.

Damus, P. S., Hicks, M. and Rosenberg, R. D. (1973). Anticoagulant action of Heparin. *Nature* **246**, 355.

Demain, A. L. (1984). Biosynthesis of β-lactam antibiotics. *Handbook of Indust. Pharmacol.* **61**(1), 189.

Deml, M., Foret, F. and Bocek, P. (1985). Electric sample splitter for capillary zone electrophoresis. *J. Chromatogr.* **320**, 159.

Drossmann, H., Luckey, J. A., Kostichka, A. J., Cunha, J. D. and Smith, L. M. (1990). High-speed separations of DNA sequencing reactions by capillary electrophoresis. *Anal. Chem.* **62**, 900.

Dwyer, J. L. (1984). Scaling up bio-product separation with high performance liquid chromatography. *Bio/Technology*, November, 957.

Ewing, A. G., Wallingford, R. A. and Olefirowicz, T. M. (1989). Capillary electrophoresis. *Anal. Chem.* **61**, 292A.

Foret, F., Szoko, E. and Karger, B. L. (1993). Trace analysis of proteins by capillary zone electrophoresis with on-column transient isotachopphoretic preconcentration. *Electrophoresis* **14**, 417.

Freitag, R., Reif, O.-W., Weidemann, R. and Kretzmer, G. (1996). Production of recombinant h-AT III with mammalian cells using capillary electrophoresis for production monitoring. *CYTO 479 Cytotechol.*, accepted.

Frenz, J., Wu, S.-L. and Hancock, W. (1989). Characterization of human growth hormon by capillary electrophoresis. *J. Chromatogr.* **480**, 379.

Fuchigami, T., Imasaka, T. and Shiga, M. (1993). Subattomole detection of amino acids by capillary electrophoresis based on semiconductor laser fluorescence detection. *Anal. Chim. Acta* **282**, 209.

Garner, M. and Chrambach, A. (1992). Resolution of circular, nicked circular and linear DNA, 4.4 kb in length, by electrophoresis in polyacrylamide solutions. *Electrophoresis* **13**, 176.

Grossmann, P. D. and Soane, D. S. (1991). Capillary electrophoresis of DNA in entangled polymer solutions. *J. Chromatogr.* **559**, 257.

Guzmann, N. A., Ali, H., Moschera, J., Iqbal, K. and Waseen Malik, A. (1991). Assessment of capillary electrophoresis in pharmaceutical applications. Analysis and quantification of a recombinant cytocine in an injected dosage form. *J. Chromatogr.* **559**, 307.

Guzmann, N. A., Moschera, J., Iqbal, K. and Waseen Malik, A. (1992). Effect of buffer constituents an the determination of therapeutic proteins by capillary electrophoresis. *J. Chromatogr.* **608**, 197.

Hebenbrock, K. (1993). *Entwicklung und Optimierung von Analysenmethoden für die Überwachung von Kultivierungen rekombinanter E. coli mit der Kapillarelektrophorese*. PhD-Thesis, Institut für Technische Chemie, University of Hannover, Germany.

Hebenbrock, K., Maschke, H. E., Schügerl, K. and Friehs, K. (1991). Plasmidanalytik mittels Kapillarelektrophorese. *Biotec Gentechnologie* **4**, 35.

Hebenbrock, K., Schügerl, K. and Freitag, R. (1993). Analysis of Plasmid-DNA and cell protein of recombinant *Escherichia coli* using capillary gel electrophoresis. *Electrophoresis* **14**, 753.

Heiger, D. N., Cohen, A. S. and Karger, B. L. (1990). Separation of DNA restriction fragments by high performance capillary electrophoresis with low and zero crosslinked polyacrylamide using continuous and pulsed electric fields. *J. Chromatogr.* **516**, 33.

Hjerten, S. and Kubo, K. (1993). A new type of pH- and detergent-stable coating for elimination of electroendosmosis and adsorption in electrophoresis. *Electrophoresis* **14**, 390.

Hobbs, J. R. (1970). Association of Clinical Pathologists Broadsheet 68.

Holzhauer-Rieger, K. (1990). *Untersuchungen bei der Fermentation von Cephalosporium acremonium mit Hilfe der Hochleistungs-Flüssigchromatographie — Einsatz von On-line- und Off-line-Verfahren*. Ph.D.-Thisis, Institut für Technische Chemie, University of Hannover, Germany.

Hoylaerts, M., Owen, W. G. and Collen, D. (1984). Involvement of heparin chain length in the heparin-catalyzed inhibition of thrombin by antithrombin III. *J. Biol. Chem.* **259**, 5670.

Hurni, W. H. and Miller, W. J. (1991). Analysis of a vaccine purification process by capillary electrophoresis. *J. Chromatogr.* **559**, 337.

Jorgenson, J. W. and Lukacs, K. D. (1981). Zone electrophoresis in open-tubular glass capillaries. *Anal. Chem.* **53**, 1298.

Judd, R. C. (1990). Peptide Mapping. In: *Methods in Enzymology*, Vol. 182, Murray, M. P. (Ed.). Academic Press, Orlando, FL.

Karube, I. (1988). Novel Immunosensor Systems. In: *Biotec, Vol. 2, Biosensors and Environmental Biotechnology*, Hollenberg, C. P. and Sahm, H. (Eds). Gustav Fischer Verlag, Stuttgart – New York, pp. 37–47.

Kasper, T. J., Melera, M., Gozel, P. and Brownlee, R. G. (1988). Separation and detection of DNA by capillary electrophoresis. *J. Chromatogr.* **458**, 303.

Kistler, P. and Nitchmann, H. S. (1962). Large scale production of human plasma fractions. *Vox Sang.* **7**, 414.

Lane, D. A. and Caso, R. (1989). Antithrombin: Structure genomic organisation, function and inherited deficiency. *Bailliere's Clin. Haematol.* **2**, 961.

Lausch, R., Scheper, Th., Reif, O.-W., Schlösser, J., Fleischer, J. and Freitag, R. (1993). Rapid capillary gel electrophoresis of protein. *J. Chromatogr. A* **654**, 90.

Li, S. F. Y. (1992). *Capillary Electrophoresis: Principles, Practice and Applications, J. Chromatogr. Library, Vol. 52.* Elsevier, Amsterdam.

Ludwig, A., Tomeczkowski, J. and Kretzmer, G. (1992). Influence of the temperature on the shear stress sensibility of adherent BHK 21 cells. *Appl. Microbiol. Biotechnol.* **38**, 323.

Mattiasson, B. and Hakanson, H. (1992). Immunochemical based assays for process control. *Adv. Biochem. Eng. Biotechnol.* **46**, 81.

Middendorf, C., Schulze, B., Freitag, R., Scheper, Th., Howald, M. and Hoffmann, H. (1993). On-line immunoanalysis for bioprocess control. *J. Biotech.* **31**, 395.

Motsch, S. R., Kleemiß, M. H. and Schomburg, G. (1991). Production and application of capillaries filled with agarose gel for electrophoresis. *High Resol. Chromatogr.* **14**, 629.

Mourey, L., Samana, J. P., Delarue, M., Choay, J., Lormeau, J. C., Petitou, M. and Moras, D. (1990). Antithrombin III: Structural and functional aspects. *Biochimie* **72**, 599.

Nathakarnkitkool, S., Oefner, P. J., Bartsch, G., Chin, M. A. and Bonn, G. K. (1992). High-resolution capillary electrophoretic analysis of DNA in free solution. *Electrophoresis* **13**, 18.

Nielsen, R. G., Rickard, E. C., Danta, P. F., Sharknas, D. A. and Sittampalam, G. S. (1991). Separation of antibody-antigen complexes by capillary electrophoresis, isoelectric focussing and high-performance size exclusion chromatography. *J. Chromatogr.* **539**, 177.

Regnier, F. E., Leesong, I. and Harmon, B. J. (1994). Electrophoretically mediated Micro-analysis (EMMA) with biological extracts. Presented at the 6th *International Symposium on HPCE*, San Diego, CA.

Reif, O.-W. (1994). *Präparative und analytische Methoden in der Hochleistungs-Membran Chromatographie und Kapillarelektrophorese.* PhD Thesis, Institut für Technische Chemie, University of Hannover, Germany.

Reif, O.-W. and Freitag, R. (1993). Characterization and application of strong ion-exchange membrane adsorbers as stationary phases in high-performance liquid chromatography of proteins. *J. Chromatogr. A* **654**, 29.

Reif, O.-W. and Freitag, R. (1994a). Comparison of membrane adsorber (MA) based multi-stage chromatographic purification schemes for the down-stream processing of recombinant h-AT III. *Bioseparation* **3**, 369.

Reif, O.-W. and Freitag, R. (1994b). Control of the cultivation process of antithrombin III and its characterization by capillary electrophoresis. *J. Chromatogr. A* **680**, 383.

Reif, O.-W., Lausch, R., Scheper, Th. and Freitag, R. (1994). FITC-labeled Protein G as an affinity ligand in affinity/immuno capillary electrophoresis with fluorescence detection. *Anal. Chem.* **66**, 4027.

Righetti, R. G., Caglio, S., Saracchi, M. and Quaroni, S. (1992). 'Laterally aggregated' polyacrylamide gels for electrophoresis. *Electrophoresis* **13**, 587.

Rosenberg, R. D. and Damus, P. S. (1973). The purification and mechanism of action of human antithrombin III-heparin cofactor. *J. Biol. Chem.* **248**, 6490.

Schultz, N. M. and Kennedy, R. T. (1993). Rapid immunoassays using capillary electrophoresis with fluorescence detection. *Anal. Chem.* **65**, 3161.

Schwartz, H. E., Ulfelder, K., Sunzeri, F. J., Busch, M. P. and Brownlee, R. G. (1991). Analysis of DNA restriction fragments and PCR products towards detection of the AIDS (HIV-1) virus in blood. *J. Chromatogr.* **559**, 257.

Stephens, A. W., Siddiqui, A. and Hirs, H. W. (1987). Expression of functional active human antithrombin III. *Proc. Natl. Acad. Sci. USA* **84**, 3885.

Sziele, D., Brüggemann, O., Döring, M., Freitag, R. and Schügerl, K. (1994). Adaptation of a microdrop injector to sampling in capillary electrophoresis. *J. Chromatogr. A* **669**, 254.

Tehrani, J., Macomber, R. and Day, L. (1991). Capillary electrophoresis: An integrated system with an unique split-flow sample introduction mechanism. *J. High Resolut. Chromatogr.* **14**, 10.

Tsuij, K. (1993). Evaluation of sodium dodecyl sulfate non-acrylamide, polymer gel-filled capillary electrophoresis for molecular size separation of recombinant bovine somatropin. *J. Chromatogr.* **652**, 139.

Verheggen, T., Beckers, J. and Everaerts, F. (1988). Simple sampling device for capillary isotachophoresis and capillary zone electrophoresis. *J. Chromatogr.* **452**, 615.

Vinther, A., Pertersen, J. and Soberg, S. (1992). Capillary electrophoretic determination of the protease Savinase in cultivation broth. *J. Chromatogr.* **608**, 205.

Wallingford, R. A. and Ewing, A. G. (1987). Characterization of a microinjector for capillary electrophoresis. *Anal. Chem.* **59**, 678.

Wallingford, R. A. and Ewing, A. G. (1988). Capillary zone electrophoresis with electrochemical detection in 12.7 μm diameter columns. *Anal. Chem.* **60**, 1972.

Wallingford, R. A. and Ewing, A. G. (1990). Capillary electrophoresis. *Adv. Chromatogr.* **29**, 1.

Wasley, L. C., Atha, D. H., Bauer, K. A. and Kaufmann, R. J. (1987). Expression and characterization of human antithrombin III synthesized in mammalian cells. *J. Biol. Chem.* **262**, 1476.

Wirth, M., Reiser, W. and Zettelmeissl, G. (1987). Recombinant animal cell lines for production of glycoproteins. In: *Modern Approaches to Animal Cell Technology*, Spier, R. E. and Griffinth, J. B. (Eds). Butterworth, p. 108.

Yim, K. W. (1991). Fractionation of the human recombinant tissue plasminogen activator (rtPA) glycoforms by high-performance capillary zone electrophoresis and capillary isoelectric focussing. *J. Chromatogr.* **559**, 401.

Zettelmeissl, G., Ragg, H. and Karges, H. E. (1987). Expression of biologically active human antithrombin III in chinese hamster ovary cells. *Bio/Technology* **5**, 720.

Zettelmeissl, G., Wirth, M., Hauser, H. and Kuepper, H. A. (1988). Efficient expression system for human antithrombin III in baby hamster kidney cells. *Behring Inst. Mit.* **82**, 26.

APPENDIX – ABBREVIATIONS

AT III	Antithrombin III
b	Bovine
bp	Base pairs
BHK	Baby hamster kidney (cells)
CE	Capillary electrophoresis
CGE	Capillary gel electrophoresis
CHO	Chinese hamster ovary (cells)
CIEF	Capillary isoelectric focusing
CITP	Capillary isotachophoresis
CZE	Capillary zone electrophoresis
E. coli	*Escherichia coli*
ELISA	Enzyme-linked immunosorbent assay
EOF	Electroendosmotic flow
FACE	Fluorescence affinity capillary electrophoresis
FCS	Fetal calf serum
FIA	Flow injection analysis
FITC	Fluoresceine isothiocyanate

FPLC	Fast protein liquid chromatography
h	Human
HPLC	High performance liquid chromatography
HPMC	Hydroxypropylmethyl cellulose
i.d.	Inner diameter
IEXMA	Ion exchanger membrane adsorber chromatography
IEP	Isoelectric point
IgG	Immunoglobulin G
MA	Membrane adsorber
MECC	Micellar electrokinetic chromatography
Mw	Molecular weight
PAA	Polyacrylamide
r	Recombinant
RP–HPLC	Reversed phase–HPLC
RSD	Relative standard deviation
SDS	Sodiumdodecylsulfate
SRID	Single radial immunodiffusion

Progress in HPLC-HPCE, Vol. 5, pp. 155–172
H. Parvez *et al.* (Eds)
© VSP 1997.

Isoform characterization, peptide mapping and amino acids analysis of monoclonal antibodies by high performance capillary electrophoresis

J. F. VEILLON, C. RAMON and N. BIHOREAU

Laboratoire Français du Fractionnement et des Biotechnologies,
3 avenue des Tropiques, BP 100, 91943 Les Ulis Cedex, France

INTRODUCTION

High performance capillary electrophoresis (HPCE) analysis is finding increased usefulness for the characterization of proteins, peptides and amino acids. This paper describes the identification of isoproteins by free solution capillary electrophoresis (FSCE) or by capillary isoelectric focusing (CIF). For a purified monoclonal antibody, both techniques revealed several peaks on the electropherograms suggesting the presence of isoforms which differ by only 0.1 pH unit. Further, to determine which of the heavy or light chains is responsible for this heterogeneity, a kinetic analysis of immunoglobulin reduction was followed.

Peptide mapping of a human anti-Rhesus(D) monoclonal antibody has also been performed by HPCE using an ion-pairing agent such as hexane sulfonic acid (HSA) to minimize the interations between the capillary wall and the solute, and to increase the resolution of tryptic peptides. After injection of 6 femtomoles of the proteolysed antibody, an efficient separation of this complex peptide mixture was obtained at 15 kV with a 30 mM sodium phosphate pH 2.5 buffer containing 100 mM HSA.

HPCE has been also used as quick screening tool for sample purity evaluation prior to sequence analysis of isolated peptides obtained after peptide mapping by reverse phase high performance liquid chromatography (RP–HPLC). For some fractions, whereas the RP–HPLC chromatograms showed a single peak characterizing a single peptide, various peaks were observed by HPCE revealing the presence of different peptides in the presumably pure fractions.

Amino acids were analyzed by micellar electrokinetic capillary chromatography using a cationic surfactant and organic modifiers in the running buffer.

This optimisation led to the separation of eighteen phenylthiocarbamoyl amino acids 14 min after injection of very low quantities (100 fmol) and volumes (4 nl) of sample.

These results indicate that HPCE is a technique of choice for structural analysis of proteins; it is a rapid, sensitive, resolutive, quantitative and low consuming method.

Capillary electrophoresis (CE) is a powerful method, complementary to conventional techniques, for the characterization of biomolecules. As described by Jorgenson and Lukacs (1983), the highly efficient separations achieved by CE, which in terms of theoretical plates number is on the order of 10^6, are due to the use of high voltages. The use of capillaries with an internal diameter reduced to $50-100$ μm is indispensable to dissipate the Joule heat generated under these conditions.

Several modes of separation based on a different principle can be performed by CE. In free solution capillary electrophoresis, the proteins or peptides injected are separated according to differences in their electrophoretic mobilities. This parameter depends on the specific mobility of each molecule and on the electroendoosmosis which is a bulk flow of liquid resulting from the effect of the electric field on the electrical double-layer adjacent to the capillary wall. As reported by Lukacs and Jorgenson (1985), the electroosmosis is directly related to the pH of the running buffer. Since separations in FSCE depend on charge-to-mass ratio differences between molecules (Grossman *et al.*, 1989), one can manipulate the properties of the solvent (pH, ionic strength or viscosity) to modify the selectivity of the separation. Considering this selectivity, FSCE is well suited for the identification of subtle differences in protein molecules in their native state. Indeed, this technique has been employed to characterize isoforms of different proteins such as purified monoclonal antibodies, recombinant erythropoietin, and tissue plasminogen activator (Compton, 1991; Tran *et al.*, 1991; Taverna *et al.*, 1992). The presence of isoproteins has been attributed by these authors to posttranslational modifications such as deamidation, sulfation, or glycosylation of the proteins.

The main problem encountered in FSCE is the adsorption of polypeptides onto the capillary wall due to coulombic interactions between positively charged proteins and negatively charged capillary wall. This adsorption prohibits proper separations and leads to unreproducible peak areas and migration times. One option for avoiding wall adsorption is to operate at a pH above the pI of the proteins present in the sample, resulting in a repulsion between the capillary wall and the negatively charged polypeptides. Under these conditions, the electroosmotic flow is high and must be equal in magnitude but opposite in sign to the electrophoretic mobility to achieve a good resolution of substances having similar mobilities (Jorgenson and Lukacs, 1983). An alternative procedure to limit wall interactions is to operate at acidic pH (pH 2). In this case, electroosmosis is practically suppressed due to the reduction of the charge of the silica wall which no longer interacts with the positively charged polypeptides. Protein–capillary interactions can also be minimized by permanent modification of the capillary wall by covalently bonded phases (Novotny *et al.*, 1990). Nevertheless, the most serious problem of the silylation-based treatments is their notorious instability. The last alternative is a dynamic deactivation of the capillary using buffer additives such as alkali metal, hydrophylic polymers, ion pairing agents, surfactants or small amines (Green and Jorgenson, 1989; Rohlicek and Deyl, 1989; Landers *et al.*, 1992; Tran *et al.*, 1991).

Peptide mapping represents another important application of HPCE to analyse the primary structure of a protein. In this case, FSCE is performed as a complementary technique to reverse-phase HPLC since it uses a different mechanism for the analytical separation of peptides. Recently, Rush *et al.* (1993) described the peptide mapping analysis of recombinant human erythropoietin by FSCE using an ion pairing agent to increase selectivity.

Since FSCE does not separate the neutral or hydrophobic molecules, micellar electrokinetic capillary chromatography (MECC) has been developed as an alternative analytical method. It is based on the differential partitioning of an analyte between an ionic micelle and the surrounding aqueous phase in an open tubullar capillary (Terabe *et al.*, 1984). Above the critical micellar concentration (CMC), the ionized monomers of the surfactant aggregate creating micelles, considered as a dynamic pseudostationary phase which interacts with the analytes. Protein sequence is determined, after repetitive Edman degradation reaction, by the characterization of the cleaved terminal derivatized amino acids generally performed by liquid chromatography separation. Among the different applications of MECC, identification of phenylthiohydantoin (PTH) amino acids has been reported by Otsuka *et al.* (1985).

The present study describes optimisation of the analytes migration conditions in the capillary to segregate by FSCE the isoforms of a purified murine antiFcγR1 monoclonal immunoglobulin. Capillary isoelectric focusing, previously described by many authors (Hjerten *et al.*, 1987; and Chen and Wiktorowicz, 1992), has also been employed to confirm protein microheterogeneity and to determine the isoelectric point of each isoform. Further, a comparative kinetic analysis of reduction of this antibody by FSCE and two-dimensional (2D) electrophoresis is presented. In the second part of this paper, we describe the optimization of experimental conditions to obtain a reproducible and characteristic peak profile of a trypsin digested anti-Rhesus(D) human monoclonal antibody with a highly sensitive FSCE method. Since it is a fast, low consuming and selective technique, CE has also been employed to evaluate the purity of the tryptic peptides obtained from reverse-phase HPLC. Before peptide sequencing, such screening analysis should save time and sample, and will increase sequencer productivity. In the last part of this study, we present a rapid and efficient separation of fmol amounts of phenylthiocarbamoyl (PTC) amino acids by MECC.

MATERIALS AND METHODS

Equipment

High performance capillary electrophoresis was carried out on a model 270 A automated system (Applied Biosystems (ABI), San Jose, CA, USA). Data were recorded and integrated using a Macintosh IIcx with a data acquisition card (version 3.5) and software (model 600, ABI). The isoelectric focusing analysis has been performed with coated capillaries (50 μm i.d. × 78 cm) purchased from ABI. Uncoated fused-silica capillaries tubing (50 μm i.d. × 72 cm) were used for the other HPCE analysis. CE column with built-in Z shaped flow cell was purchased from ABI. This high sensitivity optical cell was assembled to a specific fused silica tubing with i.d. of 75 μm and a length from the injection end to the detector of 100 cm.

Materials

The monoclonal antiFcγR1 antibody was obtained from the culture of an hybridoma cell line (clone 22) in an automated continuous system using a protein free medium (Dhainaut *et al.*, 1992). After purification, the immunoglobulin was buffer exchanged into phosphate buffered saline (PBS), pH 7.2. As described in our previous work (Vincentelli and Bihoreau, 1993), the purity of the immunoglobulin was characterized by a single band and a single peak after SDS PAGE and HPLC analysis.

The monoclonal anti-Rhesus(D) antibody was secreted by a human lymphoblastoid cell F5 (Goossens *et al.*, 1987) in a hollow-fiber cartridge in presence of a serum free medium. After the purification process, the immunoglobulin was stored in an alanine-NaCl buffer, pH 6.8, at 4°C.

Reagents

1-Hexane sulfonic acid (HSA), 2-(N-cyclohexamino) ethane sulfonic acid (CHES), anhydrous sodium dihydrogen phosphate, 3-[(3cholamidopropyl) dimethylammonio]-1-propane sulfonate (CHAPS), Tris [hydroxymethyl] aminomethane, Tris [hydroxymethyl] aminomethane hydrochloride, Nonidet NP 40 and boric acid were purchased from Sigma. Tetra ethyl ammonium bromide (TEAB), and hexadecyl-trimethyl ammonium bromide (CTAB) were products of Aldrich. Acetonitrile and trifluoroacetic acid (TFA) were from J. T. Baker. Urea, trypsin, disodium tetraborate anhydrous, ethanol and methanol were obtained from Merck. Orthophosphoric acid was supplied by Fluka. Phenylisothiocyanate (PITC), triethylamine (TEA) and amino acids standards were from Pierce. Dithiothreitol (DTT), acrylamide N,N'-methylenebisacrylamide solution (37.5 : 1), N,N,N', N'-tetraethylene diamine (TEMED), ammonium persulfate and sodium dodecyl sulfate (SDS) were purchased from Biorad. Methylcellulose and the protein standards used for capillary isoelectric focusing were products of ABI. 2D ampholytes pH 3–10 were supplied by Millipore and ampholytes pH 5–7 range were from Pharmacia. All reagents were HPLC grade or analytical quality. The buffers were passed through a 0.22 Millex filter and degassed before use.

Analytical procedures

Two-dimensional SDS-polyacrylamide gel electrophoresis. Two-dimensional gel electrophoresis was performed using for the first dimension, the Millipore investigator 2D electrophoresis system. Isoelectric focusing (IEF) was conducted in a 4% T, 2.6% C polyacrylamide tube gel (18 cm length and 1 mm i.d.). The first migration gel was prepared by mixing 5.65 ml of an acrylamide stock solution (30% acrylamide, 0.8% N,N'-methylene bis acrylamide), 350 μl of 2D carrier ampholytes and 40 μl of a 10% ammonium persulfate solution. This gel contained also 9.5 M urea, 5 mM CHAPS and 2% (v/v) Nonidet NP 40 (Lopez *et al.*, 1991). The migration was performed at 1000 V for 16 h and 2000 V for 30 min using 100 mM NaOH and 100 mM H_3PO_4 as cathode and anode buffers, respectively. When the IEF migration was achieved, the gel was incubated 2 min in a solution containing 0.3 M Tris Base, 0.075 M Tris HCl, 50 mM DTT and 3% SDS (w/v). The IEF gel was then loaded on a SDS polyacrylamide gel and the second dimension was performed as described by Laemmli (1970).

Free solution capillary electrophoresis. The capillary was conditionned by flushing with 1 M NaOH for 20 min. Before each sample injection, the capillary was rinced with 0.1 M NaOH, and equilibrated by 20 mM CHES buffer, pH 9.5, under 20″ Hg vacuum for 5 min. After sample loading at the anodic side, the migration was performed at 3 kV toward the cathode. The absorbance was monitored on line at 200 nm.

Capillary isoelectric focusing. The coated capillary was washed with methanol during 10 min, rinced with water, filled with the cathode electrolyte solution (20 mM NaOH in 0.4% methylcellulose) during 10 min and then flushed out with ampholytes, 5–7 pH range, final concentration of 1.25% in 0.4% methylcellulose) during 2.5 min (Chen and Wiktorowicz, 1992). After sample loading, the capillary was flushed by the same buffer for 2.5 min. The protein was focused at 30 kV with the anode electrolyte solution (100 mM phosphoric acid in 0.4% methylcellulose) for 6.5 min. Proteins were then eluted at 30 kV, under 5″ Hg vacuum, toward the cathode (containing 20 mM NaOH). Detection was accomplished by absorption at 280 nm.

Peptide mapping. The lyophylised anti-Rhesus(D) human monoclonal antibody (0.5 to 2 mg) was denatured in a 8 M urea, 0.4 M ammonium hydrogen carbonate solution, reduced by dithiothreitol (20 moles DTT/mole cysteine) at 50 °C for 15 min and alkylated by iodoacetamide (2 moles/mole DTT) for 15 min in darkness at room temperature. Then, the polypeptides were diluted four times in water and hydrolysed with trypsin (1/10, w/w) 18 h at 37 °C (Stone *et al.*, 1989). Before each injection, the capillary was rinsed with 0.1 M NaOH, and flushed for 5 min with a 30 mM NaH_2PO_4/H_3PO_4 running buffer, pH 2.5, containing an alkyl sulfonate: the hexane sulfonic acid (Rush *et al.*, 1993). Electrophoresis was carried out at 30 kV, at 30 °C. After migration, the peptides were monitored at 200 nm.

Reverse-phase HPLC was performed on a Waters HPLC system using a C_{18} column (0.46 × 25 cm; Vydac) with particle size of 10 μm (300 Å). The column was equilibrated with 0.1% trifluoroacetic acid (TFA) (buffer A). The non polar solvent (buffer B) was composed of 0.115% TFA in acetonitrile. The column was eluted at a flow rate of 0.8 ml/min with 0% B for 15 min after injection followed by linear gradients of 0 to 33% over a period of 265 min and of 33 to 60% during 30 min. The chromatography was performed at 30 °C and components were monitored at 214 nm.

Amino acids analysis. The amino acids were dried and derivatized for 20 min at room temperature in a solution containing phenylisothiocyanate (PITC) 10% (v/v), ethanol 70% (v/v), water 10% (v/v) and triethylamine (TEA) 10% (v/v). The PTC-amino acids generated were redried under 65 mtorr vacuum and diluted in acetonitrile (Aitken *et al.*, 1989). For HPCE analysis, the capillary was washed with 0.1 M NaOH solution and flushed by the running buffer for 5 min before sample loading. Amino acids migrated within the capillary in a 6 mM tetraborate/phosphate buffer, pH 7.1 containing 12.5 mM CTAB (Corran and Sutcliffe, 1993) with 40 mM TEAB. Under cationic micellar conditions, the electrophoresis were carried out at 30 kV toward the anode. The electropherograms were monitored at 254 nm.

RESULTS AND DISCUSSION

Protein application

SDS–PAGE analysis of the antiFcγR1 IgG showed a single Coomassie-stained band on the gel (Dhainaut *et al.*, 1992; Vincentelli and Bihoreau, 1993) indicating that the protein preparation used for the study was highly purified and homogenous. By contrast, the two-dimensional gel electrophoresis characterization of this Mr 150 000 IgG showed four bands corresponding to pI values ranging from 6.3 to 6.5 (Fig. 1). This result suggested the presence of several isoforms of the purified antibody.

pI determinations of the different IgG isoproteins were also performed by isoelectric focusing capillary electrophoresis. Figure 2 represents the electropherogram obtained after mobilization of the focused proteins. IgG was eluted

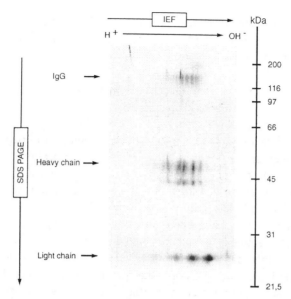

Figure 1. Two-dimensional gel electrophoresis of a partially reduced antiFcγR1 immunoglobulin. First dimension: isoelectric focusing (IEF) was performed at 1000 V during 16 h using a 3–10 pH gradient. Second dimension: SDS PAGE was conducted at 25 mA during 4 h with a 10% T polyacrylamide gel. Buffer: Tris-glycine SDS, pH 8.3. Different proteins of molecular mass ranging from 200 to 21.5 kDa were used as markers. The gel was stained in a 0.2% Coomassie brillant blue R-250 solution.

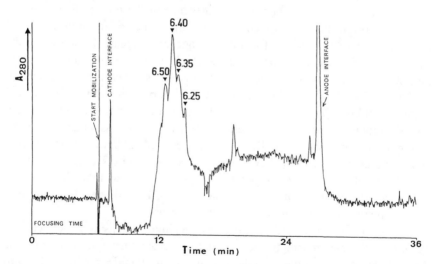

Figure 2. Capillary isoelectric focusing of the antiFcγR1 antibody. After the capillary conditioning with ampholytes 5–7 (under 20″ Hg vacuum), 4 nl of the sample (0.5 mg/ml) were injected. Conditions: capillary length, 78 cm; capillary diameter, 50 μm; prefocusing time, 6.5 min at 30 kV (400 V/cm); mobilization, 30 kV, 5″ Hg vacuum. Cathode electrolyte solution: 20 mM NaOH in 0.4% methylcellulose. Anode electrolyte solution: 100 mM H_3PO_4 in 0.4% methylcellulose. The proteins were monitored at 280 nm.

12 min after injection in between the first and last peak, which correspond
to the cathode and anode interfaces, respectively. This purified protein was
characterized by four peaks corresponding to pI values of 6.25, 6.35, 6.40
and 6.50.

Since the isoforms have the same molecular mass, they are separated in
FSCE only by their differences in charge. As reported by Landers *et al.* (1992),
isoforms separation could be achieved by modification of the running buffer
(pH and concentration) and the applied voltage. Figure 3 illustrates the overlayed electropherograms obtained at pH 11, 9.5 and 8.3 after injection of the
antiFcγR1 IgG. At pH 11 (Fig. 3A), the protein was eluted as a broad peak
without separation of the isoforms. Some sign of separation was observed at
pH 8.3 (Fig. 3B) and clearly improved using a 20 mM CHES buffer, pH 9.5
(Fig. 3C) and reducing the applied voltage to 3 kV (Fig. 3D). These results
indicated that the antibody was characterized in FSCE by different peaks confirming the presence of four isoforms. At the different pH values tested, the
protein and the inner capillary wall were both negatively charged, minimizing
the effect of solute-wall interaction. Further, the electroendoosmotic flow generated upon application of the electrical field and the electrophoretic migration
of the solutes were opposite, increasing the resolution. The separation observed
at pH 9.5 should be due to the combination of increased charge differences between the IgG isoforms and of the reduction in electrophoretic flow.

A kinetic analysis of reduction has been followed by 2D gel electrophoresis
and FSCE analysis to ascertain that the presence of these isoproteins was not
due to IgG fragments and to identify which of the heavy and light chains of
the antibody was responsible for this heterogeneity. Figure 1 shows that the
Mr 50 000 heavy chain or the Mr 25 000 light chain were characterized by
multiple bands indicating the presence of isoproteins for each reduced form of
the IgG1. The pI values of the heavy chain isoforms ranged from 6.1 to 6.4
while the different light chain polypeptides have a more basic pI ranging from
6.3 to 7. Figure 4 represents the overlayed electropherograms obtained after injection of the unreduced, partially and totally reduced antibody. These different
steps were characterized, in FSCE, by the decrease and disappearance of the
peak 2 (Fig. 4A and 4B) corresponding to the total IgG and the simultaneous
appearance of heterogenous peaks (peak 1 and 3, Fig. 4B and 4C) reflecting
the presence of the light and heavy chain isoforms.

Although the negatively charged heavy and light chains were attracted to
the anode, the electroosmotic flow, which is greater in velocity, carried the
charged species toward the cathode with an electrophoretic mobility related to
the pI. Considering the different pI values (Fig. 1), at pH 9.5 the heavy chains
were more electronegatively charged than the light chains leading to a greater
electrophoretic mobility in the capillary toward the anode and thus to a slower
global migration velocity toward the cathode as revealed on Fig. 4C (peaks 3).

These results, which are consistent with the data obtained after the 2D electrophoresis analysis (Fig. 1), validated CE as an attractive complement to

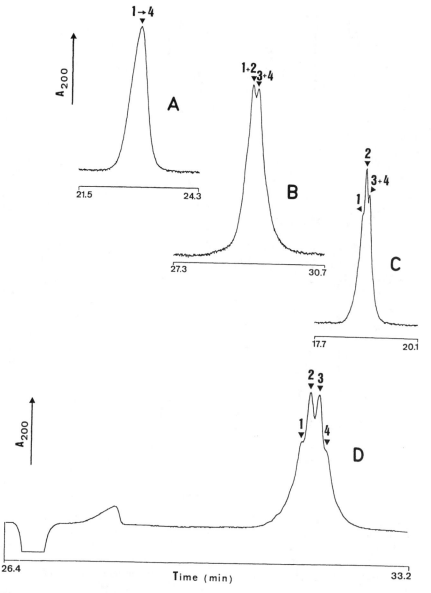

Figure 3. Effect of buffer pH and applied voltage on the separation of antiFcγR1 isoforms by FSCE. Conditions: buffer (A) 20 mM CAPS, pH 11; (B) 20 mM borate, pH 8.3; (C) 20 mM CHES, pH 9.5, at 5 kV (70 V/cm); and (D) 20 mM CHES, pH 9.5, at 3 kV (42 V/cm). The horizontal axis corresponds to the migration time and the vertical axis represents the absorbance at 200 nm.

conventional analytical techniques for the characterization of protein isoforms which differ by only 0.1 pH unit. In our previous work, F(ab')$_2$ isoforms of the same antibody have been characterized by FSCE with a detection limit of 80 amol (Vincentelli and Bihoreau, 1993).

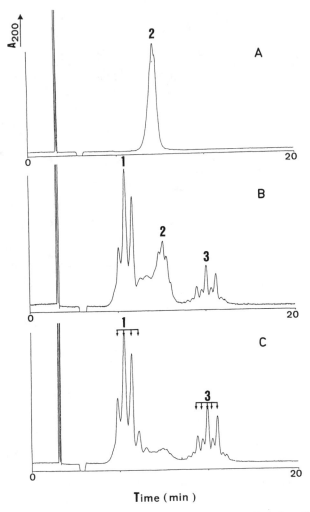

Figure 4. Electropherograms of the native (A), partially reduced (B) and totally reduced (C) antiFcγR1 antibody. The electrophoresis was performed at 30 kV (416 V/cm) for the first 2.5 min and 3 kV (42 V/cm) for the remaining 17 min using a 20 mM CHES buffer, pH 9.5. Peak 2 corresponds to the IgG isoforms while the peaks 1 and 3 represent the light and heavy chains of the antibody, respectively. The different electropherograms were monitored at 200 nm.

Peptide mapping

The goal of this study was to present capillary electrophoresis as a highly resolutive technique, complementary to reverse-phase high-performance liquid chromatography, for peptide mapping analysis. Human monoclonal anti-Rhesus(D) antibody, a protein of Mr 150 000 and with 16 disulfide linkages, served as a model for this study. Reproducibility of the electropherograms obtained from a digested polypeptide allowed the definition of a protein specific peptide map. To obtain the desired peptide in a reproducible manner after a complete digestion of the protein, any cleavage sites must be exposed (Cobb and Novotny, 1992). Thus, the antibody has been denatured and the disulfide bonds have been eliminated by reduction and alkylation prior to trypsin proteolysis.

Recently, peptide mapping analysis has been performed by CE using an ion pairing reagent such as alkyl sulfonate. The rationale for employing this type of surfactant was to increase selectivity by reducing the electrophoretic mobility of the analyte and by decreasing solute-wall interactions. Figure 5 represents the maps obtained using different concentrations of hexane sulfonic acid (HSA). In the absence of surfactant in the running buffer (Fig. 5A) the tryptic peptides were not clearly separated. By contrast, in presence of 50 mM HSA (Fig. 5B), the mobility of the peptides decreased and the resolution was improved. The total protease digestion of the reduced and alkylated antibody should generate 64 peptides according to the theoretical amino acid composition. Despite the complexity of the mixture, addition of 100 mM HSA (Fig. 5C) led to the identification of an equivalent number of peaks which are nearly baseline resolved. The separation took 35 min. Figure 5C shows that the electropherogram is segregated into two portions; in the first one we observed an abundance of sharp peaks (detected between 6 and 12 min) with slight overlaps while in the second portion, broader peaks were detected (between 12 to 35 min). This difference could be explained either by interactions of the last eluted peptides with the silanols of the capillary wall or by postranslational modifications, such as glycosylation, of these specific peptides as described by Rush *et al.* (1993). To improve resolution in the first portion of the map, the applied voltage has been reduced to 15 kV. We observed in Fig. 5D that the selectivity was enhanced by decreasing the proper velocity of each peptide. The HPCE sensitivity has been assessed in the peptide mapping analysis using a high sensitivity optical cell. Figure 6 presents the resolution of a characteristic peptide map obtained after injection of 6 fmol (2 nl) of tryptic digested immunoglobulin.

Complementarity between HPLC and CE was also achieved for purity screening of peptides, subsequent to HPLC fractionation, and prior to peptide sequencing. The different tryptic peptides obtained after digestion of the anti-Rhesus(D) monoclonal antibody have been isolated by RP–HPLC (Fig. 7). Considering the great number of peptides, a slow acetonitrile gradient (slope of 0,12%/min) was used to attain a good resolution. Nevertheless, under these conditions, the separation took 270 min.

After this preparative chromatography, a comparative analytical study of each eluted peptides has been performed by CE and HPLC. The efficiency of this purity screening method is presented in Fig. 8 for three different fractions (fr1, fr2 and fr3, Fig. 7) which were characterized by a single symmetrical peak on the HPLC profile (Figs 8A, 8B and 8C). The electropherograms obtained after injection in the capillary of fraction 1 (Fig. 8D) also showed a single peak confirming the purity of the sample. By contrast, the heterogeneity observed in Figs 8E and 8F suggested the presence of different 'contaminating' peptides

J. F. Veillon et al.

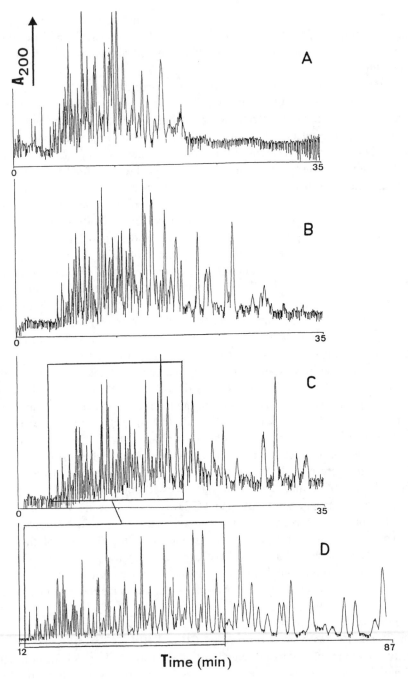

Figure 5. Peptide mapping of the human monoclonal anti-Rhesus(D) antibody by FSCE: Effects of the HSA concentration and applied voltage on peptide resolution. Conditions: capillary length, 72 cm, capillary diameter, 50 μm; field, 30 kV (416 V/cm); running buffer, A: 30 mM sodium phosphate/phosphoric acid, pH 2.5, containing B: 50 mM HSA, C: 100 mM HSA and D: 100 mM HSA at 15 kV (208 V/cm). Detection was accomplished by using UV absorption at 200 nm.

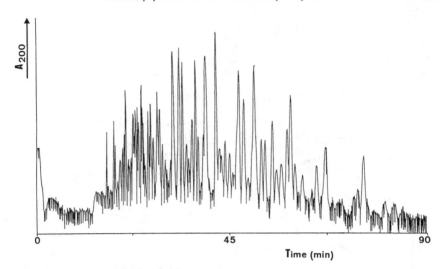

Figure 6. Detection limits for the tryptic peptides of the human monoclonal anti-Rhesus(D) antibody using a high sensitivity optical cell. Peptides were separated with a phosphate buffer containing 100 mM HSA, pH 2.5, at 30 kV. Capillary: 130 cm × 75 μm i.d. Detection: UV absorption at 200 nm.

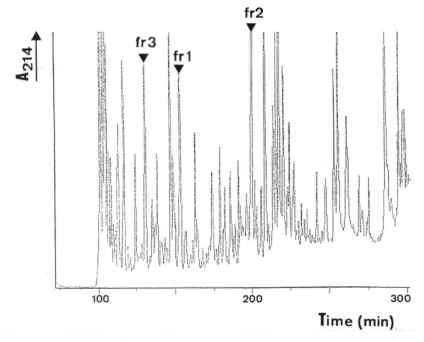

Figure 7. Separation of tryptic peptides of monoclonal anti-Rhesus(D) antibody by reversed-phase HPLC. Chromatographic conditions: column, 218TP104 (4.6 × 250 mm, Vydac); flow rate, 0.8 ml/min; detector, 214 nm wavelength; buffer A, 0.1% TFA in water; buffer B, 0.115% TFA in acetonitrile; gradient, 0–33% B in 265 min, 33–60% B in 30 min. Different fractions (fr1, fr2, and fr3) have been collected for FSCE characterization.

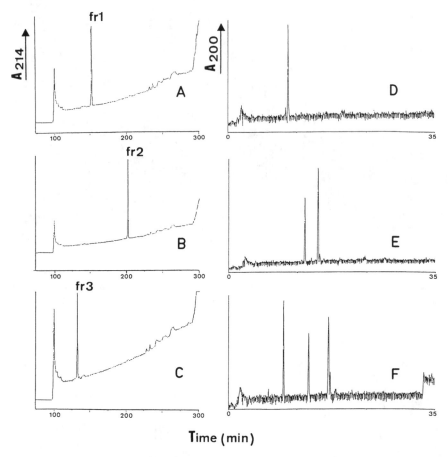

Time (min)

Figure 8. Comparative screening of the tryptic peptides by RP–HPLC and FSCE. Fractions 1, 2 and 3 eluted from the preparative RP–HPLC (Fig. 7) have been separately analysed by RP–HPLC (A, B, C) and FSCE (D, E, F). RP–HPLC conditions: same as in Fig. 7, injection volume, 200 µl. FSCE conditions: same as in Fig. 5B, injection volume, 40 nl. Chromatograms and electropherograms were monitored at 214 and 200 nm, respectively.

in fractions 2 and 3. These results indicated that FSCE, based on a mechanism of separation different from that of RP–HPLC, is a powerful method to obtain complementary information about sample purity. In this case, a second preparative HPLC separation method using different conditions had to be performed.

Amino acids analysis

Among the different amino acids analyses performed by micellar electrokinetic capillary chromatography using various derivatizing agents (Liu *et al.*, 1988; Novotny *et al.*, 1990; Terabe *et al.*, 1993), separation of PTH-amino acids has been demonstrated by Otsuka *et al.* (1985) using SDS or dodecyltrimethylammonium bromide (DTAB). In our study, the effect of the addition of organic

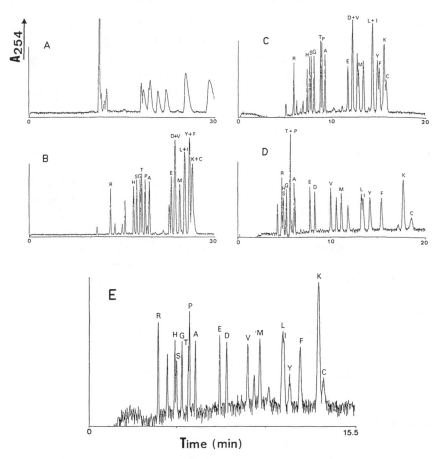

Figure 9. Micellar capillary electrophoresis separation of 17 PTC-amino acids with CTAB. Migration conditions: field, 15 kV (208 V/cm); buffer, 6 mM tetraborate/phosphate buffer, pH 7.1 containing A: 1 mM CTAB; B: 10 mM CTAB. Amino acids separation was then conducted at 30 kV, for 30 min, with C: 10 mM CTAB, D: 10 mM CTAB, 40 mM TEAB and E: 12.5 mM CTAB, 40 mM TEAB. Detection limit was determined by injection of 100 fmol of each residue.

modifiers to a cetyltrimethyl ammonium bromide (CTAB) micellar solution was determined to improve resolution of 17 PTC-amino acids. Figure 9A shows that under the CMC (1 mM CTAB), the PTC-amino acids were not resolved while in presence of 10 mM CTAB (above the CMC), 15 peaks were detected (Fig. 9B). An increase of the applied voltage to 30 kV (Fig. 9C) led to a separation between lysine and cysteine and between phenylalanine and tyrosine. No significant separation effect was observed after addition of modifiers such as TMAB or methanol while, as presented in Fig. 9D, with 40 mM TEAB, aspartic acid and valine were clearly separated. Figure 9E shows that, by using a tetraborate-phosphate, pH 7.1, buffer containing 12.5 mM CTAB and 40 mM TEAB, the 17 PTC-amino acids were resolved, with an overlap between leucine and isoleucine. Under these conditions, PTC-tryptophan, which had been injected separately, eluted between phenylalanine and lysine (data not shown). Separation of the derivatized residues took only 14 min with a detection limit of 100 fmol (S/N = 3). This sensitivity could be improved 30-fold using the high sensitivity optical cell. Detection of subfemtomole quantities (0.5 fmol) of PTH-amino acids have been reported by Waldron and Dovichi (1992) using a thermooptical absorbance detector.

CONCLUSION

Capillary electrophoresis is probably the most successful analytical method developed during the last several years for the separation and the characterization of proteins. It provides quantitative information and reveals subtle differences in protein molecules in their native state. Further, the high resolution, the short time of analysis and the low sample quantities required identifies capillary electrophoresis as a powerful complementary method to conventional techniques for the studies of the primary structure of proteins such as peptide mapping. For these structural analyses, conclusive results have been recently reported by coupling capillary electrophoresis with mass spectrometry (Niessen *et al.*, 1993). This technique will be a method of choice for protein characterization since it combines on-line the highly efficient separations in capillary electrophoresis to the ultrasensitive detection and the precise identification of a mass spectrometer.

REFERENCES

Aitken, A., Geisow, M. J., Findlay, J. B. C., Holmes, C. and Yarwood, A. (1989). Peptide preparation and characterization. In: *Protein Sequencing. A Practical Approach*, Findlay, J. B. C. and Geisow, M. J. (Eds). IRL Press, Oxford, pp. 43–68.

Chen, S-W. and Wiktorowicz, J. E. (1992). Isoelectric focusing by free solution capillary electrophoresis. *Anal. Biochem.* **206**, 84–90.

Cobb, K. A. and Novotny, M. V. (1992). Peptide mapping of complex proteins at the low-picomole level with capillary electrophoretic separations. *Anal. Chem.* **64**, 879–886.

Compton, B. J. (1991). Electrophoretic mobility modeling of proteins in free zone capillary electrophoresis and its application to monoclonal antibody microheterogeneity analysis. *J. Chromatogr.* **559**, 357–366.

Corran, P. H. and Sutcliffe, N. (1993). Identification of gonadorelin (LHRH) derivatives: comparison of reversed-phase high performance liquid chromatography and micellar electrokinetic chromatography. *J. Chromatogr.* **636**, 87–94.

Dhainaut, F., Bihoreau, N., Metereau, J. L., Lirochon, J., Vincentelli, R. and Mignot, G. (1992). Continuous production of large amounts of monoclonal immunoglobulins in hollow fibers using protein-free medium. *Cytotechnology* **10**, 33–41.

Goossens, D., Champomier, F., Rouger, P. and Salmon, C. (1987). Human monoclonal antibodies against blood antigens: preparation of series of stable EBV-immortalized B-clones producing high levels of antibody of different isotypes and specificities. *J. Immunol. Methods* **101**, 193–220.

Green, J. S. and Jorgenson, J. W. (1989). Minimizing adsorption of proteins on fused silica in capillary zone electrophoresis by addition of alkali metal salts to the buffers. *J. Chromatogr.* **478**, 63–70.

Grossman, P. D., Colburn, J. C., Lauer, H. H., Nielsen, R. G., Riggin, R. M., Sittampalam, G. S. and Rickard, E. C. (1989). Application of free-solution capillary electrophoresis to the analytical scale separation of proteins and peptides. *Anal. Chem.* **61**, 1186–1194.

Hjerten, S., Elenbring, K., Kilar, F., Liao, J., Chen, A. J. C., Siebert, C. J. and Zhu, M. (1987). Carrier-free zone electrophoresis, displacement electrophoresis and isoelectric focusing in a high performance electrophoresis apparatus. *J. Chromatogr.* **403**, 47–61.

Jorgenson, J. W. and Lukacs, K. D. A. (1983). Capillary zone electrophoresis. *Science* **222**, 266–272.

Laemmli, U. K. (1970). Cleavage of structural proteins during the assembly of the head of bacteriophage T4. *Nature* **277**, 680–688.

Landers, J. P., Oda, R. P., Madden, B. J. and Spelsberg, T. C. (1992). High-performance capillary electrophoresis of glycoproteins: the use of modifiers of electroosmotic flow for analysis of microheterogeneity. *Anal. Biochem.* **205**, 115–124.

Liu, J., Cobb, K. A. and Novotny, M. V. (1988). Separation of precolumn ortho-phthalaldehyde-derivatized amino acids by capillary zone electrophoresis with normal and micellar solutions in the presence of organic modifiers. *J. Chromatogr.* **468**, 55–65.

Lopez, M. F., Patton, W. F., Utterback, B. L., Chung-Welch, N., Barry, P., Skea, W. M. and Cambria, R. P. (1991). Effect of various detergents on protein migration in the second dimension of two-dimensional gels. *Anal. Biochem.* **199**, 35–44.

Lukacs, K. D. and Jorgenson, J. W. (1985). Capillary zone electrophoresis: effect of physical parameters on separation efficiency and quantitation. *J. High Resolut. Chromatogr. Chromatogr. Commun.* **8**, 407–411.

Niessen, W. M. A., Tjaden, U. R. and van der Greef, J. (1993). Capillary electrophoresis-mass spectrometry. *J. Chromatogr.* **636**, 3–19.

Novotny, M. V., Cobb, K. A. and Liu, J. (1990). Recent advances in capillary electrophoresis of proteins, peptides and amino acids. *Electrophoresis* **11**, 735–749.

Otsuka, K., Terabe, S. and Ando, T. (1985). Electrokinetic chromatography with micellar solutions separation of phenylthiohydantoin-amino acids. *J. Chromatogr.* **332**, 219–226.

Rohlicek, V. and Deyl, Z. (1989). Simple apparatus for capillary zone electrophoresis and its application to protein analysis. *J. Chromatogr.* **494**, 87–99.

Rush, R. S., Derby, P. L., Strickland, T. W. and Rohde, M. F. (1993). Peptide mapping and evaluation of glycopeptide microheterogeneity derived from endoproteinase digestion of erythropoietin by affinity high-performance capillary electrophoresis. *Anal. Chem.* **65**, 1834–1842.

Stone, K. L., LoPresti, M. B., Myron Crawford, J., DeAngelis, R. and Williams, K. R. (1989). Enzymatic digestion of proteins and HPLC peptide isolation. In: *A Practical Quide to Protein and Peptide Purification for Microsequencing*, Matsudaira, P. T. (Ed.). Academic Press, San Diego, pp. 31–47.

Taverna, M., Baillet, A., Biou, D., Schlüter, M., Werner, R. and Ferrier, D. (1992). Analysis of carbohydrate-mediated heterogeneity and characterization of N-linked oligosaccharides of glycoproteins by high performance capillary electrophoresis. *Electrophoresis* **13**, 359–366.

Terabe, S., Otsuka, K., Ichikawa, K., Tsuchiya, A. and Ando, T. (1984). Electrokinetic separation with micellar solutions and open-tubular capillaries. *Anal. Chem.* **56**, 111–113.

Terabe, S., Miyashita, Y., Ishihama, Y. and Shibata, O. (1993). Cyclodextrin-modified micellar electrokinetic chromatography: separation of hydrophobic and enantiomeric compounds. *J. Chromatogr.* **636**, 47–55.

Tran, A. D., Park, S., Lisi, P. J., Huyhh, O. T., Ryall, R. R. and Lane, P. A. (1991). Separation of carbohydrate-mediated microheterogeneity of recombinant erythropoietin by free solution capillary electrophoresis: effects of pH, buffer type and organic additives. *J. Chromatogr.* **542**, 459–471.

Vincentelli, R. and Bihoreau, N. (1993). Characterisation of each isoform of a F(ab′)$_2$ by capillary electrophoresis. *J. Chromatogr.* **641**, 383–390.

Waldron, K. C. and Dovichi, N. J., (1992). Sub-femtomole determination of phenylthiohydantoin-amino acids: capillary electrophoresis and thermooptical detection. *Anal. Chem.* **64**, 1396–1399.

APPENDIX: ABBREVIATIONS

CE	capillary electrophoresis
CHES	2-(N-cyclohexamino)ethane sulfonic acid
CMC	critical micellar concentration
CTAB	hexadecyl trimethyl ammonium bromide
FSCE	free solution capillary electrophoresis
HPCE	high performance capillary electrophoresis
HSA	hexane sulfonic acid
IEF	isoelectric focusing
MECC	micellar electrokinetic capillary chromatography
PTC	phenylthiocarbamoyl
PTH	phenylthiohydantoin
RP–HPLC	reverse phase high performance liquid chromatography
SDS	sodium dodecyl sulfate
SDS	sodium dodecyl sulfate
SDS PAGE	SDS protein analysis gel electrophoresis
TEAB	tetra ethyl ammonium bromide

Progress in HPLC-HPCE, Vol. 5, pp. 173–197
H. Parvez *et al.* (Eds)
© VSP 1997.

Capillary zone electrophoresis and isotachophoresis of biologically active peptides

VÁCLAV KAŠIČKA and ZDENĚK PRUSÍK

Institute of Organic Chemistry and Biochemistry, Academy of Sciences of the Czech Republic, Flemingovo 2, 166 10 Prague 6, Czech Republic

INTRODUCTION

High-performance electromigration separation techniques, capillary zone electrophoresis (CZE) and capillary isotachophoresis (CITP), are presented as powerful and useful techniques in the field of analysis and preparation of biologically active peptides. They can provide fast and accurate information on peptide purity and they can be used for modelling and evaluation of preparative peptide separations. They are becoming a recognized complement and/or counterpart of high-performance liquid chromatography.

The rules for rational selection of experimental conditions of CZE separation of peptides are given and suitable methods of peptide detection in CZE and CITP are described.

Some examples of applications of CZE and CITP to peptide purity determination are demonstrated and different ways of peptide purity evaluation are presented.

Based on the correlation between CZE and free-flow zone electrophoresis (FFZE) a procedure for conversion of analytical capillary separations into preparative separations realized by FFZE is described and its application to analysis and preparation of synthetic biopeptides is shown.

Use of CITP for determination of low-molecular-mass ionic admixtures in peptide preparations is presented.

Peptides represent a large and complex group of biologically active substances. Their occurrence and function in nature are extremely variable and of vital importance. Peptides act as hormonal, neuronal and immunity regulators, co-enzymes or enzyme inhibitors, drugs, toxins, antibiotics, etc. Consequently, any high resolution method applicable to their efficient separation and characterization is of value for those chemists, biologists and all other specialists who are dealing with peptide research and applications. For that reason, the use of capillary zone electrophoresis and isotachophoresis for peptide analysis, preparation and physico-chemical characterisation is an area of great importance.

Natural peptides might be composed of over 20 amino acid species linked by peptide bonds and sometimes also cross-linked by the disulphide bridges whereas noncoded amino acid residues and various types of crosslinking may occur in synthetic peptides. A huge amount of variation in amino acid sequences in a polypeptide chain is the cause of an extraordinary heterogeneity of peptides. They differ in their size — depending on the number of linked amino acid residues, their relative molecular mass can range from a few hundreds for oligopeptides containing 2–10 amino acid residues up to several thousands for polypeptides containing tens of amino acid residues. Electric charge and steric arrangement of a peptide depend not only on the type and number of amino acid residues but also on their sequence in the peptide chain. All these differences result in different electrophoretic mobilities of peptides allowing thus the use of high performance electromigration methods — capillary zone electrophoresis (CZE) and capillary isotachophoresis (CITP) — for peptide separation.

SELECTION OF SEPARATION CONDITIONS

General

When selecting suitable conditions for the CZE and ITP separations of peptides the general rules for CZE (Grossman, 1992; Li, 1992; Foret *et al.*, 1993) or for ITP (Everaerts *et al.*, 1976; Boček *et al.*, 1988) should be followed and in addition, the following specific properties of peptides should be taken into account.

1. Amphoteric character.
2. Diversity of amino acid composition and sequences resulting in a wide range of solubilities and mobilities, including relatively low mobilities and solubilities.
3. Biological activity, chemical and temperature lability.
4. Complexity of peptide mixtures from biological material, mutual interactions.
5. Adsorption of peptides to the fused silica.

The influence of these factors on the selection of parameters of CZE separation is described below; the selection of parameters of ITP separation can be found in our previous papers (Kašička and Prusík, 1989, 1991).

Composition and pH of background electrolyte

Peptides are amphoteric (poly)electrolytes or (poly)ampholytes. Their effective (net) charge (sum of all charges including their sign) is strongly dependent on pH because of the presence of various amounts of different types of ionogenic groups. Consequently, for the rational selection of pH and composition of background electrolyte (BGE) it is advantageous to know the pH dependence of the effective charge and specific charge of peptides to be analyzed by CZE.

Therefore, we have developed a procedure and computer program for calculation of these dependences. It allows us to calculate the effective and specific charges of any peptide, the amino acid sequence of which is known and if the pK values of the ionogenic groups present are known or at least can be estimated. The program is based on the mathematical model of the acid-base equilibria of a general ampholyte. It is described in detail elsewhere (Kašička and Prusík, 1989); here, only the main points will be mentioned.

Let us consider a peptide P with maximal (positive) charge M and with minimal (negative) charge N. Let $P(J)$ be the ionic form of peptide P with charge J and $K(J)$ be the apparent dissociation constant of the equilibria between the ionic forms $P(J + 1)$ and $P(J)$, i.e.

$$K(J) = c_{P(J)}c_H/c_{P(J+1)}, \qquad (1)$$

where $c_{P(J)}$ and $c_{P(J+1)}$ are the equilibrium concentrations of peptide ionic forms $P(J)$ and $P(J + 1)$ and c_H is the equilibrium concentration of hydrogen ions. The molar fraction of component $P(J)$, $D_{P(J)}$, referred to the total concentration, c_P, of peptide P is defined as

$$D_{P(J)} = c_{P(J)}/c_P. \qquad (2)$$

It was derived that $D_{P(J)}$ can be expressed as a function of c_H with parameters $K(J)$, N, M, i.e.

$$D_{P(J)} = f(c_H; [K(J), N, M]), \quad J = N \ldots M. \qquad (3)$$

If molar fractions of all ionic forms $P(J)$ are calculated, then the effective charge of peptide P, $z_{P,ef}$, can be determined:

$$z_{P,ef} = \sum J D_{P(J)}, \quad J = N \ldots M. \qquad (4)$$

From the effective charge, $z_{P,ef}$, the specific charge, $z_{P,sp}$, can be calculated, i.e. the charge referred to unit relative molecular mass, M_P, of peptide P:

$$z_{P,sp} = z_{P,ef}/M_P. \qquad (5)$$

The relationship between effective mobility of peptide P, $m_{P,ef}$, and its effective charge, $z_{P,ef}$, and relative molecular mass, M_P, is described by Offord's equation (Offord, 1966) derived empirically from peptide separations by paper zone electrophoresis:

$$m_{P,ef} = k z_{P,ef}/(M_P)^{2/3}, \qquad (6)$$

where k is an empirical constant.

The validity of Offord's relation was confirmed also for free-solution CZE by Rickard *et al.* (1991), whereas Compton (1991) prefers the form:

$$m_{P,ef} = A z_{P,ef} / \left[B(M_P)^{1/3} + C(M_P)^{2/3} \right]. \tag{7}$$

Other forms were suggested by Grossman *et al.* (1989):

$$m_{P,ef} = A \log(z_{P,ef} + 1)/n^B, \tag{8}$$

(n is the number of amino acid residues) and by Cifuentes and Poppe (1994):

$$m_{P,ef} = A \log(1 + B \cdot z_{P,ef})/(M_P)^C. \tag{9}$$

In all these equations A, B, C are empirical constants dependent on the electrolyte system used.

The calculated dependences of effective and specific charges of simple peptides of diglycine and triglycine on pH are shown in Fig. 1.

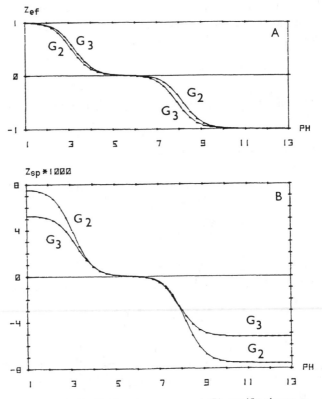

Figure 1. Dependence of (A) effective charge, z_{ef}, and (B) specific charge, z_{sp}, of diglycine, Gly_2, and triglycine, Gly_3, on pH.

It is obvious that from these dependences useful information for the selection of pH and composition of BGE can be obtained, e.g. the regions of minimal (negative) and maximal (positive) charges, regions of strong or weak dependence of effective and specific charge on pH and the isoionic (isoelectric) point (pI) can be determined.

The pH of BGE should be at least 1–2 units distant from pI since close to pI both electrophoretic mobility and solubility of peptide are relatively low. The pH of BGE should be taken from the region where specific charge is greater than 2×10^{-4}–5×10^{-4}. With respect to the fact that effective mobility is directly proportional to the effective charge and indirectly proportional to the relative molecular mass (see (6)–(9)), it is obvious that the difference of peptide mobilities will be maximal at the same region where the difference of specific charges is maximal.

For the given example of Gly_2 and Gly_3 it is evident that a suitable pH for their separation is either in the low acid region (pH < 3) or at alkaline pH (pH > 9). On the basis of these calculations, and with respect to other aspects (see later on), the 0.5 mol/l acetic acid, pH 2.5, was chosen as BGE for the separation of these two peptides which are often used as test mixture in our laboratory. A record of the CZE separation of these two peptides and electroosmotic flow marker, phenol, is in Fig. 2.

The calculation of the pH dependence of specific charge for a more complicated example, synthetic hexapeptides of growth hormone releasing peptide (GHRP), is shown in Fig. 3. Curve 1 is calculated for His^1-GHRP (sequence His-D-Trp-Ala-Trp-D-Phe-Lys.NH$_2$, $N = 0$, $M = 3$, $M_r = 873.1$, pK(2) = 5.8, pK(1) = 7.8, pK(0) = 10.4) and curve 2 for Tyr^1-GHRP (sequence Tyr-D-

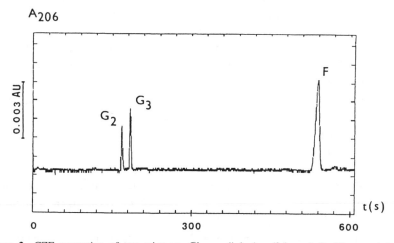

Figure 2. CZE separation of test mixture: Gly_2 – diglycine (0.3 mg/ml), Gly_3 – triglycine (0.3 mg/ml), F – phenol (0.1 mg/ml). BGE: 0.5 mol/l acetic acid, pH 2.5. Capillary: i.d. 0.05 mm, effective length 220 mm, total length 310 mm. Voltage 9.0 kV, current 10.1 µA, temperature, ambient 23 °C. A_{206} – absorbance at 206 nm, t – migration time.

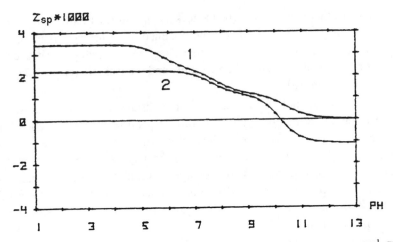

Figure 3. Calculated dependence of specific charge, z_{sp}, of synthetic peptides His[1]-GHRP (curve 1) and Tyr[1]-GHRP (curve 2) on pH. Peptide sequences and other parameters are given in the text.

Trp-D-Ala-Trp-D-Phe-Lys.NH$_2$, $N = -1$, $M = 2$, $M_r = 899.1$, pK(1) = 7.6, pK(0) = 10.0, pK(−1) = 10.4).

From the course of these dependencies it follows that the suitable region for the separation is at acid pH and that for pH < 4.5 the difference of specific charge (effective mobility) is independent of pH. Separation at high pH (pH > 11) should be from the point of view of different specific charges and is also possible but it cannot be recommended because of hydrolysis of amide groups at high pH.

The accuracy of these calculated dependencies is negatively influenced by the fact that the average (estimated) pK values of the ionogenic group were used instead of individual constants of these groups in the given peptide. For that reason the conclusions derived from the calculated dependencies can be used only as a first approximation of the suitable pH which can be further optimized by experimental tests.

Recently a similar strategy for the prediction of peptide charge and mobility from the peptide sequence has been published (Cifuentes and Poppe, 1994) which tries to predict more precise values of individual pK values of ionogenic groups present in the peptide chain.

According to the predicted suitable pH for peptide separation the composition of BGE is selected. The main constituent(s) of BGE should possess buffering capacity at given pH, i.e. their pK should fulfil the conditions pH$_{BGE}$ = pK ± 0.5.

Of course, pH of BGE must be selected not only according to pH dependence of effective and specific charge of peptides to be separated, but also with respect to their solubility, chemical and thermal lability and biological activity. These aspects are discussed in more detail in the following chapter.

Factors influencing selection of background electrolyte

Optimal pH region and generally the optimal composition of BGE should fulfil the following conditions:

a) sufficient solubility of analyzed peptides (at least about 1 mmol/l),

b) chemical stability of peptides including some of their labile groups (amides, disulphides, etc.),

c) preservation of biological activity (if peptides are used for biological tests after their CZE separation),

d) effective mobilities of peptides and their relative differences should be sufficiently high for their separation (*ca.* $m_{ef} > 1 \times 10^9 \text{ m}^2 \text{V}^{-1} \text{s}^{-1}$, $\Delta m_{ef} / m_{ef} > 0.02$),

e) adsorption of peptides to capillary wall is suppressed.

Some rules for the selection of suitable composition and pH of BGE with respect to these demands are briefly given further.

Solubility. Peptide solubility can be affected by a number parameters:

a) *pH.* The pH of BGE should be at least 1–2 pH units distant from the isoelectric point since the solubility of peptide is relatively lower at pI and its close vicinity.

b) *Ionic strength and the buffer constituents.* Composition of BGE for CZE has much less limitation than for ITP. The composition of the BGE is relatively free: it is possible to use relatively concentrated solutions of acids and bases (e.g. 0.1 M phosphoric acid, 0.5 M acetic acid, 0.1 M Tris) which is favourable for peptide solubilization. Higher ionic strength can contribute to better solubility of peptides but too conductive buffers should be avoided since great Joule heat would be generated in these buffers at the high voltages generally applied in CZE.

c) *Solubilizing additives.* Different types of additives can be added to the BGE to improve the peptide solubilization, chaotropic agents, as urea and its derivatives (Nielsen and Rickard, 1990), nonionogenic and zwitterionic detergents (Swedberg, 1990; Greve *et al.*, 1994), and cyclodextrins and micelle forming compounds (Liu *et al.*, 1990). These agents should be added to BGE very carefully since they can also change or reverse the electroosmotic flow, they can form complexes or ion pairs with peptides or they can change the separation mechanism from charge/size-based separation into hydrophobicity based separation, e.g. by micellar electrokinetic chromatography (Terabe, 1989). This technique is suitable for separation of uncharged peptides or peptides with the same or similar specific charge, i.e. peptides containing the same charged amino acids and similar non-charged amino acids (Yashima *et al.*, 1992). CZE separations are mostly performed in aqueous buffers but organic solvents such as methanol and acetonitrile can be also added to BGE (Johansson *et al.*, 1991).

Biological activity, chemical and thermal lability. Biological activity (e.g. hormonal or immunochemical) of peptides is connected with a certain conformation of the polypeptide molecule which is stable only under defined conditions. Consequently, if the biological activity of a peptide is to be preserved even after the peptide CZE separation, then this separation must be performed under such conditions where denaturation of the peptide does not occur, i.e. composition, pH, solvents and additives of BGE have to be chosen with respect to this demand.

Chemical lability of some groups in the peptide molecule has to be taken into account, e.g. peptides containing amides, disulphide and sulphhydryl groups should not be analyzed at high pH because of hydrolytic and trans-sulphidation reactions of these groups at this pH.

In addition, the electric input power should be sufficiently low (*ca.* lower than 2 W per meter of capillary length) and active removal of Joule heat by flowing air or circulating liquid is recommended in order to prevent thermal denaturation of separated polypeptides.

DETECTION

Conductivity detection

Similar to other charged compounds, peptides can be detected on the basis of different electric conductivity of the peptide zone in comparison with the surrounding media, i.e. with the background electrolyte in CZE or with neighbouring zones in CITP. Whereas this type of universal detection is commonly used in CITP, often complementary with specific UV detection, the conductivity detection of peptides in CZE is not commonly used.

The most frequent application of conductivity detection occurs in CITP analyses of complex reaction products resulting from peptide synthesis, where both peptides and low molecular ionic admixtures are present. As the length of a zone in the ITP regimen is directly proportional to the quantity of the ion species and the quality is characterized by the height of the signal, simple comparison of a length of zones with and without specific detection signal (mostly UV absorption signal) easily detects the peptide zone among other ion species zones.

Reproducibility of the signal height of the conductivity detector with contact electrodes may be distorted by interaction of the electrode with the peptide. Especially peptides containing cysteine and histidine residues may interfere with the contact electrodes. No such problems occur when ITP contactless high frequence conductivity detectors are used (Gaš *et al.*, 1989).

Mass spectrometric detection

Mass spectrometry (MS) represents an ideal detection technique for CZE because of its universality, sensitivity and selectivity. Mass spectrometry combined with high resolution capillary electrophoresis is becoming of importance for peptide mapping study of peptides — protein fragments from specific enzymatic hydrolysis. Electrospray ionisation is the preferred mode of coupling the CE apparatus with a mass analyser. In such an arrangement, the best signal-to-noise ratio was found when acidic buffers of low ionic strength for CE of peptides were applied (Moseley *et al.*, 1992).

If the electrospray ionisation process is incompatible with CE separation conditions, matrix assisted laser desorption-ionisation (MALDI) may be used as an alternative means of introducing peptide to the mass analyser. Keough *et al.* (1992) reported a 250 fmol detection limit for smaller peptides and 100 fmol off-line detection limit for α-lactalbumin. The MS detection enables us to detect and characterize absolutely also the nonpeptidic component attached to the peptide chain.

Light absorption detection

Universal UV absorption detection. Each peptide molecule absorbs light energy in the short wavelength UV region. The typical structure occurring in each peptide molecule is the peptide bond CO−NH which strongly absorbs UV light in the region near to 200 nm. For practical reasons i.e. for UV light absorption by the background electrolyte, some limitations in background electrolyte composition arise near the 200 nm region. A better situation and sufficient sensitivity was found in the wavelength region 205−214 nm sometimes up to 220 nm. The UV absorption of the peptide zone in this spectral region characterizes the peptide bond quantitatively, i.e. the longer the peptide, the higher the signal response at equal molar concentration. Below 205 nm, the sensitivity of detection of a peptide bond rises considerably, but the majority of buffers are inappropriate. Near 200 nm, phosphate buffers can still be successfully applied. For polypeptides separated in the presence of synthetic water-soluble polymers, a wavelength of 230 nm provides a good compromise between detectability and peak-area analysis accuracy on one side and stronger light absorption of the background electrolyte on the other side.

Specific UV absorption detection. Peptides containing aromatic amino acid residues can be detected most conveniently by UV light absorption at 275−280 nm. The highest absorption coefficient is characteristic of tryptophane and tyrosine residues, where significant changes of the absorption coefficient influenced by pH occur. Less sensitive is the detection of the phenylalanine residue. Some ITP apparatuses are equipped with a monochromatic UV absorption detector at 254 nm. Although the absorption coefficients and hence the resulting detectability of aromatic residue containing peptides at 254 nm is lower, higher steady state concentrations of a peptide zone in ITP regimen usually compensate for this drawback.

Light emission detection

Laser-induced fluorescence detection (LIF) of native peptides. The native fluorescence of peptides excited by UV light in the region 200–300 nm is exclusively dependent on aromatic amino acid residues. The optimal excitation wavelength was found at 275 nm. The tryptophane residue obviously plays the dominating role. The quantum yield of fluorescence of tryptophane is positively influenced by a hydrophilic microenvironment. The fluorescence intensity of tyrosine is about two orders of magnitude weaker. Almost negligible is the fluorescence of the phenylalanine residue.

Availability of deep UV lasers makes possible very efficient excitation of native fluorescence of peptides containing tryptophane and tyrosine residues. This LIF approach greatly simplifies the preseparation sample handling. Low nanomolar up to picomolar detection limits for polypeptides were reached in CZE (Lee and Yeung, 1992). The detection limits for peptides are at least two orders of magnitude less when compared with UV absorption at 214 nm (Chan et al., 1993).

Laser-induced fluorescence detection (LIF) of labelled peptides. Both precolumn and postcolumn labelling of peptides may be used for enhancement of sensitivity in CZE separations. Naturally, differences in the extent of the precolumn incorporation of the tags causing rise of multiple peaks after CE must be taken into account. Often precolumn derivatisation reagents reacting with the aminogroup were used, such as 5-dimethyl-aminonaphthalene-1-sulphonyl (dansyl) chloride, naphthalene-2,3-dicarboxaldehyde (NDA), fluorescein isothiocyanate (FITC), etc. Subzeptomole detection limits for CE of FITC derivatives of peptides were reported. CE peptide mapping of 360 attomole of tryptic HSA digest involving selective derivatization of the arginine-containing peptides with benzoin was performed by Cobb and Novotny (1992).

Nonfluorescent reagents such as fluorescamine and *o*-phthalaldehyde (OPA) may be used both for precolumn and postcolumn derivatization under mild conditions. Lower resolution of the CE zones as a result of postcolumn detection may be expected.

In spite of some above mentioned disadvantages of LIF of labelled peptides, LIF detection in general seems to be the most sensitive detection mode at the present time.

ANALYSIS OF PEPTIDE PRODUCTS

General

Qualitative and quantitative analysis of synthetic peptides and peptides isolated from natural material is the most frequent application of CZE and ITP in peptide chemistry. Especially, the need for a purity test of synthetic peptides has arisen in many areas in recent years.

In biochemistry, synthetic peptides are used as substrates and inhibitors of enzymes in the elucidation of the mechanism of their catalytic effect; in the investigation and modelling of the interactions of antigens with antibodies, hormones with receptors, and of peptides and proteins with nucleic acids; in the mapping of antigenic determinants (epitopes) of proteins, and in the study of the dependence of secondary and tertiary structure of a peptide chain on amino acid sequence. The use of synthetic peptides in pharmaceutical research and in human and veterinary medicine is also widespread. In the food industry, peptides are used as sweeteners and additives.

In a majority of these applications of synthetic, natural or biotechnologically prepared peptides, CZE and CITP can be used as sensitive control methods for the determination of their purity, or as control methods of the efficiency of the individual methods used for their purification. CZE and CITP provide rapid and accurate qualitative and quantitative data about the peptide preparations.

Evaluation of peptide purity

Peptide purity and content in the sample can be quantified in several ways depending on whether the standard of the given peptide is available or not.

If the standard peptide P is available, then the quantity of this peptide in the analyzed sample can be determined absolutely by the calibration curve method or by the method of internal standard addition, i.e. by comparison of migration times, peak heights or peak areas in CZE, or by comparison of relative step height and step length in ITP. If the standard is not available, which is the case for peptides synthesized or isolated for the first time, the relative evaluation of peptide purity has to be used.

ITP degree of purity. The ITP degree of purity of peptide P, $p_{P,ITP}$, is defined as the ratio of zone length of the peptide P itself, l_P, to the total zone length, l_T, of all UV-positive zones on the isotachopherogram of a preparation of peptide P (see Fig. 4). The total zone length, l_T, is given by the sum of all (n) zones present on the ITP-gram (including the zone of peptide P itself):

$$p_{P,ITP} = l_P / l_T = l_P / \sum_i l_i, \quad i = 1 \ldots n. \tag{10}$$

The zone lengths of possible impurities from electrolyte system must not, of course, be included in the total length, l_T.

Application of CITP to the determination of the purity of a polypeptide isolated from natural material, bull seminal trypsin inhibitor, BUSI II (Kašička and Prusík, 1989) is shown in Fig. 4.

ITP analysis was performed in a cationic electrolyte system (leading electrolyte 0.01 M NaOH adjusted with acetic acid to pH 4.8, 0.02% polyvinylalcohol, terminating electrolyte 0.01 mol/l BALA, adjusted with acetic acid

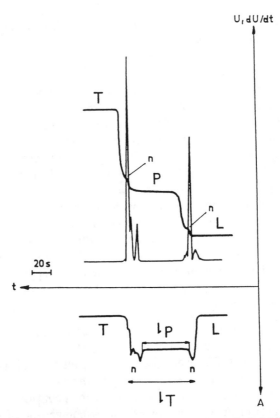

Figure 4. Determination of degree of purity of bull seminal trypsin isoinhibitor, BUSI II, by CITP. l_P – zone length of pure peptide; l_T – total length of all UV-positive zones; L, (T) – zone of leading (terminating) electrolyte; n – nonidentified admixtures. A – absorbance at 254 nm, U – voltage of potential gradient detector, t – time. For experimental details see the text.

to pH 4.5) in a home-made ITP analyser (Kašička and Prusík, 1989). It is equipped with a teflon capillary, i.d. 0.45 mm, length 230 mm and two potential gradient detectors and a UV-photometric detector at 254 nm. Sample concentration was 4.5 mg/ml and applied sample volume was 2 μl.

The advantage of ITP is that the concentration of analyte in its zone is regulated by Kohlrausch function and the zone length is directly proportional to the analyte amount in the sample. Due to this fact, the ITP degree of purity can be approximately identified with the molar fraction of the analyte (peptide P) in the sample (peptide preparation). This approximation is the more precise the closer are the charges and the effective mobilities of the components present in the sample which is often the case for synthetic peptide preparations where the admixtures are similar to the main synthetic product.

CZE degree of purity. The CZE degree of purity of peptide P, $p_{P,CZE,h}$ ($p_{P,CZE,A}$), is defined as the ratio of the peak height (area) of peptide P itself to the sum of heights (areas) of all (n) peaks present on the CZE-gram of the given preparation of peptide P:

$$p_{P,CZE,h} = h_P / \sum_i h_i, \quad i = 1 \ldots n, \tag{11}$$

$$p_{P,CZE,A} = A_P / \sum_i A_i, \quad i = 1 \ldots n. \tag{12}$$

Application of CZE to the determination of peptide purity is demonstrated in Fig. 5. It shows CZE analysis of preparation of pig insulin isolated from natural material (Barth, 1994). In addition to the main peak of insulin (I), one admixture with lower mobility and three admixtures with higher mobilities are present. Height-based purity degree of this preparation is 75.1% and area based peak purity is 85.8%.

It must be emphasized that both ways of expression of CZE purity degree can be used only as approximate and relative criteria of peptide purity since the molar absorption coefficients of the individual sample components may be generally very different. However, in the case of synthetic peptide preparations, most admixtures will have a similar structure to that of the main synthetic product and consequently, also, similar absorption coefficients.

The peak height-based CZE purity degree is recommended to be used for CZE-grams with great differences in migration times and with similar peak shape and width.

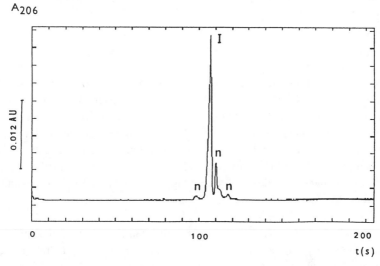

Figure 5. Determination of degree of purity of pig insulin by CZE. Sample concentration: 0.8 mg/ml. BGE: 0.05 M borate, adj. with NaOH to pH 9.1. Voltage 10.0 kV, current 22.5 μA. Capillary: the same as in Fig. 2. I – peak of insulin, n – nonidentified admixtures, A_{206} – absorbance at 206 nm, t – time.

The area-based CZE purity degree should be used for evaluation of CZE-grams where the peaks of some components are broader not because of slow migration through the detector but due to adsorption or electromigration dispersion.

Only illustrative applications of CZE and CITP to analysis of peptide products were shown here. More examples can be found in several reviews and books (e.g. Novotny *et al.*, 1990; Kašička and Prusík, 1991; Colburn, 1992; Li, 1992; Foret *et al.*, 1993; Monnig and Kennedy, 1994).

CONVERSION OF ANALYTICAL CAPILLARY SEPARATIONS INTO PREPARATIVE SCALE

General

The advantages of CZE in the analysis of peptides were demonstrated in the previous section. In this field, CZE is already accepted as a recognized counterpart and/or complement to what, until now, was the most widely spread separation technique — reversed-phase high-performance liquid chromatography (RP–HPLC).

However, in the field of peptide preparation, the application potential of CZE is relatively limited. The lower potential is caused by two facts:

1. More complicated adaptation of CE systems from analytical to preparative scale due to the fact that both ends of the capillary are dipped into the buffer in the electrode compartments and electric field is applied during the whole of the experimental time;

2. Low preparative capacity of capillary systems.

Whereas the first problem was solved by several special procedures for fraction collections, e.g. by interruption of electric field and changing of electrode vessels and using pressure or electroosmotic flow for sample zone collection (Banke *et al.*, 1991; Cheng *et al.*, 1992), the second problem, limited preparative capacity, is fundamental. Because of the small dimensions of the capillary separation compartment (typical i.d. is 0.050–0.1 mm) only small amounts of peptides (less than 1 µg) can be isolated from capillary systems. These amounts are sufficient only for a limited number of applications, e.g. for further characterization of isolated peptides by sequencing, amino acid analysis, MS-analysis or for very sensitive enzymatic and immunochemical tests in biochemistry, molecular biology and medicine.

The preparative capacity of capillary systems can be only partially increased by the increasing of inner diameter of the capillary because of dramatic loss of separation efficiency in capillaries with i.d. greater than 0.200–0.3 mm. In the wide bore capillaries, the Joule heat is much more poorly removed from the separation compartment, which leads to radial temperature gradient and broadening of sample zones.

Partial increasing of preparative capacity can be achieved in rectangular cross-section capillary columns with the dimensions 0.05×1 mm (Tsuda et al., 1990) or by a microconcentric capillary column (Fujimoto et al., 1994).

Free-flow zone electrophoresis (FFZE)

Several order enlargement of preparative capacity is achieved if electrophoretic separation is realized in the continuous free-flow arrangement in the flow-through electrophoretic chamber (Prusík, 1974, 1979; Wagner et al., 1984). In this instrumental format, the capacity can be increased up to hundreds of milligrams per hour.

The principle of free-flow zone electrophoresis (Hannig and Heidrich, 1989; Wagner and Heinrich, 1990; Roman and Brown, 1994) is as follows (see Fig. 6). The separation compartment is formed by two parallel glass plates. The background electrolyte (carrier buffer) is continuously laminary flowing in the narrow gap (0.5 mm) between these two plates. Sample solution is also continuously introduced into the background electrolyte as a narrow zone. Electric field is applied perpendicular to the direction of laminar flow of background electrolyte and sample, and causes the different deflection of the sample components depending on their effective mobilities. Cations are deflected to cathode, anions to anode and non-charged components will move in the straight direction if no electroosmotic flow occurs in the chamber. At the outlet side of the chamber the sample components are continuously collected in the fraction collector.

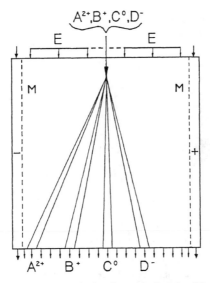

Figure 6. Cross-section of the flow-through chamber and principle of free-flow zone electrophoresis. A, B, C, D – sample components, E – background electrolyte, M – ion exchange membranes separating the electrode compartments.

Correlation of CZE and FFZE

CZE and free-flow zone electrophoresis (FFZE) represent two modes of the same separation principle. Both of them are performed in the carrier-less medium with the same background electrolyte. Consequently, a direct correlation exists between these two methods, which can be utilized for conversion of capillary microscale separation into preparative ones. The correlation of CZE and FFZE and the procedure for conversion of CZE to FFZE was described in more detail elsewhere (Prusík *et al.*, 1990; Kašička *et al.*, 1992); here, only the main points will be mentioned.

First, suitable experimental conditions are developed for CZE separation of given peptide preparation (composition and pH of BGE, voltage, capillary length, etc.). Then, under the same conditions, the separation of a test mixture, containing charged standard components S (diglycine and triglycine) and the non-charged standard component N (phenol) is performed by CZE (see Fig. 2) and by FFZE (see Fig. 7). From the CZE and FFZE separations of the test mixture the ratio, r, of electrophoretic velocities in CZE and FFZE is determined.

$$r = v_{ep,FFZE,S}/v_{ep,CZE,S}. \tag{13}$$

Further procedure is based on the assumption that r is approximately constant for different charged components separated by CZE and FFZE under the same separation conditions. This means that if this coefficient is once determined for the standard component S, it can be used for calculation of electrophoretic velocity of peptide P or any other component in FFZE if their electrophoretic

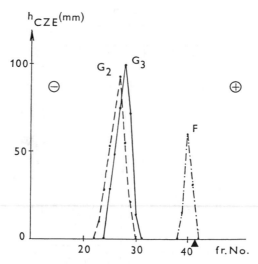

Figure 7. Separation of standard mixture by FFZE. Gly_2 – diglycine (15 mg/ml), Gly_3 – triglycine (15 mg/ml), F – phenol (5 mg/ml). BGE: 0.5 mol/l acetic acid. Flow-through time 31 min, sample flow rate 1.5 ml/h, voltage 3000 V, current 120 mA, temperature $-3\,°C$. h_{CZE} – peak height of analyte in CZE, fr. no. – fraction number of FFZE, ▲ – sample inlet position.

velocities in CZE were measured.

$$v_{ep,FFZE,P} = r\, v_{ep,CZE,P}. \tag{14}$$

From the calculated electrophoretic velocity of peptide P the resulting migration distance, d_r, of different sample components can be predicted and their separability can be estimated:

$$d_{r,P} = v_{ep,FFZE,P}\, t_f + d_{eo}, \tag{15}$$

where t_f is flow-through time of FFZE and d_{eo} is contribution of electroosmotic flow to the resulting distance, which is determined from the FFZE experiment with the non-charged component.

According to these predicted migration distances, the separation conditions of FFZE, namely voltage and flow-through time, can be optimized.

Applications of CZE and FFZE to peptide analysis and preparation

Combined application of CZE and FFZE to peptide separation is demonstrated by analysis and preparation of some biologically active peptides synthesized in our Institute.

Growth hormone releasing peptide, GHRP, a hexapeptide of the sequence His-D-Trp-Ala-Trp-D-Phe-Lys.NH$_2$ was synthesized by the solid phase method and purified by gel permeation chromatography on Sephadex G-10 and then analyzed by CZE (see Fig. 8). The electropherogram shows a relatively high content of the main synthetic product (the highest peak) but some admixtures

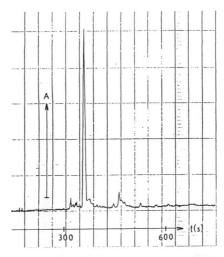

Figure 8. CZE analysis of synthetic growth hormone releasing peptide. Sample concentration: 2 mg/ml. BGE: 0.5 mol/l acetic acid. Capillary: i.d. 0.05 mm, effective length 500 mm, total length 650 mm. Voltage 16.5 kV, current 8.0 μA. A – absorption at 206 nm, t – time.

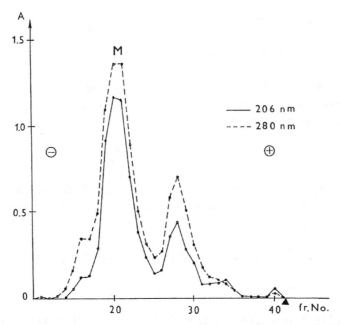

Figure 9. FFZE separation of growth hormone releasing peptide.   Sample concentration: 20 mg/ml. BGE: 0.5 mol/l acetic acid. Flow-through time 31 min, sample flow rate 1.5 ml/h, voltage 2700 V, current 106 mA, temperature $-2\,^\circ$C. M – main synthetic product, A – absorbance, fr. no. – fraction number, ▲ – sample inlet position.

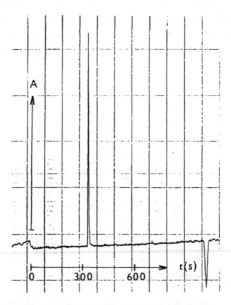

Figure 10. CZE analysis of fraction 20 of FFZE separation of GHRP presented in Fig. 9. Aliquot of fraction 20 was directly applied to CZE analysis. Conditions of CZE are the same as in Fig. 8. A – absorption at 206 nm, t – time.

with both higher and lower mobilities are still present. For that reason, this peptide preparation was subjected to preparative FFZE separation. The record of this separation is shown in Fig. 9.

The qualitative similarity of separation profile of both CZE and FFZE reflects the correlation of these two methods. The separation power of FFZE is of course much lower and the contaminants with higher and lower mobilities are not separated so distinctly and completely as in CZE, but at least from some fractions pure peptide can be obtained. It is shown in Fig. 10 where CZE analysis of fraction 20 is presented. The CZE purity degree is approaching 100% in this case.

Another example is CZE analysis of crude synthetic preparation of [D-Tle^{2,5}]-dalargin, a synthetic hexapeptide with the sequence Tyr-D-Tle-Gly-Phe-D-Tle-Arg which is an analogue of dalargin – an enkephalin type peptide with opiate activity. CZE-gram of the crude synthetic product (see Fig. 11) shows that in addition to the main synthetic product M, there are many other admixtures present in this preparation. For better orientation some of them are indicated, the fastest one by F, the slowest charged component by S and the non-charged component by N.

The lyophilizate of crude synthetic product was dissolved in 0.5 mol/l acetic acid, pH 2.5 (40 mg/ml) and applied to FFZE separation. The record of this separation is shown in Fig. 12.

Qualitative similarity of CZE and FFZE separation of this product can be seen from the comparison of Fig. 11 and Fig. 12. In addition to the main peak M, the peaks of fastest, slow and uncharged components (F, S, N) can be identified in both records. Obviously, a better separation of sample components

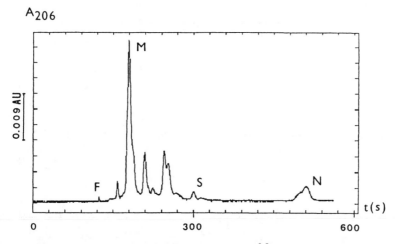

Figure 11. CZE analysis of crude synthetic product of [D-Tle^{2,5}]-dalargin. Sample concentration 1.5 mg/ml. BGE: 0.5 mol/l acetic acid. Capillary: i.d. 0.05 mm, effective length 200 mm, total length 310 mm. Voltage 9.0 kV, current 10.1 µA. F – fastest component, M – main synthetic product, S – slow component, N – noncharged component, A_{206} – absorbance at 206 nm, t – time.

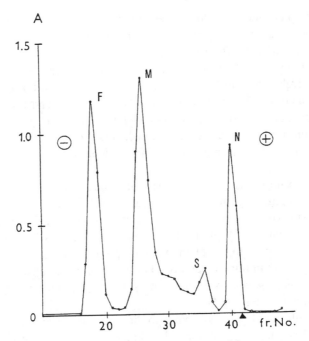

Figure 12. FFZE separation of crude synthetic product of [D-Tle^{2,5}]-dalargin. Sample concentration 38 mg/ml. BGE: 0.5 mol/l acetic acid. Flow-through time 31 min, sample flow rate 1.5 ml/h, voltage 3000 V, current 120 mA, temperature $-3\,^{\circ}$C. F – fastest component, M – main synthetic product, S – slow component, N – noncharged component, A_{280} – absorbance at 280 nm, fr. no. – fraction number, ▲ – sample inlet position.

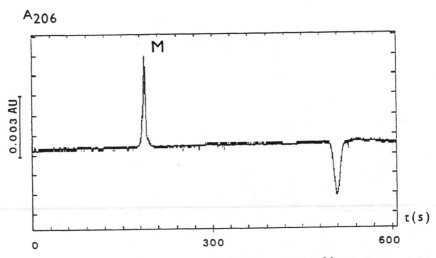

Figure 13. CZE analysis of fraction 26 of FFZE separation of [D-Tle^{2,5}]-dalargin presented in Fig. 12. Aliquot of FFZE fraction was directly applied to CZE analysis. Experimental conditions are the same as in Fig. 11, M – main synthetic product.

is achieved in CZE than in FFZE, especially in the region between peaks M and S. However, it is important that the main synthetic product (peak M) is relatively well separated from the other synthetic admixtures. This was confirmed by CZE analysis of the fractions representing peak M (see the single peak on CZE-gram of fraction 26 presented in Fig. 13).

The above described procedure, i.e. a combination of CZE and FFZE, was applied to analysis and preparation of other biopeptides, e.g. analogues of GHRP and its fragment (Kašička *et al.*, 1994) and analogues of LHRH (Prusík *et al.*, 1991).

A combination of CZE and FFZE represents a systematic approach to analysis and preparation of synthetic peptides.

CZE is used for analysis of peptide preparations and for the development of optimal separation conditions on a microscale. Then, based on the correlation between CZE and FFZE, the optimized conditions of analytical CZE separation are converted into preparative separations realized by FFZE.

FFZE allows preparative separations of peptides with the capacity of about 100 mg/h. The advantage of this method is that it works continuously, in a free solution under mild conditions where the biological activity of peptides is retained and the loss of material is minimized.

The purity of peptides prepared by FFZE and/or other methods is tested by CZE.

ION ANALYSIS OF PEPTIDES BY CITP

In the course of peptide synthesis and purification, some organic and inorganic acids (e.g. formic, acetic, trifluoroacetic, hydrochloric, hydrofluoric, *p*-toluensulphonic) and basic compounds (e.g. ammonia, alkylammonium derivatives, triethylamine, pyridine, piperidine) are used. Often, at least some of these agents remain as contaminants of peptide preparations. Due to the toxicity of some of these agents, their content in a peptide preparation has to be controlled before they are used in biological tests and in human and veterinary medicine.

The determination of these ionic admixtures in peptide preparations can be advantageously performed by CITP (Janssen and van Nispen, 1984). CITP instruments are mostly equipped with on-column universal conductivity detection by which all types of ions can be detected, i.e. simultaneous determination of cations separated in cationic ITP regimen or simultaneous determination of anions separated in anionic ITP regimen can be achieved.

We have developed a method for ITP determination of anions (fluorides, acetates, trifluoroacetates) in peptide pharmaceuticals, derivatives of oxytocin, vasopressin and LHRH. Analysis is performed in anionic ITP regimen with the following electrolyte system: leading electrolyte 0.01 mol/l HCl, 0.02 mol/l histidine, pH 5.9, terminating electrolyte 0.01 mol/l sodium glutamate, pH 6.7.

Quantitative evaluation of anion contents is performed by the calibration curve method, i.e. the dependence of zone length on the anion amount is first

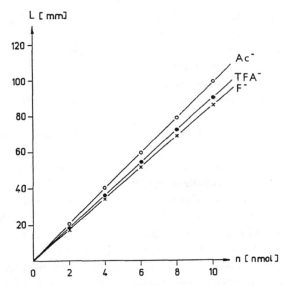

Figure 14. Calibration curves for ITP determination of anionic admixtures in peptide preparations. F^- – fluorides, TFA^- – trifluoroacetates, Ac^- – acetates; L – length of ITP zone, n – sample amount.

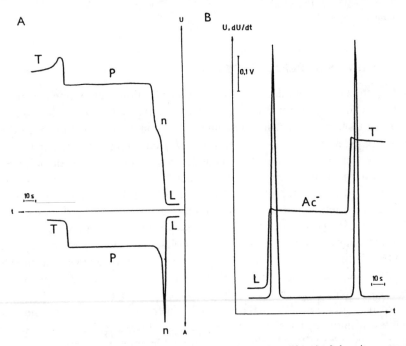

Figure 15. ITP analysis of pharmaceutical preparation of Adiuretin ([8-D-Arg]-deaminovasopressin). A: analysis of peptide in cationic mode, B: determination of anionic admixtures in anionic mode. L (T) – leading (terminating) electrolyte, P – peptide zone, n – nonidentified admixture, Ac^- – acetates, U – voltage of potential gradient detector, A – absorbance at 254 nm, t – time.

measured (see Fig. 14) and then from this curve and from the zone length of anion in the peptide sample the amount of anion in this peptide preparation is determined.

Example of ITP analysis of pharmaceutical preparation of Adiuretin ([8-D-Arg]-deaminovasopressin) in both cationic and anionic ITP regimen is shown in Fig. 15.

In the cationic regimen, the purity of the peptide itself was tested and in the anionic regimen, the content of acetic acid was determined. As can be seen from the record, no other anionic admixture was found in this preparation.

The determination of acetates or chlorides in peptide preparations is of considerable practical importance since many biologically active peptides are stored and biologically and pharmacologically tested as acetates or chlorides and for the full characterization of these preparations the quantity of these anions has to be determined.

The advantage of ITP in ion analysis of peptides is the simple preparation of the sample (only weighing and dissolution), low sample consumption (0.2–0.3 mg), short time of analysis (10–15 min), high precision (relative error 1–2%) and simultaneous determination of either anions or cations.

ACKNOWLEDGEMENTS

The financial support of this work by Grant Agency of the Academy of Sciences of the Czech Republic, grant no. 45 511 and by the Grant Agency of the Czech Republic, grant no. 203/93/0718 and grant no. 203/94/0698 is acknowledged.

The authors thank Mrs V. Lišková for her technical assistance in experimental work and for the help in preparation of this paper.

REFERENCES

Banke, N., Hansen, K. and Diers, I. (1991). Detection of enzyme activity in fractions collected from free solution capillary electrophoresis of complex samples. *J. Chromatogr.* **559**, 325–335.

Barth, T. (1994). Unpublished results.

Boček, P., Deml, M., Gebauer, P. and Dolník, V. (1988). *Analytical Isotachophoresis*. Verlag Chemie VCH, Weinheim.

Chan, K. C., Janini, G. M., Muschik, G. M. and Issaq, H. J. (1993). Pulsed UV laser-induced fluorescence detection of native peptides and proteins in capillary electrophoresis. *J. Liquid. Chromatogr.* **16**, 1877–1890.

Cheng, Y. F., Fuchs, M., Andrews, D. and Carson, W. (1992). Membrane fraction collection for capillary electrophoresis. *J. Chromatogr.* **608**, 109–116.

Cifuentes, A. and Poppe, H. (1994). Simulation and optimization of peptide separation by CZE. *J. Chromatogr. A.* **680**, 321–340.

Cobb, K. A. and Novotny, M. (1992). Peptide mapping of complex proteins at low-picomole level with capillary electrophoretic separations. *Anal. Chem.* **64**, 879–886.

Colburn, J. C. (1992). Capillary electrophoresis separations of peptides: practical aspects and applications. In: *Capillary Electrophoresis: Theory and Practice*, Grossman, P. D. and Colburn, J. C. (Eds). Academic Press, San Diego, pp. 237–271.

Compton, B. J. (1991). Electrophoretic mobility modeling of proteins in free zone capillary electrophoresis and its application to monoclonal antibody microheterogeneity analysis. *J. Chromatogr.* **559**, 357–366.

Everaerts, F. M., Beckers, J. L. and Verheggen, T. P. E. M. (1976). *Isotachophoresis — Theory, Instrumentation and Applications.* Elsevier, Amsterdam.

Foret, F., Křivánková, L. and Boček, P. (1993). *Capillary Zone Electrophoresis.* Verlag Chemie, Weinheim.

Fujimoto, C., Matsui, H., Sawada, H. and Jinno, K. (1994). The use of microconcentric column in capillary electrophoresis. *J. Chromatogr. A.* **680**, 33–42.

Gaš, B., Zuska, J. and Vacík, J. (1989). Measurement of limiting mobilities by capillary isotachophoresis with a constant temperature at the site of detection. *J. Chromatogr.* **470**, 69–78.

Greve, K. F., Nashabeh, W. and Karger, B. L. (1994). Use of zwitterionic detergents for the separation of closely related peptides by capillary electrophoresis. *J. Chromatogr. A.* **680**, 15–24.

Grossman, P. D. (1992). Background concepts. In: *Capillary Electrophoresis*, Grossman, P. D. and Colburn, J. C. (Eds). Academic Press, San Diego, pp. 3–43.

Grossman, P. D., Colburn, J. C. and Lauer, H. H. (1989). A semiempirical model for the electrophoretic mobilities of peptides in free-solution CE. *Anal. Biochem.* **179**, 28–33.

Hannig, K. and Heidrich, H. K. (1989). *Free-Flow Electrophoresis.* GIT, Darmstadt.

Janssen, P. S. L. and van Nispen, J. W. (1984). Isotachophoretic determination of anions and cations in peptides. *J. Chromatogr.* **287**, 166–175.

Johansson, I. M., Huang, E. C., Henion, J. D. and Zweigrnbaum, J. (1991). CE — atmospheric pressure ionisation MS for characterisation of peptides. *J. Chromatogr.* **554**, 311–327.

Kašička, V. and Prusík, Z. (1989). Isotachophoretic analysis of peptides. *J. Chromatogr.* **470**, 209–221.

Kašička, V. and Prusík, Z. (1991). Application of capillary isotachophoresis in peptide analysis. *J. Chromatogr.* **569**, 123–174.

Kašička, V., Prusík, Z. and Pospíšek, J. (1992). Conversion of capillary zone electrophoresis to free-flow zone electrophoresis using a simple model of their correlation. *J. Chromatogr.* **608**, 13–22.

Kašička, V., Prusík, Z., Smékal, O., Hlaváček, J., Barth, T., Weber, G. and Wagner, H. (1994). Application of capillary and free-flow zone electrophoresis and isotachophoresis to the analysis and preparation of the synthetic tetrapeptide fragment of growth hormone releasing peptide. *J. Chromatogr. B.* **656**, 99–106.

Keough, T., Takigiku, R., Lacey, M. P. and Purdon, M. (1992). Matrix assisted laser desorption mass spectrometry of proteins isolated by capillary zone electrophoresis. *Anal. Chem.* **64**, 1594–1600.

Lee, T. T. and Yeung, E. S. (1992). High sensitivity laser-induced fluorescence detection of native proteins in capillary electrophoresis. *J. Chromatogr.* **595**, 319–325.

Li, S. F. Y. (1992). *Capillary Electrophoresis.* Elsevier, Amsterdam.

Liu, J., Cobb, K. A. and Novotny, M. V. (1990). CE separation of peptides using micelle-forming compounds and cyclodextrins as additives. *J. Chromatogr.* **519**, 189–197.

Monnig, C. A. and Kennedy, R. T. (1994). Capillary electrophoresis. *Anal. Chem.* **66**, 280R–314R.

Moseley, M. A., Jorgenson, J. W., Shabanowitz, J., Hunt, D. F. and Tomer, K. B. (1992). Optimization of capillary zone electrophoresis/electrospray ionization parameters for mass spectrometry and tandem mass spectrometry analysis of peptides. *J. Am. Soc. Mass Spectrom.* **3**, 289–300.

Nielsen, R. G. and Rickard, E. C. (1990). Method optimization in CZE analysis of hGH tryptic digest fragments. *J. Chromatogr.* **516**, 99–114.

Novotny, M. V., Cobb, K. A. and Liu, J. (1990). Recent advances in capillary electrophoresis of proteins, peptides and amino acids. *Electrophoresis* **11**, 735–749.

Offord, R. E. (1966). Electrophoretic mobilities of peptides on paper and their use in the determination of amide groups. *Nature* **211**, 591–593.

Prusík, Z. (1974). Free-flow electromigration separations. *J. Chromatogr.* **91**, 867–872.

Prusík, Z. (1979). Continuous flow-through electrophoresis. In: *Electrophoresis, Part A: Techniques*, Deyl, Z. (Ed.). Elsevier, Amsterdam, pp. 229–251.

Prusík, Z., Kašička, V., Mudra, P., Štěpánek, J., Smékal, O. and Hlaváček, J. (1990). Correlation of capillary zone electrophoresis with continuous free-flow zone electrophoresis: Application to the analysis and purification of synthetic growth hormone releasing peptide. *Electrophoresis* **11**, 932–936.

Prusík, Z., Kašička, V., Weber, G. and Pospíšek, J. (1991). Correlation of capillary electrophoresis with continuous free-flow electrophoresis systems and their application in peptide separations: Analysis and preparation of LHRH derivative. In: *Elektrophorese Forum '91*, Radola, B. J. (Ed.). Technische Universität München, pp. 201–206.

Rickard, E. C., Strohl, M. M. and Nielsen, R. G. (1991). Correlation of electrophoretic mobilities from capillary electrophoresis with physicochemical properties of proteins and peptides. *Anal. Biochem.* **197**, 197–207.

Roman, M. C. and Brown, P. R. (1994). Free-flow electrophoresis as a preparative separation technique. *Anal. Chem.* **66**, 86A–94A.

Swedberg, S. A. (1990). Use of non-ionic and zwitter-ionic surfactants to entrance selectivity in HPCE. *J. Chromatogr.* **503**, 449–452.

Takigiku, R., Keough, T., Lacey, M. P. and Schneider, R. E. (1990). CZE with fraction collection for desorption MS. *Rapid Commun. Mass Spectrom.* **4**, 24–29.

Terabe, S. (1989). Electrokinetic chromatography: an interface between electrophoresis and chromatography. *Trends Anal. Chem.* **8**, 129–134.

Tsuda, T., Sweedler, J. V. and Zare, R. N. (1990). Rectangular capillaries for CZE. *Anal. Chem.* **62**, 2149–2152.

Wagner, H., Mang, V., Kessler, R. and Speer, W. (1984). A free-flow system for preparative separations. In: *Analytical and Preparative Isotachophoresis*, Holloway, C. J. (Ed.). Walter de Gruyter, Berlin, pp. 347–356.

Wagner, H. and Heinrich, J. (1990). Free-flow electrophoresis for the separation and purification of biopolymers. In: *Modern Methods in Protein and Nucleic Acid Research*, Tschesche, H. (Ed.). Walter de Gruyter, Berlin, pp. 69–97.

Yashima, T., Tsuchiya, A., Morita, O. and Terabe, S. (1992). Separation of closely related large peptides by micellar electrokinetic chromatography with organic modifiers. *Anal. Chem.* **64**, 2981–2984.

APPENDIX — ABBREVIATIONS

BALA	beta-alanine
BGE	background electrolyte
CE	capillary electrophoresis
CZE	capillary zone electrophoresis
CITP	capillary isotachophoresis
FFZE	free-flow zone electrophoresis
GHRP	growth hormone releasing peptide
HPCE	high-performance capillary electrophoresis
HPLC	high-performance liquid chromatography
HSA	human serum albumin
ITP	isotachophoresis
LIF	laser induced fluorescence
MS	mass spectrometry

HPCE IN DRUG ABUSE
AND DRUG INTERACTIONS

Progress in HPLC-HPCE, Vol. 5, pp. 201–254
H. Parvez *et al.* (Eds)
© VSP 1997.

Analysis of drugs of abuse by micellar electrokinetic capillary chromatography

V. CRAIGE TRENERRY[1] and ROBERT J. WELLS[2]

[1]*Australian Government Analytical Laboratories, 338–340 Tapleys Hill Road, Seaton, SA 5023, Australia*
[2]*Australian Government Analytical Laboratories, 1 Suakin Street, Pymble, NSW 2073, Australia*

INTRODUCTION

Micellar Electrokinetic Capillary Chromatography (MEKC) is a rapidly expanding branch of Capillary Electrophoresis (CE) which provides not only the outstanding separation efficiency of CE but also offers the advantage that neutral molecules may be separated. Furthermore, several new parameters such as surfactant type, organic modifier and ion pairing reagent can be effectively utilised to allow fundamental alteration in many separation systems.

In 'traditional' CE, ionisable substances are separated by a combination of their electrophoretic mobility coupled with the endoelectroosmotic flow in the system. Both of these parameters are pH dependent. The ideal pH for electrophoretic separation of a mixture is in a range near the pK_a values of the target analytes whereas the electroosmotic flow of a system is very low at pH 2 but increases rapidly with increasing pH.

Therefore, as a first approximation, the most important factor in 'traditional' CE is the choice of the buffer pH and the buffering capacity of the buffer at that pH.

Various buffer additives are capable of significantly changing this relatively simple picture. Thus the neutral eight unit macrocyclic polysaccharide, β-cyclodextrin, forms weak inclusion complexes with many aromatic compounds. Whilst complexed to β-cyclodextrin, the effective molecular weight of a simple aromatic compound is increased by between two and four times whilst the charge on the molecule remains constant. This results in a significant decrease in electrophoretic mobility of the compound and this decrease in mobility is greatest for those compounds for which the stability constant of complex formation with β-cyclodextrin is the greatest.

Addition to the buffer of either anionic or cationic substances, such as hexane-sulfonate or tetrabutylammonium salts, which are capable of forming reversible ion pairs with analytes of interest, also effectively alter the net average molecular weight of the analyte. In this case, however, the formation of the reversible ion pair also alters the net average charge of the ionised analyte and therefore separations by CE are strongly influenced by both the concentrations of the additive and the buffer pH.

Of course, both types of buffer additive can be used in combination to give separations which are now dependent on several competing mechanisms. Theoretical modelling of such systems becomes increasingly difficult whilst it works reasonably well in simple systems.

Addition of organic solvents to buffers also affects separations. Not only do solvents alter the viscosity and dielectric constant of buffers but they can also modify the effective electrical charge on the analyte resulting in alteration of electrophoretic properties. In general, polar solvents which can interact with the capillary wall slow the endoelectroosmotic flow when added to a buffer and this flow can be more readily correlated to the viscosity of the pure solvent rather than the viscosity of the buffer. In the case of alcohols, this is presumably because the alcohol interacts strongly with the capillary wall and is the dominant factor in determining the properties of the double layer which in turn regulates electroosmotic flow. Diols are even more effective in slowing down the flow. Although protic solvents decrease the electroosmotic flow according to their viscosity, the same is not necessarily true of aprotic solvents. Acetonitrile forms very non-viscous mixtures with buffers and electroosmotic flow is affected minimally by addition of this modifier whereas dimethylsulfoxide (DMSO) has more effect and sometimes acetone has an even greater one, despite the fact that acetone is 10 times less viscous than DMSO in the pure form.

Addition of micellar substances

When anionic or cationic micellar substances, most commonly sodium dodecylsulfate (SDS) or cetyltrimethylammonium bromide (CTAB), are added to the buffer used in separation, another separation mechanism comes into play. SDS is a charged molecule which, above the critical micelle concentration of 5×10^{-4} M, forms molecular aggregations of up to 50 molecules which can encapsulate smaller molecules. This encapsulation – de-encapsulation process is extremely rapid because of the small molecular dimensions of the surfactant micelles and, in the time frame of CE, may be regarded as an instantaneous molecular equilibrium.

The relative time a molecule spends micelle-bound rather than in solution will depend on the relative affinity of a particular substance for organic or for aqueous media. Thus, polar substances such as methanol and sugars will have little affinity with lipophilic micelles; they will move with the electroosmotic flow (EOF) of the system and can be used as markers for the 'void volume' of the system. However, non-polar, non-charged substances such as anthracene or Sudan III which have little affinity for water will spend most of their time associated with micelles and their migration time will approximate to that of micelles. Most substances will therefore have a CE residence time within a window between the emergence of methanol or Sudan III. Exceptions to this are charged molecules which are hydrophilic and have positive electrophoretic mobility. These substances will appear before the emergence of the neutral 'marker' substance.

MEKC is a potentially 'universal' method for the separation of low molecular weight compounds (its use for macromolecular compounds is very limited because large molecules are not incorporated into micelles). It may also be used in conjunction with other methods of modifying CE separations. For instance, an extensive survey has been published where cyclodextrins, optically active ion pairing compounds, surfactants and organic modifiers were used in conjunction to effect direct chiral separation. Although the authors did not seek or claim optimisation, it was clear from the results that some or all of the additives contributed to separations and the relative contribution of each additive was not immediately discernible. The cauldron of buffer mixture additives used in many separations in this paper did not, however, appear to decrease separation significantly indicating that the many reversible equilibria operating during the separations were *all* extremely fast.

To summarise the variables that can be used in MEKC, we can draw up the following buffer modifications that can be utilised:

change pH	alter electroosmotic flow
	alter electromobility of analytes
add neutral complexing agents	alter electromobility
add ion pair substances	alter electromobility
add organic modifier	alter electroosmotic flow
	alter electromobility

All this is additional to the effects that can be obtained by alteration of the micelle-forming surfactant. Anionic surfactants such as SDS produce micelles which migrate in the opposite direction to the EOF during a separation. Because the use of cationic surfactant CTAB reverses the direction of EOF when used, it also possesses an electromobility in the opposite direction to the EOF. However, we have found in practice that at constant pH, compounds usually have a significantly faster migration time when CTAB is used as surfactant than with SDS. This probably reflects the comparative lower stability of CTAB-analyte micelles over SDS-analyte micelles.

It is clear that many factors can affect the stability of micelle-analyte agglomerates, one of which is the use of organic modifiers in the buffer which not only slow the EOF but have a profound effect on separations.

Finally, it must be mentioned that there has been a strong preference in CE to work either below pH 3 where electroosmotic flow is very low or above pH 8 where the change of electroosmotic flow with pH is small. At pHs between these values, the rapid change of electroosmotic flow with pH make the design of a stable reproducible system using uncoated capillaries very problematical.

The above summary of factors affecting separations in CE are dealt with in greater detail in several reviews of CE which have recently appeared (Li, 1992; Kuhn and Hoffstetter-Kuhn, 1993; Weinberger, 1993). We have given this brief summary here to explain some of the rationale used in the development of new quantitative procedures described later.

Aims of present work

Our aim in the present work was to develop a universal CE method whereby any one of a number of classes of drugs of abuse and associated impurities could be quantitatively analysed under the same electrophoretic conditions. In MEKC all substances should migrate within a window between the emergence of an unretained polar substance such as methanol and the emergence of the micelle. Therefore this aim did not appear unreasonable. What had to be established, however, was the quantitative reliability and repeatability of results obtained.

In the event, we have found that heroin and cocaine seizures can be conveniently analysed under identical electrophoretic conditions whereas buffer modifications are required to obtain adequate peak shapes and quantitative data in the case of opium alkaloids and, more particularly, in the case of amphetamines.

Finally we report a CE method to differentiate dextro- and levorphanol derivatives. In this case, dextrophan analogues are commonly used as antitussive agents whereas their levorotatory counterparts are powerful narcotic analgesics.

MATERIALS AND METHODS

Reagents

Purified drug standards were obtained from the Curator of Standards, Australian Government Analytical Laboratories, 1 Suakin Street, Pymble, NSW 2073, Australia. Sodium dodecyl sulfate (SDS) was obtained from E. Merck Pty Ltd. Kilsyth, Victoria, 3137, Australia. Cetyltrimethylammonium bromide (CTAB) was obtained from Sigma Chemical Company, St. Louis, MO 63178, USA. All other chemicals and solvents were of AR grade or HPLC grade and were used without further purification.

Micellar electrokinetic capillary electrophoresis

Qualitative work descibed later was performed with 60–75 cm × 50–75 μm i.d. fused silica capillary tubes (Isco Inc., Lincoln, Nebraska), with an effective length of 35–50 cm to the detector window. Quantitative work was performed with fused silica capillaries with the following dimensions — heroin assays:

72 cm × 75 μm i.d., effective length 50 cm; cocaine and amphetamine assays: 65 cm × 75 μm i.d., effective length 40 cm; opiate assays: 70 cm × 50 μm i.d., effective length 45 cm.

An Isco Model 3140 Electropherograph (Isco Inc.) was used for all of the analyses. For quantitative work, the instrument operated at −15 kV and at a temperature of 30 °C for heroin, cocaine and amphetamines and at −25 kV and a temperature of 28 °C for opiate alkaloids. The sample solutions were loaded onto the capillary under vacuum (Vacuum level 2, 5.0 kPa s for heroin and cocaine and Vacuum level 2, 10.0 kPa s for amphetamines and opiate alkaloids for the Isco 3140 Electropherograph).

It has been found that substance elution times in all cases studied are dependent not only on the particular batch or brand of fused silica capillary but also, to some extent, on the make of the instrument. It has not proved possible to transfer a method from one instrument to another and obtain identical resident times, even using the same capillary. Relative resident times remain unchanged. Therefore, *the use of an internal standard in all this work is essential.*

Procedure for capillary preparation and handling

The capillary was filled with 1 M NaOH and allowed to stand for 1 h. This solution was replaced with 0.1 M NaOH, allowed to stand for a further hour and washed with deionised water before filling with the running buffer. For CTAB buffers, the capillary was cleaned by washing with 0.1 M HCl for 10 min, followed by successive washings with deionised water, 0.1 M NaOH and deionised water before refilling with buffer. This was done on a weekly basis. For both qualitative and quantitative analyses, the capillary was flushed with running buffer for 2 min between runs.

Preparation of buffers

Heroin assays. 0.1 M SDS buffers were prepared by diluting of a stock solution of SDS (0.2 M) with an equal volume of a 1:1 mixture of KH_2PO_4 (0.02 M) and sodium tetraborate (0.02 M). 0.05 M CTAB buffers were prepared by dissolving 1.83 g CTAB in 100 ml of a 1:1 mixture of KH_2PO_4 (0.02 M) and sodium tetraborate (0.02 M). The pH of the solution was adjusted to 8.6 with 2.5% phosphoric acid or 0.05 M NaOH solution as required. Operating buffers were prepared by adding 2.5 ml of acetonitrile to 22.5 ml of the CTAB buffer, filtering the solution through a 0.45 μm PTFE filter disc and degassing with sonication before use.

Cocaine assays. CTAB buffers (0.05 M) were prepared as described for heroin, except that 1.875 ml of acetonitrile was used for the final solution.

Amphetamine assays. 0.025 M CTAB buffers were prepared by dissolving 0.92 g CTAB in 100 ml of 0.01 M sodium tetraborate. The pH of the solution is adjusted to 11.5 with 1 M NaOH solution. The running buffer was prepared by adding 88 ml of this solution to 1 ml of ethanolamine and 11 ml of dimethyl-sulphoxide (DMSO). The solution was mixed thoroughly and filtered through a 0.45 μm PTFE filter disc before use.

Crude morphine, poppy straw and opium assays. 0.05 M CTAB buffers were prepared by dissolving 0.92 g CTAB in 50 ml of a 1:1 mixture of 0.02 M KH_2PO_4 and 0.02 M sodium tetraborate. The pH of the solution was adjusted to 8.6 with 2.5% phosphoric acid or 0.01 M NaOH as required. Operating buffers were prepared by adding 2.5 ml of dimethylformamide to 22.5 ml of the CTAB buffer and the solution filtered through a 0.45 μm PTFE filter disc before use.

Detection

Heroin and related compounds were detected at 280 nm at 0.01 AUFS, cocaine and its derivatives were detected at 230 nm at 0.05 AUFS, amphetamines and related compounds were detected at 254 nm at 0.01 AUFS and the opiate alkaloids detected at 254 nm at 0.005 AUFS. Morphinans were detected at 200 nm and 0.01 or 0.02 AUFS. Electropherograms were recorded and processed with the ICE Data Management and Control Software supplied with the model 3140 Electropherograph.

Standards: Qualitative MEKC work

Stock solutions were prepared at 1 mg/ml by dissolving the substances in 0.01 M HCl. For MEKC with SDS buffers, the stock solutions were diluted with 1:1 phosphate-borate buffer and water to pH 7. For MEKC with CTAB buffers, the stock solutions were diluted with water only. The solutions were refrigerated until required.

Quantitative CE work

Heroin assays. Stock solutions of morphine, O^6-monoacetylmorphine and acetylcodeine, caffeine and theophylline were prepared daily at 1 mg/ml by dissolving the substances in 0.01 M HCl. Pholcodine was prepared in the same way at 3 mg/ml.

Heroin standard. Standard solutions were prepared by dissolving a known weight of heroin hydrochloride (15–20 mg) in 2 ml of pholcodine solution (3 mg/ml), adding suitable amounts of morphine, O^6-monoacetylmorphine and acetylcodeine stock solutions and diluting to 10 ml with 0.01 M HCl. The solutions were filtered through a 0.45 μm cellulose acetate filter and refrigerated before use. Heroin gave linear detector response between 0.1 and 3 mg/ml whilst O^6-monoacetylmorphine and acetylcodeine were linear from 0.1 to at least 0.8 mg/ml. A typical standard solution contains 0.1 mg/ml morphine, 0.1 mg/ml O^6-monoacetylmorphine, 0.1 mg/ml acetylcodeine and 1.5 mg/ml heroin and 0.6 mg/ml pholcodine.

Caffeine standard. The standard solution was prepared by mixing 1 ml of caffeine solution (1 mg/ml) with 1 ml of theophylline solution (1 mg/ml) and diluting to 10 ml with 0.01 M HCl. Caffeine gave linear detector response up to 0.2 mg/ml.

Cocaine assays. Stock solutions of benzoylecgonine and *trans*-cinnamoylcocaine were prepared daily at 0.2 mg/ml by dissolving the substances in 0.01 M HCl. Phocodine was prepared in the same way at 3 mg/ml. The solutions were refrigerated until required. Standard solutions were prepared daily by dissolving a known weight of cocaine (approximately 12 mg) in 5 ml of pholcodine solution (3 mg/ml), adding suitable amounts of benzoylecgonine and *trans*-cinnamoylcocaine solutions and diluting to 25 ml with 0.01 M HCl. The solutions were filtered through a 0.45 μm cellulose acetate filter and refrigerated before use. Cocaine gave linear detector response between 0.1 and 0.6 mg/ml whilst benzoylecgonine and *trans*-cinnamoylcocaine were linear to 0.25 mg/ml and 0.2 mg/ml respectively. A typical standard solution contains 0.5 mg/ml cocaine, 0.02 mg/ml benzoylecgonine, 0.02 mg/ml *trans*-cinnamoylcocaine and 0.6 mg/ml pholcodine.

Amphetamine assays. Stock solutions of amphetamines and related substances were prepared weekly at 1 mg/ml by dissolving the substances in 0.01 M HCl. Caffeine was used as the internal standard and prepared in the same way at 1 mg/ml. The solutions were refrigerated before use. The stock solutions were diluted with deionised water and filtered through a 0.45 μm cellulose acetate filter disc before analysis. A typical standard solution contains 0.4 mg/ml of amphetamine (or related substances) and 0.1 mg/ml of caffeine. The detector response was linear to 1.6 mg/ml for the amphetamines.

Crude morphine, poppy straw and opium assays. Stock solutions of morphine were prepared weekly at 3 mg/ml by dissolving the substances in 0.01 M HCl. Stock solutions of the minor alkaloids were prepared in the same way at 1 mg/ml. Pholcodine was used as the internal standard and prepared in the same way at 4 mg/ml. The solutions were refrigerated before use. The stock solutions were diluted with deionised water and filtered through a 0.45 μm cellulose acetate filter disc before analysis. The detector response was linear to 1 mg/ml for morphine and codeine, and to 0.2 mg/ml for oripavine, thebaine, papaverine and narcotine. A typical standard solution contains 1.0 mg/ml of morphine, 0.1 mg/ml codeine, 0.02 mg/ml oripavine, 0.02 mg/ml thebaine, 0.02 mg/ml papaverine, 0.02 mg/ml narcotine and 0.4 mg/ml of pholcodine.

Samples

Samples of illicit heroin, cocaine and amphetamines were actual seizures made by the Australian Federal Police. Some of the heroin samples were prepared from crude morphine produced from Australian grown *Papaver somniferium* poppies from a previous study. The poppy straw was produced from Australian grown *Papaver somniferium* poppies and kindly supplied by Glaxo, Australia. The opium and opium dross was supplied by Curator of Standards, Australian Government Analytical Laboratories, 1 Suakin Street, Pymble, NSW 2073, Australia.

Heroin samples. Sample solutions were prepared by mixing a weighed amount of illicit substance (20–70 mg) with 2 ml pholcodine solution (3 mg/ml) and diluting the solution to 10 ml with 0.01 M HCl. The mixture was sonicated for 2 min, mixed thoroughly and filtered through a 0.45 µm cellulose acetate filter disc and refrigerated before use.

Cocaine samples. Sample solutions were prepared by mixing a known weight of the seizure or sample (15–25 mg) with 5 ml of pholcodine solution (3 mg/ml) and diluting to 25 ml with 0.01 M HCl. The mixture was sonicated for 2 min, mixed thoroughly and filtered through a 0.45 µm cellulose acetate filter disc and refrigerated before use.

Amphetamine samples. Sample solutions were prepared by mixing a weighed amount of sample (10–50 mg) with 1 ml of caffeine solution and diluting to 10 ml with 0.01 M HCl. The mixture was sonicated for 2 min, mixed thoroughly, filtered through a 0.45 µm cellulose acetate filter disc and refrigerated before use.

Crude morphine, opium and poppy straw preparations. The crude morphine solutions were prepared by mixing a weighed amount of sample (10–50 mg) in with 1 ml of pholcodine solution (4 mg/ml) and diluting the solution to 10 ml with 0.01 M HCl (unless otherwise specified in the text).

Opium and opium dross were extracted using the procedure of Srivastava and Maheshawari (1985). Essentially, 1 g of opium was extracted with 20 ml of 2.5% acetic acid and then made to volume (100 ml) with water. For the opium dross, 1 g of sample was extracted with 2.5% acetic acid (5 × 20 ml) and the solutions combined. The solutions were filtered and 20 ml added to 60 ml of water. The pH of the solution was adjusted to 9.2 with concentrated ammonia solution and the solution was then extracted with chloroform (6 × 50 ml). The chloroform was dried over sodium sulphate and the solvent removed *in vacuo* with a rotary evaporator. The residue was dissolved in 5 ml of 0.05 M HCl. 1 ml of this solution was diluted to 5 ml with deionised water for HPLC analysis. The same dilution was made for MEKC analysis, except that pholcodine was added at a final concentration of 0.4 mg/ml.

The poppy straw was extracted with 1) lime-water, 2) 5% acetic acid solution and 3) Soxhlet extraction with 30% ethanol in chloroform.

1) Lime-water extraction (Dainis and Ronis, 1983). 100 ml of deionised water was added to 8 g of finely milled poppy straw and 2 g of calcium hydroxide and the mixture shaken vigorously for 25 min. The mixture was filtered and 20 ml was diluted with 60 ml of deionised water. The pH of the solution was adjusted to 9.2 with 10% acetic acid and then extracted with 25% ethanol in chloroform (5 × 50 ml). Care had to be exercised to avoid the formation of emulsions. The combined extracts were washed with deionised water (2 × 30 ml) and then dried with sodium sulphate. The solvent was removed under vacuum with a rotary evaporator. The residue was dissolved in 25 ml of 0.05 M HCl. 1 ml of this stock solution was diluted to 5 ml with deionised water for HPLC analysis. For MEKC analysis, 9 ml of the stock solution was diluted with 1 ml of internal standard solution and filtered through a 0.8 μm cellulose acetate filter before analysis.

2) 5% acetic acid extraction (Dainis and Ronis, 1983). 100 ml of 5% acetic acid was added to 8 g of finely milled poppy straw and the mixture shaken vigorously for 25 min. The mixture was filtered and 20 ml was diluted with 60 ml of deionised water. The pH of the solution was adjusted to 9.2 with concentrated ammonia solution and then extracted with chloroform (5 × 50 ml). The combined extracts were washed with deionised water (2 × 30 ml) and then dried with sodium sulphate. The solvent was removed under vacuum with a rotary evaporator. The residue was dissolved in 25 ml of 0.05 M HCl. 1 ml of this stock solution was diluted to 5 ml with deionised water for HPLC analysis. For MEKC analysis, 9 ml of the stock solution was diluted with 1 ml of internal standard solution and filtered through a 0.8 μm cellulose acetate filter before analysis.

3) 30% ethanol/chloroform extraction. 8 g of finely milled poppy straw was extracted with 30% ethanol/chloroform in a Soxhlet extractor for 6 h. The final volume of solvent after extraction was 130 ml. 25 ml was removed and taken to dryness with a rotary evaporator. The residue was dissolved in 10 ml 0.05 M HCl. 2 ml of this stock solution was diluted to 5 ml with deionised water for HPLC analysis. For MEKC analysis, 2 ml of the stock solution was added to 0.5 ml of internal standard solution and the solution made to 5 ml with deionised water. The solution was filtered through a 0.8 μm cellulose acetate filter before analysis.

High performance liquid chromatography

Heroin assays. The sample solutions were prepared as described in the above section for the MEKC analyses. No internal standard was used for the HPLC analyses. The analyses were performed with a model 501 HPLC pump, model 710B WISP and a model 490 programmable multiwavelength UV detector using a microporasil column (3.9 × 300 mm) equipped with a silica precolumn (Waters Chromatography Division of Millipore, Milford, MA, USA) with a mobile phase consisting of trimethylpentane, diethyl ether, methanol and diethylamine in the ratio 156:64:30:1 with a flow rate of 1.2 ml/min. The compounds were detected at 280 nm at 0.2 AUFS, and the data analysed with a HP3350A Laboratory Data System (Hewlett-Packard Company, Palo Alto, CA, USA).

Crude morphine, poppy straw and opium preparations. Sample preparation as for MEKC assays. The analyses were performed with a model 501 HPLC pump, model 712 WISP and a model 484 tunable absorbance UV detector using a 250 mm × 4.6 mm stainless steel column filled with 10 µm Spherisorb CN packing material equipped with a CN pre-column (Waters Chromatography Division of Millipore, Milford, MA 01757, USA). A mobile phase consisting of a 100:10:5 mixture of 1% ammonium acetate (adjusted to pH 5.8 with 10% acetic acid), acetonitrile and dioxan at a flow rate of 1.5 ml/min was used for the analyses. The compounds were detected at 254 nm at 0.05 AUFS, and the data analysed with a HP3350A Laboratory Data System (Hewlett-Packard Company).

Gas chromatography

Cocaine assays. Sample solutions were prepared by mixing a known weight of sample (approximately 15 mg) with 10 ml of eicosane solution (1 mg/ml) and diluting to 25 ml with a mixture of $CH_3OH : CHCl_3$ (10:90). The mixture was sonicated for 2 min, mixed thoroughly, filtered through a 0.45 µm PTFE filter disc and refrigerated before use. The analyses were performed with a Varian 3400 gas chromatograph equipped with a Varian 8100 autosampler and a flame ionisation detector (Varian Associates Inc., Walnut Creek, CA, USA), using a 12QC2/BP5 0.25 fused silica capillary column (SGE Pty Ltd., Ringwood, Victoria, Australia) operating in the split injection mode. Hydrogen was used as the carrier gas at a pressure of 12 psi and a split ratio of approx. 50:1. The injector temperature was 220°C and detector temperature was maintained at 280°C. After injection the column temperature was programmed at 10°C/min from 150° to 230°C and held at the final temperature for 2 min. The data were analysed with a HP 3350A Laboratory Data System (Hewlett-Packard Company).

Amphetamine assays. Sample solutions were prepared by dissolving a known weight of substance (20–70 mg) in 2 ml of deionised water, basifying the solution with 5 drops of concentrated ammonia solution and extracting with methylene chloride (2 × 3 ml, 1 × 2 ml). The organic solution was dried over sodium sulphate, filtered and added to 1 ml of diphenylamine (4 mg/ml) in methylene chloride and made to 10 ml with methylene chloride. The analyses were performed with a Varian 3400 gas chromatograph equipped with a Varian 8100 autosampler and a flame ionisation detector (Varian Associates Inc.), using a 25QC2/BP1 0.25 fused silica capillary column (SGE Pty Ltd.), operating in the split injection mode. Hydrogen was used as the carrier gas at a pressure of 12 psi and a split ratio of approx. 50:1. The injector temperature was 230°C and detector temperature was maintained at 280°C. After injection the column temperature was maintained at 80°C for two minutes followed by temperature programming from 80 to 200°C at 10°C/min and holding at the final temperature for 2 min. The data were analysed with a HP 3350A Laboratory Data System (Hewlett-Packard Company).

DISCUSSION

Pioneering work on the separation of drugs of abuse by CE was reported by Weinberger and Lurie (1991). Repetition of the conditions of Weinberger and Lurie gave satisfactory separation of all heroin and cocaine alkaloids with some later emerging peaks showing broadening. The attempted use of these conditions, however, for routine quantitative analysis did not yield reproducible results. We therefore investigated alternative buffer systems which would allow reproducible and reliable quantitation of illicit drug seizures with the same degree of analytical confidence as currently utilised validated methods.

Heroin (Trenerry *et al.*, 1994a)

Although the separation of drugs by the Weinberger and Lurie buffer system was excellent (Fig. 1), the relatively long analysis time was a disadvantage. Furthermore, it did not prove feasible to run many samples reproducibly without extensive and frequent capillary maintenance. We therefore sought a buffer system which was more robust for multiple samples and delivered faster separation times.

The repeatability of migration times and some peak shapes were distinctly improved by substitution of dimethyl sulfoxide (DMSO) for acetonitrile. Although addition of either 10 or 15% DMSO to the SDS-phosphate-borate buffer system gave satisfactory separations, the 10% DMSO modified buffer gave most consistent retention times. The separation of common heroin constituents with buffer modified with DMSO shown in Fig. 2 demonstrates that although reproducibility was satisfactory, analysis times were unacceptably long.

Buffer systems containing cetyltrimethylammonium bromide (CTAB) in place of SDS gave much faster separation times with both DMSO and acetonitrile as buffer additives. The polarity of the electropherograph was reversed for these experiments and the order of separation of some heroin constituents varied from the SDS system. Optimum heroin constituent separations were obtained with 10% acetonitrile employed as micelle modifier. A typical separation is shown in Fig. 3. Use of CTAB concentrations lower than 0.05 M gave inferior

V. C. Trenerry and R. J. Wells

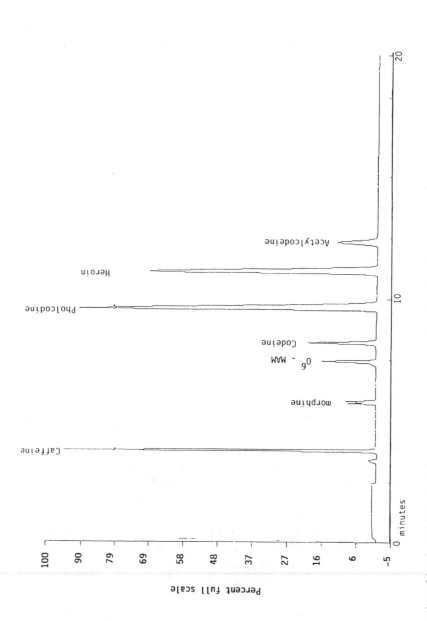

Figure 1. Separation of heroin and related substances using a mixture of 15% CH₃CN and 85% of a buffer consisting of 0.1 M SDS and a 1:1 mixture of 0.01 M KH₂PO₄ and 0.01 M sodium tetraborate, pH 8.6. The *x*-axis gives migration time in minutes.

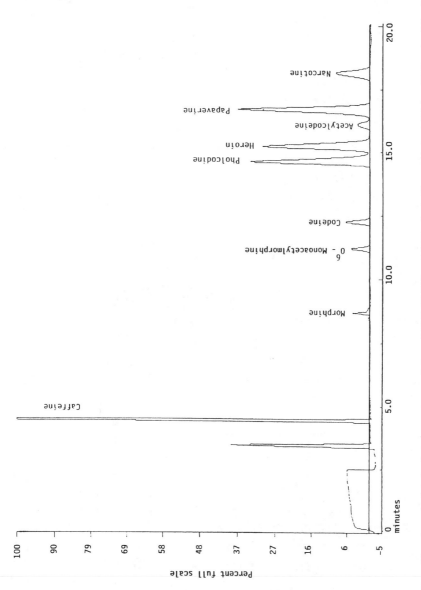

Figure 2. Separation of heroin and related substances using a mixture of 10% DMSO and 90% of a buffer consisting of 0.1 M SDS and a 1:1 mixture of 0.01 M KH$_2$PO$_4$ and 0.01 M sodium tetraborate, pH 8.6. The x-axis gives migration time in minutes.

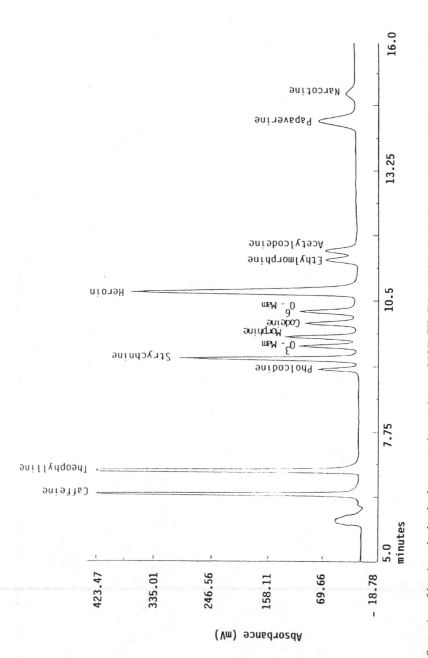

Figure 3. Separation of heroin and related substances using a mixture of 10% CH₃CN and 90% of a buffer consisting of 0.05 M CTAB and a 1:1 mixture of 0.01 M KH₂PO₄ and 0.01 M sodium tetraborate, pH 8.6. The *x*-axis gives migration time in minutes.

separations whilst the use of higher concentrations of CTAB gave no experimental advantage. Under optimised conditions shown in Fig. 3, excellent and reproducible separations of all alkaloids of interest were obtained in 13 min at pH 8.6.

The investigation of optimum pH for heroin separation produced somewhat surprising results. While the separation of all analytes was relatively constant between pH 8.4 and 8.8, it changed for some analytes above and below these pH values. The variations at pHs lower than 8.4 could be explained by greater degree of change of electroosmotic flow with change of pH at these lower pH values giving an inherently less stable system. The relative migration time changes of analyses, some significant, at pHs higher than 8.8 (Fig. 4) is partially, but not totally, explained by different pK_a values of various alkaloids.

Stability of all analytes, particularly heroin, in the optimised pH 8.6 buffer system was satisfactory under conditions used to prepare samples and also to the slightly alkaline buffer used for electrophoretic separation. No observable analyte decomposition occurred in 20 sample runs conducted overnight (*ca.* 7 h total run time).

The linearity of all common substances found in heroin seizures was established for the concentration ranges employed. Thus, heroin gave linear detector response between 0.1 and 3 mg/ml whilst O^6-monoacetylmorphine and acetylcodeine were linear from 0.1 to at least 0.8 mg/ml. The stronger UV absorbers, narcotine and papaverine, were linear to at least 0.2 mg/ml which is well within the range expected for the presence of these alkaloids in actual seizures of heroin. Caffeine, however, only gave linear response to 0.2 mg/ml and theophylline was required as internal standard to allow accurate quantitation of caffeine.

A feature of many south-east Asian seizures of heroin are their high purity. Many assay well over 90% heroin with relatively minor quantities of O^6-monoacetylmorphine (MAM, 0.2–3%) and acetylcodeine (AC, 1–6%) as the only major impurities. It appeared, therefore, prudent to use two internal standards in the operational method, one at a high level for heroin quantitation and one at a lower level for the quantitation of MAM and AC. Pholcodine and ethylmorphine were potential internal standards but it was found that equally satisfactory quantitation of both MAM and AC could be obtained using the high level pholcodine internal standard as using the low level ethylmorphine standard (pholcodine to ethylmorphine ratio, 10:1). As a consequence of this investigation shown in Table 1 only one internal standard, pholcodine, was included in the optimised routine method.

A comparison was carried out between the quantitative MEKC method for heroin discussed above with the HPLC procedure routinely used for analysis of heroin seizures. The results of this comparison are shown in Table 2 where 10 representative heroin seizures were each analysed nine times by both the MEKC and HPLC procedures. The different methods give similar but not identical results. There is an apparent slight systematic difference in absolute

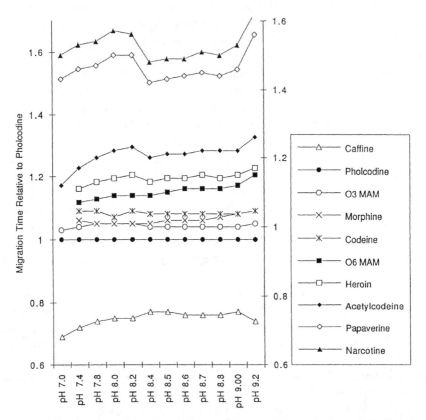

Figure 4. Migration times of heroin and related substances relative to the internal standard phol-codine using a mixture of 10% CH_3CN and 90% of a buffer consisting of 0.05 M CTAB and a 1:1 mixture of 0.01 M KH_2PO_4 and 0.01 M sodium tetraborate, pH 8.6. Results show the change in migration times with variations of buffer pH.

Table 1.

Comparison of the use of the major internal standard (pholcodine) and the minor internal standard (ethylmorphine) for the calculation of minor constituents of heroin seizures

Alkaloid	Morphine (%)		O^6-MAM (%)		DAM (%)		AC (%)	
Sample	Phol	Eth-M	Phol	Eth-M	Phol	Eth-M	Phol	Eth-M
1-68	Absent		1.0	1.0	90.6	90.9	3.9	3.9
1-68R	Absent		1.2	1.2	88.7	87.0	4.0	3.9
85/752	1.7	1.7	7.6	7.6	54.1	54.0	2.5	2.5
85/709	Absent		3.1	3.3	13.0	13.9	0.2	0.2
85/878	1.6	1.7	17.9	18.2	Absent		1.3	1.3
85/1141	Absent		3.5	3.7	74.9	78.5	8.8	9.2
85/1219	Absent		1.0	0.9	42.2	39.1	5.3	4.9
85/1222	Absent		1.0	0.9	14.1	13.7	2.0	2.0

O^6-MAM = O^6-monoacetylmorphine, DAM = diacetylmorphine (heroin), AC = acetyl-codeine, Phol = pholcodine, Eth-M = ethylmorphine.

quantitative values between the two methods. The percentage coefficient of variation (% CV) of the HPLC method is often, but not always, slightly better than that of the MEKC method. However, the % CV of the MEKC procedure is perfectly satisfactory for routine heroin analysis. The performance of the CE method in multi-laboratory Quality Assurance Studies where MEKC, HPLC and GC methods were used showed MEKC to be quantiatively comparable and more than competitive in regard to ease of use and speed of analysis. The comparative results of Quality Assurance comparisons are discussed in a later section.

The power of the MEKC method over the HPLC procedure was exemplified by comparison of the analytical results acquired from three complex samples containing non-opiate constituents. The MEKC results for these samples are shown in Table 3. Sample 92/10072 contained methaqualone and phenobarbitone in addition to the constituents shown in Table 3, the electropherogram of this seizure is shown in Fig. 5.

Table 2.

Comparison of CZE and HPLC quantitation for several heroin seizures relative to pholcodine as internal standard showing % Coefficients of Variation (% CV) between the methods

Alkaloid	Morphine (%)		O^6-MAM (%)		DAM (%)		AC (%)	
Sample	CZE	HPLC	CZE	HPLC	CZE	HPLC	CZE	HPLC
91/267	Absent		2.2	1.9	74.6	76.1	6.5	6.5
% CV			4.1	2.1	0.7	1.6	2.8	1.3
1-52	2.8	3.0	24.6	26.2	43.0	43.9	3.3	3.7
% CV			1.9	0.1	2.3	0.2	3.6	0.3
1-68	Absent		1.3	0.6	86.7	86.6	4.0	4.4
% CV			3.1	11.7	0.9	1.0	2.3	1.1
1-65	Absent		3.3	2.8	78.7	81.9	4.7	5.0
% CV			3.0	1.4	1.7	0.4	3.4	0.8
85/1141	Absent		3.5	3.2	73.4	73.1	8.5	8.5
% CV			1.7	3.8	0.5	1.4	1.1	1.1
85/1219	Absent		0.9	0.7	36.7	33.2	4.0	3.8
% CV			6.7	7.1	0.7	0.8	3.3	1.3
85/1222	Absent		1.0	1.1	13.7	13.8	2.2	2.2
% CV			5.0	4.5	2.6	3.1	3.6	4.5
85/10218	Absent		0.2	0.3	26.1	27.1	2.7	2.7
% CV			10.0	16.6	1.3	3.1	3.0	2.2
85/10220	Absent		0.5	0.6	27.3	26.8	2.9	2.8
% CV			10.0	5.0	0.8	0.6	4.1	1.4
S91/274	Absent		2.3	2.0	75.9	78.7	6.6	6.8
% CV			4.8	1.5	1.3	0.4	1.7	0.9

O^6-MAM = O^6-monoacetylmorphine, DAM = diacetylmorphine (heroin), AC = acetylcodeine. Average of 9 determinations for each sample.

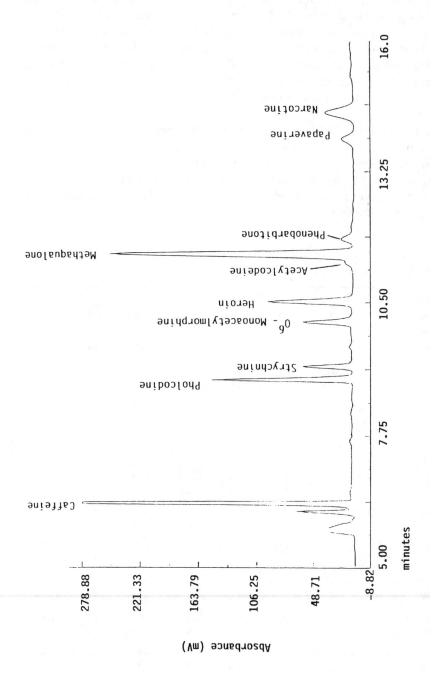

Figure 5. Separation of a complex Asian heroin using a mixture of 10% CH₃CN and 90% of a buffer consisting of 0.05 M CTAB and a 1:1 mixture of 0.01 M KH₂PO₄ and 0.01 M sodium tetraborate, pH 8.6. The x-axis gives migration time in minutes.

Table 3.

Quantitative results of 3 complex seizures containing non-opiate additives

Sample	CAF (%)	STY (%)	MOR (%)	MAM (%)	DAM (%)	AC (%)	PAP (%)	NAR (%)
85/878	48.0	ND	1.6	17.9	ND	1.3	ND	ND
92/92	18.0	ND	ND	1.0	26.7	1.7	0.7	0.6
92/10072	8.0	2.2	0.2	6.9	17.3	1.1	0.8	8.0

CAF = caffeine, STY = strychnine, MOR = morphine, MAM = O^6-monoacetylmorphine, DAM = diacetylmorphine (heroin), AC = acetylcodeine, PAP = papaverine, NAR = narcotine; ND – not determined.

The HPLC procedure is unable to compete with CE in the routine analysis of complex mixtures for the following reasons:

Caffeine and narcotine co-elute

Acetylcodeine and papaverine co-elute

O^3- and O^6-monoacetylmorphines co-elute

Codeine peak interferes with O^6-monoacetylmorphine peak

Morphine and strychnine co-elute

HPLC quantitation is suitable for routine samples that contain only acetylcodeine, heroin and O^6-monoacetylmorphine (and possibly caffeine and morphine). However, TLC is necessary to ascertain if other compounds are present before quantitation is carried out and if certain other compounds are shown to be present then HPLC quantitation is not possible for several combinations.

This is clearly demonstrated in the seizures in Table 3. For S92/92 or S92/10072 (brown powders; possibly of south-west Asian origin), HPLC cannot give quantitative results for caffeine, narcotine, papaverine or acetylcodeine. Similarly, for some south-east Asian preparations containing strychnine, the quantitation could be incorrect if morphine is present. Therefore, a preliminary TLC screening is essential. MEKC however, separates all of these compounds. It is still recommended to screen for other compounds by TLC before quantitative analysis by MEKC particularly to search for substances absorbing below 280 nm in the UV region.

Cocaine (Trenerry *et al.*, 1994b)

Repetition of the MEKC conditions which we had developed for the quantitation of heroin seizures was also satisfactory for the separation of cocaine and related substances. However, a slight variation in the concentration of the organic modifier, acetonitrile, from 10% to 7.5% and a change of UV detector monitoring wavelength from 280 to 230 nm gave the optimum compromise in selectivity, sensitivity and separation of cocaine and its analogues. An electropherogram of a standard mixture of cocaine and related substances is shown in Fig. 6, and Fig. 7 shows the relative movement of these standards during 20 replicate injections. Therefore, this buffer mixture not only gives excellent separations but is also suitable for routine work involving sequential injection of many samples.

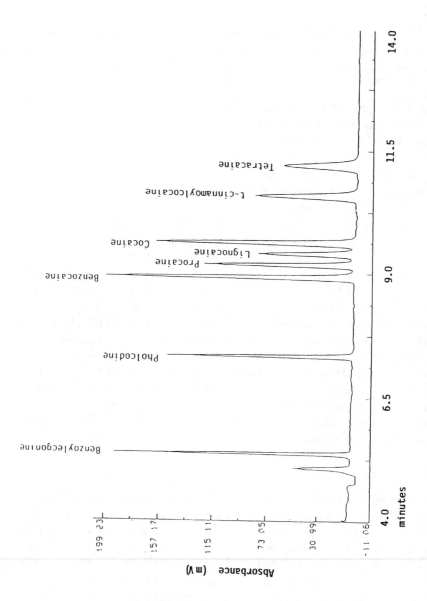

Figure 6. The optimum separation of cocaine and related substances by MEKC using a mixture of 7.5% CH₃CN and 92.5% of a buffer consisting of 0.05 M CTAB and a 1:1 mixture of 0.01 M KH₂PO₄ and 0.01 M sodium tetraborate, pH 8.6. The x-axis gives migration time in minutes.

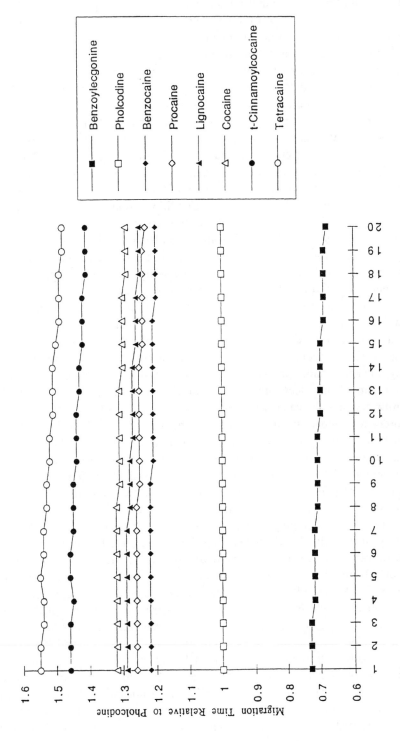

Figure 7. The retention times of cocaine and related substances relative to the internal standard pholcodine on repetitive injection of a standard mixture 20 times.

It was found that cocaine and related compounds were apparently stable to hydrolysis in the buffers used for both sample preparation and for MEKC runs. For instance, there was no measurable conversion of cocaine to benzoylecgonine during the multiple injection run performed to obtain the data for Fig. 7.

The conditions required to obtain the optimum separations of cocaine and related substances (shown in Fig. 6) are slightly different from those found to give optimum separation of heroin constituents. However, it should be noted that both routine heroin and cocaine seizures can be satisfactorily performed under the same MEKC conditions previously discussed for heroin with the exception that cocaine detection is conducted at 230 rather than 280 nm. Indeed, pholcodine can be used as internal standard for both groups of substances. Comparison electropherograms for cocaine and heroin constituents run under the same conditions are shown in Fig. 8. Conditions used in acquiring data for Fig. 8 were optimal for separation of the various cocaine related substances and not those previously reported for heroin. The MEKC method for heroin can be routinely used for quantitative work for cocaine. However, baseline separation of procaine and lignocaine is no longer possible but such a separation is infrequently required for analysis of routine illicit cocaine samples.

Detector response for various analytes were determined by plotting the concentration of each analyte against the ratio of peak area of the analyte to the peak area of the internal standard (pholocodine) maintained at constant concentration. Cocaine and tetracaine were linear up to analyte concentrations of 0.6 mg/ml; benzocaine, procaine and lignocaine up to analyte concentrations of 0.7 mg/ml; benzoylecgonine up to analyte concentrations of 0.25 mg/ml; and *trans*-cinnamoylcocaine up to analyte concentrations of 0.2 mg/ml.

In order to compare quantitative results obtained by MEKC with those acquired by a routinely used validated GC method (Noggle and Clark, 1982), cocaine seizures of different age, origin and cocaine content were analysed by each technique. The outcome of this comparison is shown in Table 4. Determinations were performed 9 times on each sample to ensure the stability of the MEKC system and the repeatability of the method. Although the comparative data were encouraging, it appeared that cocaine contents determined by MEKC were usually slightly lower than those from GC. However, the coefficient of variation between each method was comparable. A comparison of a typical electropherogram with a gas chromatogram of the same seizure is shown in Fig. 9a and 9b. Whilst separation times are comparable, benzoylecgonine is readily detected and quantitated by MEKC but is not detected under GC conditions. Ecgonine is not found by either method. Attempts to use 200 or 210 nm for the detection of ecgonine were not successful and it is very likely it is too hydrophilic to be retained by micelle entrainment. The low levels of *trans*-cinnamomylcocaine detected by GC and benzoylecgonine and *trans*-cinnamoylcocaine detected by MEKC could not be reliably quantitated relative to the internal standard which was present in concentrations appropriate to the accurate quantitation of cocaine which was the dominant constituent

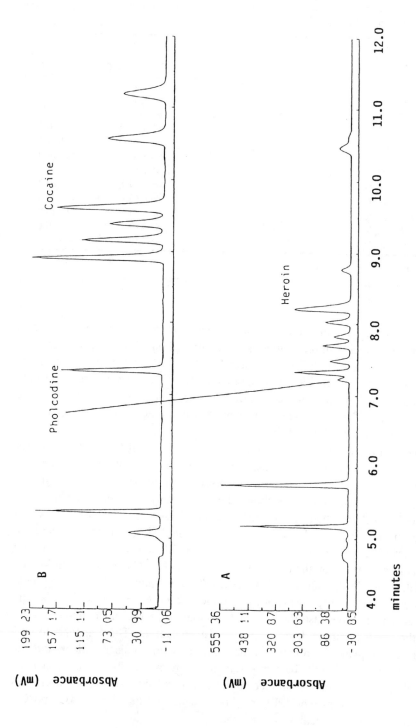

Figure 8. Comparison of the separation of A: heroin and related substances and B: cocaine and related substances using the same conditions as in Fig. 6. The *x*-axis gives the migration time in minutes. The *y*-axis is the UV absorption at 280 and 230 nm, respectively.

Table 4.

Comparison of quantitative results for laboratory prepared cocaine standards and australian federal police seizures by MEKC and GC

Alkaloid	Benzoylecgonine %	Cocaine %	Cocaine %
Sample	MEKC (% CV)	MEKC (% CV)	GC (% CV)
Standard 1	5 (3.3)	95 (2.4)	
Standard 2	4 (2.8)	96 (0.6)	
Standard 3	3 (3.2)	97 (2.0)	
Check sample*	0.6 (8)	82.6 (1.1)	86.0 (0.5)
Seizure C97/07		87.7 (1.1)	86.1 (1.3)
Seizure 1190/61		84.5 (0.8)	89.2 (0.9)
Seizure C92/572		41.7 (2.1)	37.2 (1.6)
Seizure sub-sample 1		86.7 (0.9)	88.8 (0.7)
Seizure sub-sample 2		86.9 (0.7)	88.3 (1.5)
Seizure sub-sample 3		85.6 (0.5)	89.2 (0.8)
Seizure sub-sample 4		83.5 (0.4)	87.9 (0.8)
Seizure sub-sample 5		88.7 (0.6)	90.7 (0.4)

Mean from 9 replicate analyses.

*The 'Check sample' is a carefully homogenised seizure used for Routine Quality Assurance purposes.

in all seizures. The addition of minor amounts of theophylline as an internal standard to check sample solutions containing about 0.6% benzoylecgonine and 3.8% trans-cinnamoylcocaine gave % CVs of 8% when these substances were quantitated. The electropherograms of two different seizures using both pholcodine and theophylline as internal standards are shown in Fig. 9a. The 'extra' peak in this check sample (obtained from an actual cocaine seizure) is almost certainly cis-cinnamoylcocaine.

Table 5 shows further comparisons of quantitation by MEKC with results previously obtained by other methods. The first two columns of Table 5 document the quantitative values obtained in the present work for single analyses of four cocaine seizures and the values obtained by a routinely used GC procedure. The third column shows the cocaine content reported for these same samples which had been previously analysed by using a validated HPLC method. There are distinct differences between the results with MEKC and GC values showing a much closer correlation than those acquired by the HPLC method. Both MEKC and GC gave similar but non-identical results in our hands. To ensure comparability of results between methods in Table 5, each MEKC analysis was only undertaken once because the only data available for the other methods were also the result of single determinations on each sample. There appears to be a difference in quantitation between the methods, with no discernible systematic bias evident, which will be addressed in future by results comparison from interlaboratory Quality Assurance studies. The reasons for major differences between results from the HPLC method and those from the MEKC and GC procedures is not immediately apparent.

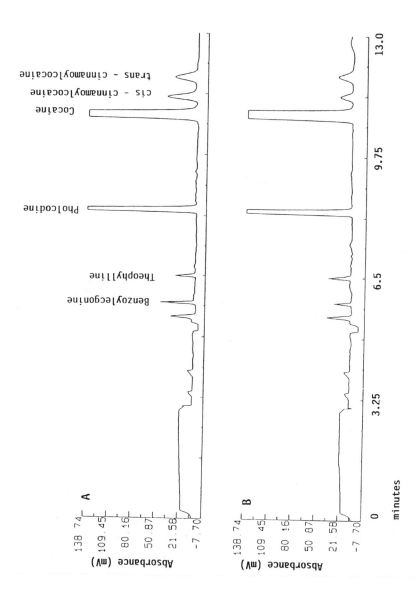

Figure 9a. MEKC separation of A: a typical cocaine seizure containing *cis*- and *trans*-cinnamoylcocaine and B: MEKC separation of a second typical cocaine seizure containing *cis*- and *trans*-cinnamoylcocaine. Both theophylline and pholcodine were added as internal standards. Conditions as in Fig. 6.

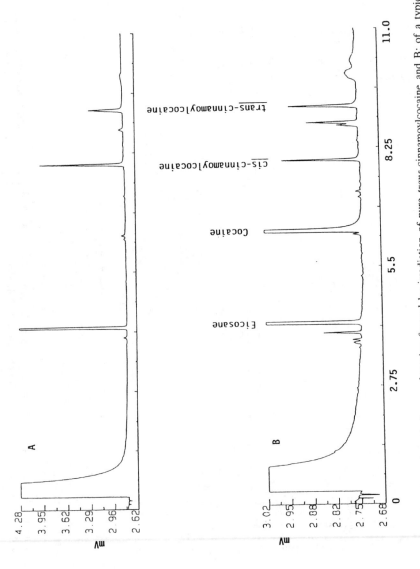

Figure 9b. GC separation of A: *cis-* and *trans-*cinnamoylcocaine formed by irradiation of pure *trans-*cinnamoylcocaine and B: of a typical cocaine seizure containing *cis-* and *trans-*cinnamoylcocaine. Eicosane was added as the internal standard.

Table 5.

Comparison of quantitative values obtained for
cocaine seizures by MEKC, GC and HPLC

Sample	Cocaine (%)		
	MEKC	GC	HPLC
93/002	29.4	28.9	29.0
93/003	69.7	68.0	72.0
93/004	82.2	79.7	88.0
92/938	71.7	69.8	76.0

Results are from a single analysis of each
sample by each method.

Quality assurance studies

Quantitative results acquired using the MEKC methodology developed for heroin and cocaine constituents have been found to be comparable with the currently used routine operational methods for each class of drug in both quantitative values obtained and coefficient of variation on repetition. Furthermore, the MEKC methods were used in interlaboratory Analytical Quality Assurance Studies on heroin and cocaine constituents amongst ten Australian Commonwealth and State Laboratories and found to give reliable results within two standard deviations of the whole set and with no significant systematic bias evident. However, these Quality Assurance studies conducted between Australian laboratories suggests that there is a significant systematic difference in the quantitation of cocaine obtained by HPLC or GC procedures respectively with HPLC results always higher. CE results more closely resemble those obtained by HPLC. The reasons for such differences are being investigated further. The great advantage of the two MEKC methods developed for heroin and cocaine lay, however, in their speed (heroin analysis time half that of the HPLC method) and the ability to analyse both drug classes using identical conditions with the exception of the wavelength used for UV detection. Furthermore, the excellent resolving power of MEKC allowed detection of additives and cutting agents in illicit seizures which were not detected by or interfered with the analysis by our standard laboratory operation methods.

Amphetamines (Trenerry *et al.*, 1995a)

The routine analysis of amphetamines and related stimulants such as ephedrine by the heroin–cocaine method discussed above was unsatisfactory. Some amphetamine derivatives could be analysed in the buffer system developed for heroin–cocaine. However, the detection and quantitation of the wide range of amphetamine derivatives required was totally unacceptable and totally new MEKC conditions were required. The new MEKC system which was subsequently developed gave excellent and reproducible separation of the amphetamine derivatives ephedrine, norephedrine, pseudoephedrine, pseudonorephedrine, amphetamine, methamphetamine, methylenedioxyamphetamine, methylenedioxymethamphetamine, 4-methoxyamphetamine, 3,4-dimethoxyamphetamine and 4-bromo-2,5-dimethoxyamphetamine.

Separation of a mixture of norephedrine, pseudoephedrine, ephedrine, amphetamine and methamphetamine with the pH 8.6 CTAB buffer used previously and modified by addition of 10% acetonitrile gave satisfactory separations and peak shapes (albeit rather broad) at low analyte concentrations. However, at higher analyte concentrations peak symmetry was lost and all peaks had a broad saw tooth shape as shown in Fig. 10. Although both separations and peak shapes were also satisfactory at low analyte concentrations when the organic modifier was either 10% dimethylformamide or 10% dimethyl sulfoxide, the same problem of broad sawtooth shaped peaks was again apparent at higher analyte concentrations.

The change of buffer from phosphate/borate to borate at pH 10.5 with 0.05 M CTAB as micellar additive and acetonitrile as micelle modifier also gave satisfactory separations. Figure 11 shows the separation of a series of norephedrines, ephedrines and amphetamines together with methylenedioxyamphetamine, methylenedioxymethamphetamine and pholcodine. Pholcodine was used as the internal standard for these experiments. Again this system caused peak tailing, particularly in the pseudoephedrine series and *d*-pseudonorephedrine appeared as a grossly distorted peak. The separations, peak widths and, in particular, the peak shape of pseudonorephedrine improved as the buffer pH was raised to 11.5 which, together with the use of acetonitrile as micelle modifier, appeared optimal for this particular system as illustrated in Fig. 12. However, accurate quantitative work could not be achieved with this buffer mixture.

Halving the concentration of CTAB to 0.025 M and replacing the 7.5% acetonitrile with 20% DMSO further improved the peak shapes of pseudonorephedrine and pseudoephedrine and the overall separation of the components (Fig. 13). However, when standard mixtures of norephedrine, pseudonorephedrine, ephedrine, pseudoephedrine, amphetamine, methylenedioxyamphetamine, methamphetamine and methylenedioxymethamphetamine of 0.1, 0.25 and 0.5 mg/ml respectively were run nine times each, the linearity of the detector response for pseudonorephedrine and pseudoephedrine in particular were totally unsatisfactory. The peak shapes of pseudonorephedrine and pseudoephedrine changed dramatically after nine runs for the 0.1 mg/ml and 0.5 mg/ml solutions. Accordingly, the percentage coefficient of variation (% CV) was poor for pseudoephedrine at low concentrations and poor for pseudonorephedrine at higher concentrations (Fig. 14). Table 6 shows the % CV for each substance run nine times with *p*-aminobenzoic acid as the internal standard.

The inclusion of β-cyclodextrin into the pH 11.5 borate-CTAB buffer did give some peak sharpening and indications of better resolution. However, standard mixtures of norephedrine, pseudonorephedrine, ephedrine and pseudoephedrine gave totally unsatisfactory reproducibility and % CV in repeated injections at lower analyte concentrations. Dimethyl-β-cyclodextrin was also unsuitable as

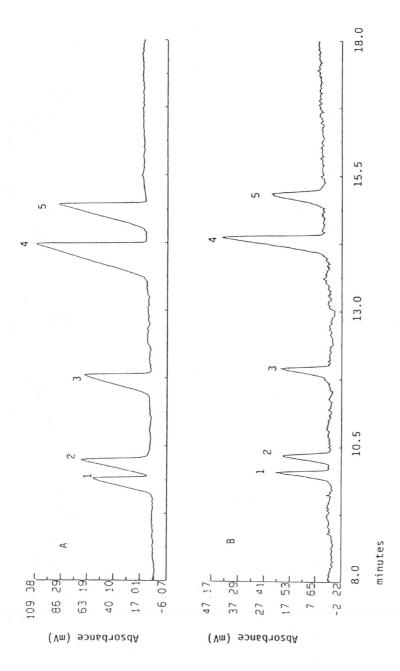

Figure 10. Partial electropherogram showing the separation and peak shapes of norephedrine (1), pseudoephedrine (2), ephedrine (3), amphetamine (4) and methamphetamine (5) for A: a concentration of 1.0 mg/ml of each analyte and B: a concentration of 0.2 mg/ml of each analyte using a mixture of 10% CH$_3$CN and 90% of a buffer consisting of 0.05 M CTAB and a 1:1 mixture of 0.01 M KH$_2$PO$_4$ and 0.01 M sodium tetraborate, pH 8.6.

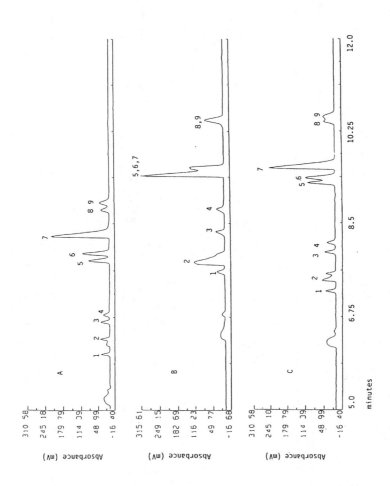

Figure 11. Partial electropherograms showing the separation and peak shapes of norephedrine (1), pseudonorephedrine (2), ephedrine (3), pseudoephedrine (4), amphetamine (5), methylenedioxyamphetamine (6), pholcodine (7), methamphetamine (8) and methylenedioxymethamphetamine (9) for (A) using a mixture of 7.5% CH$_3$CN and 92.5% of a buffer consisting of 0.05 M CTAB and 0.01 M sodium tetraborate, pH 10.5, (B) using a mixture of 7.5% CH$_3$CN and 92.5% of a buffer consisting of 0.05 M CTAB and 0.01 M KH$_2$PO$_4$, pH 10.5 and (C) using a mixture of 7.5% CH$_3$CN and 92.5% of a buffer consisting of 0.05 M CTAB and a 1:1 mixture of 0.01 M KH$_2$PO$_4$ and 0.01 M sodium tetraborate, pH 10.5. Pholcodine was added as the internal standard.

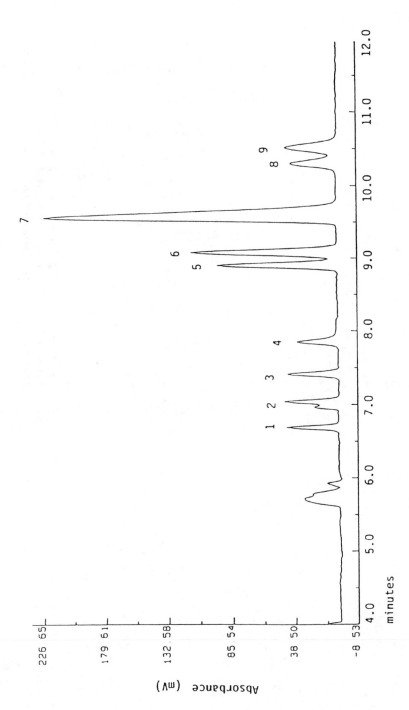

Figure 12. Partial electropherogram showing the separation and peak shapes of norephedrine (1), pseudonorephedrine (2), ephedrine (3), pseudoephedrine (4), amphetamine (5), methylenedioxyamphetamine (6), pholcodine (7), methamphetamine (8) and methylenedioxymethamphetamine (9) using a mixture of 7.5% CH₃CN and 92.5% of a buffer consisting of 0.05 M CTAB and 0.01 M sodium tetraborate, pH 11.5. Pholcodine was added as the internal standard.

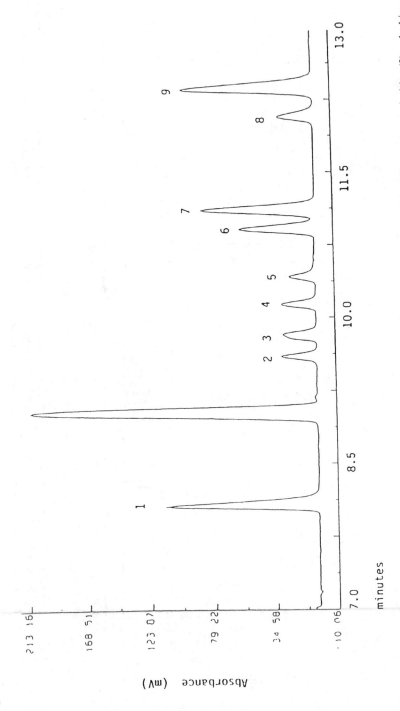

Figure 13. Partial electropherogram showing the separation and peak shapes of *p*-aminobenzoic acid (1), norephedrine (2), pseudonorephedrine (3), ephedrine (4), pseudoephedrine (5), amphetamine (6), methylenedioxyamphetamine (7), methamphetamine (8) and methylenedioxymethamphetamine (9) using a mixture of 20% DMSO and 80% of a buffer consisting of 0.025 M CTAB and 0.01 M sodium tetraborate, pH 11.5. *p*-aminobenzoic acid was added as the internal standard.

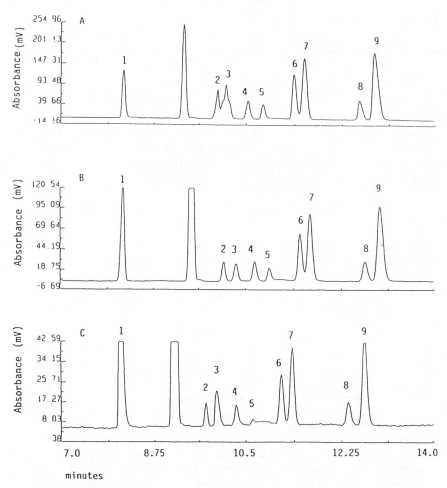

Figure 14. Partial electropherograms showing the separation and peak shapes of *p*-aminobenzoic acid (1), norephedrine (2), pseudonorephedrine (3), ephedrine (4), pseudoephedrine (5), amphetamine (6), methylenedioxyamphetamine (7), methamphetamine (8) and methylenedioxymethamphetamine (9) for A: run number 9 of a concentration of 0.5 mg/ml of each analyte, B: run number 9 of a concentration of 0.25 mg/ml of each analyte and C: run number 9 of a concentration of 0.1 mg/ml of each analyte using a mixture of 20% DMSO and 80% of a buffer consisting of 0.025 M CTAB and 0.01 M sodium tetraborate, pH 11.5. *p*-aminobenzoic acid was added as the internal standard.

a buffer additive in solving the problems with pseudonorephedrine and pseudoephedrine.

Ethanolamine containing buffers

The failure of more 'traditional' buffer mixtures to give a stable MEKC system for the whole range of amphetamines of interest led to the investigation of less conventional mixtures. Pseudoephedrine and pseudonorephedrine, the substances which had caused the major problems, are 2-aminoethanol derivatives.

Table 6.
Percentage coefficients of variation for amphetamines (internal standard: *p*-aminobenzoic acid)

Conc. (mg/ml)	NE	PNE	E	PE	A	MDA	MA	MDMA
0.1	7.7	6.0	6.9	39.1	4.1	3.0	4.6	2.8
0.25	4.6	4.1	2.8	13.0	3.4	3.4	2.4	2.0
0.5	19.3	32.2	3.4	3.2	2.0	1.8	3.1	2.1

The % CVs for norephedrine (NE), pseudonorephedrine (PNE), ephedrine (E), pseudoephedrine (PE), amphetamine (A), methylenedioxyamphetamine (MDA), methamphetamine (MA), and methylenedioxymethamphetamine (MDMA) for a repetitive series of injections of standard solutions analysed nine times. Buffer: 20% DMSO in 0.01 M sodium tetraborate and 0.025 M CTAB adjusted to pH 11.5 with 1 M NaOH.

Table 7.
Retention times for amphetamines by MEKC

Run no.	Elution times								
	Caf.	NE	PNE	E	PE	A	MDA	MA	MDMA
1	7.24	1.04	1.08	1.14	1.19	1.29	1.34	1.51	1.59
4	7.24	1.04	1.08	1.13	1.18	1.28	1.33	1.49	1.56
8	7.33	1.03	1.08	1.13	1.18	1.28	1.32	1.49	1.55
12	7.47	1.04	1.08	1.13	1.18	1.28	1.33	1.49	1.56
16	7.55	1.04	1.08	1.13	1.18	1.29	1.33	1.50	1.60
20	7.68	1.04	1.08	1.13	1.18	1.27	1.33	1.51	1.60

The retention times (min) of caffeine (Caf), norephedrine (NE), pseudonorephedrine (PNE), ephedrine (E), pseudoephedrine (PE), amphetamine (A), methylenedioxyamphetamine (MDA), methamphetamine (MA) and methylenedioxymethamphetamine (MDMA) for a repetitive series of injections of a standard solution. The elution time of caffeine, the internal standard is given together with the elution times of all other drugs relative to caffeine. Buffer: 1% ethanolamine and 11% DMSO in 0.01 M sodium tetraborate and 0.025 M CTAB adjusted to pH 11.5 with NaOH.

The errant MEKC behaviour exhibited by these substances might be expected to be suppressed if the MEKC buffer system was 'flooded' by a compound with the same structural features. This approach proved immediately successful and a stable and reproducible buffer system was rapidly developed which enabled not only the base line separation of all amphetamines of interest but gave satisfactory detector linearity and % CV on repeated injections for all compounds of interest, the problematical pseudoephedrines in particular. DMSO was found to be the micelle modifier of choice and the optimised MEKC system for amphetamine separation and quantitation was 0.025 M CTAB in a 0.01 M borate buffer modified with 11% DMSO and 1% ethanolamine and adjusted to pH 11.5 with 1 M NaOH. The electropherogram of a standard mixture of amphetamine run in this buffer is shown in Fig. 15. The reproducibility of the system over 20 runs was excellent and the variation in elution time of a number of amphetamine derivatives is shown in Table 7. Table 8 shows the % CV for each

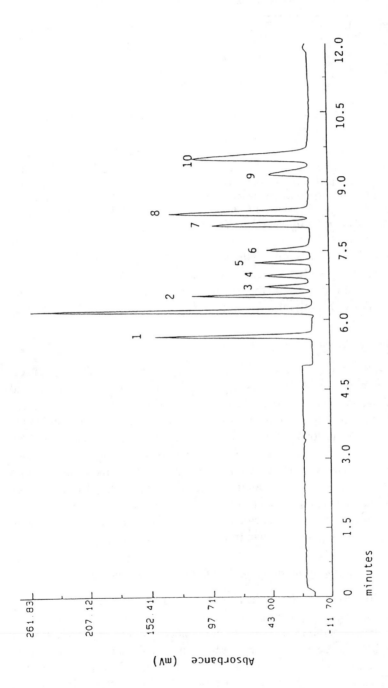

Figure 15. Electropherogram showing the separation and peak shapes of *p*-aminobenzoic acid (1), caffeine (2), norephedrine (3), pseudonorephedrine (4), ephedrine (5), pseudoephedrine (6), amphetamine (7), methylenedioxyamphetamine (8), methamphetamine (9) and methylenedioxymethamphetamine (10) using a mixture of 1% ethanolamine, 11% DMSO and 80% of a buffer consisting of 0.025 M CTAB and 0.01 M sodium tetraborate, pH 11.5. *p*-aminobenzoic acid and caffeine were added as the internal standards.

Table 8.

Percentage coefficients of variation for amphetamines (internal standard: caffeine)

Conc (mg/ml)	NE	PNE	E	PE	A	MDA	MA	MDMA
0.1					2.0	3.0		
0.2	2.5	3.1	2.7	2.1	1.0	0.9		
0.25							3.4	1.6
0.4					1.4			
0.8	0.5	1.6	0.8	2.1		1.0	1.1	2.0
1.6	0.8	0.9	0.7	1.3		1.9	1.1	1.1
1.7					1.4			
3.3					1.4			

The % CVs for norephedrine (NE), pseudonorephedrine (PNE), ephedrine (E), pseudoephedrine (PE), amphetamine (A), methylenedioxyamphetamine (MDA), methamphetamine (MA) and methylenedioxymethamphetamine (MDMA) for a repetitive series of injections of a standard solution 7 times. Buffer: 1% ethanolamine and 11% DMSO in 0.01 M sodium tetraborate and 0.025 M CTAB adjusted to pH 11.5 with NaOH.

substance run seven times with caffeine as the internal standard. The detector response was linear within the concentration ranges shown in Table 8 and a study without the use of an internal standard demonstrated linearity at least up to concentrations of 2 mg/ml.

Less commonly encountered amphetamines

Although less commonly encountered as seizures in Australia, 4-methoxy-methamphetamine (4-MMA), 3,4-dimethoxyamphetamine (3,4-DMA) and 4-bromo-2,5-dimethoxyamphetamine (4-Br-2,5-DMA) are, on occasion, encountered as illicit drugs. All three substances behave well in the ethanolamine buffer system. Detector linearity was excellent for all three compounds and % CV of quantitation after nine replicate injections was satisfactory as shown in Table 9. 3,4-DMA co-eluted with ephedrine and 4-MMA co-eluted with MDA. 4-Bromo-methoxyamphetamine appeared after MDMA. The substances which co-elute in this system can be readily differentiated by their different UV spectra if a diode array UV detector is used to record results. The separation of methoxyamphetamines is shown in Fig. 16.

Comparison of quantitation of amphetamines by MEKC and GC

It now remained to establish that the new quantitative MEKC method developed for amphetamines was not only robust and repeatable but gave comparable results to the GC method which had been previously validated. The % CV for the GC method was determined for the same eight amphetamine analogues that were evaluated in the MEKC method. As with the MEKC method, the % CV was calculated from seven replicate injections. Results are shown in Table 10. Comparison of the % CV for amphetamine by the GC method from Table 10 which those for the MEKC method shown in Table 8 shows that the results for

Table 9.

Percentage coefficients of variation for less common amphetamines

Conc (mg/ml)	3,4-DMA	4-MMA	4-Br-2,5-DMA
0.1	2.2	3.2	5.8
0.25	1.4	1.8	4.7
0.5	1.4	1.7	3.5

The % CVs for methoxymethamphetamine (4-MMA) 3,4-dimethoxyamphetamine (3,4-DMA) and 4-bromo-2,5-dimethoxyamphetamine (4-Br-2,5-DMA) for a repetitive series of injections of a standard solution 7 times. Buffer: 1% ethanolamine and 11% DMSO in 0.01 M sodium tetraborate and 0.025 M CTAB adjusted to pH 11.5 with NaOH.

Table 10.

Percentage coefficients of variation for gas chromatographic analysis of amphetamines

Conc (mg/ml)	A	MA	PNE	NE	PE	E	MDA	MDMA
0.2	1.0	1.0	4.2	4.3	6.1	4.9	1.9	1.3
0.8	0.8	0.5	0.6	0.6	0.8	2.3	0.4	0.7
1.6	0.3	1.1	0.5	0.4	0.5	0.4	0.4	0.4

The % CVs for the GC analysis of amphetamine (A), methamphetamine (MA), pseudonorephedrine (PNE), norephedrine (NE), pseudoephedrine (PE), ephedrine (E), methylenedioxyamphetamine (MDA), and methylenedioxymethamphetamine (MDMA) for a repetitive series of injections of a standard solution 7 times. Column: 25QC2/BP 1 0.25, Carrier gas: H_2, Split injection ratio: 50 : 1.

MEKC are similar to those for GC. The large % CVs for the GC procedure at low concentration coincided with poor peak shape. Both the peak shape and the % CV improved at the two higher concentrations. The GC determination requires the amphetamine salts to be converted to the free base before injection onto the column. This adds an extra step and therefore extra uncertainty to that method. A further drawback of the GC method was the co-elution of ephedrine and pseudoephedrine. A GC chromatogram of standards is shown in Fig. 17.

Results of the analysis of a series of illicit amphetamine and methamphetamine seizures by both MEKC and GC, each repeated 7 times, are reported in Table 11.

Table 12 shows the comparison of quantitative results from MEKC and GC in the present work and the GC results obtained in early 1990 for seven separate bags of an ephedrine seizure.

Both MEKC and GC give similar but non-identical results. There appears to be a difference in quantitation between the two methods, with no discernible systematic bias evident. Future interlaboratory Quality Assurance studies using both methods in parallel will be required to establish reliability and trueness of results for separate methods.

V. C. Trenerry and R. J. Wells

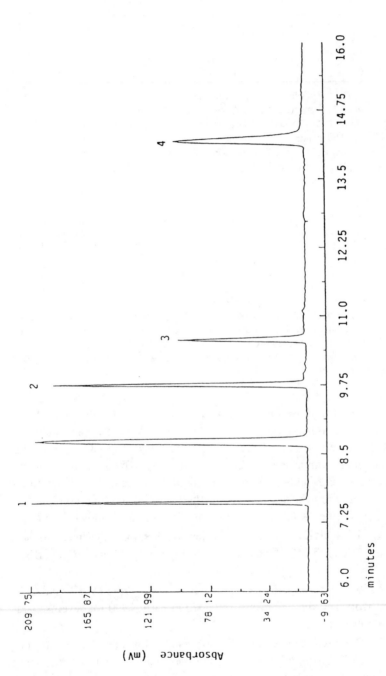

Figure 16. Partial electropherogram showing the separation and peak shapes of *p*-aminobenzoic acid (1), 3,4-dimethoxyamphetamine (2), 4-methoxymethamphetamine (3) and 4-bromo-2,5-dimethoxyamphetamine (4), using a mixture of 1% ethanolamine, 11% DMSO and 88% of a buffer consisting of 0.025 M CTAB and 0.01 M sodium tetraborate, pH 11.5. *p*-Aminobenzoic acid was added as the internal standard.

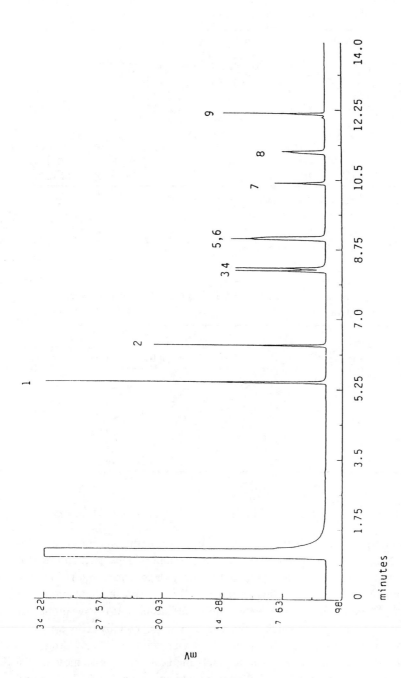

Figure 17. Gas chromatogram showing the separation of amphetamine (1), methamphetamine (2), pseudonorephedrine (3), norephedrine (4), ephedrine (5), pseudoephedrine (6), methylenedioxyamphetamine (7), methylenedioxymethamphetamine (8) and diphenylamine (9). Diphenylamine was added as the internal standard.

Table 11.

Comparison of quantitative results and the % CVs from MEKC and GC for a number of illicit amphetamine and methamphetamine seizures

Seizure	Conc. of A (%) (% CV)		Conc. of MA (%) (% CV)		Other (% CV)	
	MEKC	GC	MEKC	GC	MEKC	GC
Q92/684	23.7 (2.7)	22.0 (3.1)				
Q92/703	23.3 (2.1)	24.8 (0.4)				
93/126			7.8 (1.4)	8.9 (0.4)		
93/117			5.3 (4.0)	5.9 (0.4)		
93/118			6.0 (2.3)	5.9 (0.6)		
Q9033/3			19.7 (0.8)	20.7 (2.8)	57 (1.1)	66 (2.3)[1]
90/54	4.4 (2.4)	4.8 (1.0)			18.4 (1.3)	18.7 (2.9)[2]

A = amphetamine, MA = methamphetamine.
[1] shown to be pseudoephedrine by MEKC.
[2] shown to be ephedrine by MEKC.

Table 12.

Comparison of quantitative results from MEKC and GC in the present work and the GC results obtained in early 1990 for seven separate bags of an ephedrine seizure

Sample	Sub-sample	% Ephedrine		
		MEKC (1993)	GC (1993)	GC (1990)
90/3	Bag 1	69.2	73.9	80.3
	Bag 2	75.8	80.9	74.2
	Bag 3	78.1	76.7	79.1
	Bag 4	76.2	76.9	81.7
	Bag 5	74.3	78.3	84.3
	Bag 6	78.8	79.1	71.1
	Bag 7	72.2	77.3	76.2

Chiral separation of optical isomers of racemethorphan and raceorphan

Dextromethorphan, a synthetic analogue of codeine, is the *d*-enantiomer of 3-methoxy-N-methyl-morphinan. It is an effective and widely used anti-tussive agent with low toxicity and low potential for drug dependancy and is found in a large number of proprietory cough remedies. By contrast, the *l*-enantiomer of 3-methoxy-N-methylmorphinan (levomethorphan) has no anti-tussive properties but is a narcotic analgesic which is not available commercially. However, *l*-3-hydroxy-N-methylmorphinan (levorphanol) is a currently marketed drug used in the relief of pain. It is a powerful narcotic analgesic, five times more potent than morphine. Use of suitably labelled material has shown that 43% of an oral dose of dextromethorphan is excreted in urine within 24 h. It is known to be metabolised by both O- and N-demethylation. Metabolism by O-demethylation metabolic pathway leads to the production of d-3-hydroxy-N-methylmorphinan (Benson *et al.*, 1953).

Although the use of preparations containing dextromethorphan by athletes is allowed by the International Olympic Committee, the use of levorphanol is expressly banned and detection in routine urinary screening of athletes would lead to disqualification and the implementation of disciplinary action. Thus direct differentiation between the presence of an allowed substance dextrorphan [a metabolite of dextromethorphan] and levorphanol (a banned substance) in the urine of an athlete depends on the ability of the analyst to separate these two enantiomers.

Current analytical methods can not simultaneously separate the enantiomers of 3-methoxy-N-methylmorphinan or 3-hydroxy-N-methylmorphinan; thus it is only possible to distinguish between an athlete who has been taking a cough syrup from one who has used the banned substance levorphanol by the preparation of an diastereoisomeric derivative of 3-hydroxy-N-methylmorphinan. The presently used method for differentiation depends on the detection of the antitussive d-3-methoxy-N-methylmorphinan with the assumption that any co-occuring 3-hydroxy-N-methylmorphinan detected is a metabolite derived from dextromethorphan. Thus anyone taking dextromethorphan and levorphanol in combination may escape detection using this criterion.

Enantiomeric separations by MEKC

Initial attempts to obtain separation of the enantiomers of both 3-methoxy-N-methylmorphinan and 3-hydroxy-N-methylmorphinan explored the use of α-, β- and γ-cyclodextrins in association with either sodium dodecyl sulfate (SDS) or cetyltrimethylammonium bromide (CTAB). Although partial separations could be obtained with some cyclodextrin containing buffers, no complete separation of enantiomers was attained. Addition of the usual organic modifiers used in MEKC (methanol, acetonitrile etc.) also gave unsatisfactory separations. Similar lack of success resulted from the use of dimethylated or trimethylated β-cyclodextrins.

Some partial successes obtained with certain β-cyclodextrin systems encouraged further exploration of different organic modifiers to enhance chiral differentiation. The water solubility of β-cyclodextrin is enhanced by the addition of up to 30% ethanol or propanol, This enhancement has been ascribed to the formation of a loose host-guest complex between the short chain alcohol and β-cyclodextrin (Bender and Komiyama, 1978). This work was repeated and it was confirmed that β-cyclodextrin was over twice as soluble in 30% ethanol or 1-propanol in buffer containing sodium dodecyl sulfate as it was in water (Bender and Komiyama, 1978). By contrast, the solubilty of β-cyclodextrin in 30% aqueous solutions of methanol or 1-butanol was lower than in pure water. We reasoned that not only would ethanol or 1-propanol enhance the solubility of β-cyclodextrin but they could also increase the selectivity of host-guest complex formation between β-cyclodextrin and other substances.

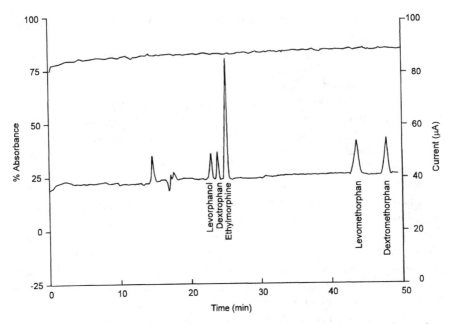

Figure 18. Separation of the standard mixture of 0.2 ppm levorphanol, dextrophan, levomethorphan and dextromethorphan together with 0.5 ppm ethyl morphine.

These expectations were realised with the excellent separation of morphinan diastereoisomers using 60 mM β-cyclodextrin dissolved in a running buffer containing 20% 1-propanol as an organic modifier. Furthermore, excellent separations of the enantiomers of both 3-methoxy-N-methylmorphinan and 3-hydroxy-N-methylmorphinan were obtained with 1-propanol modified buffered β-cyclodextrin-SDS mixtures as shown in Fig. 18. No enantiomeric separations were obtained using either methanol or butanol as the organic modifier. The separation efficiency of the column was maintained for at least 18 replicate injections with only capillary washing being required between injections. Day-to-day reproducibility was excellent if the capillary was washed with 1 M sodium hydroxide and water before use.

In order to accurately assess recoveries of analytes from urine and to obtain reliable quantitative data, an internal standard was required. Ethylmorphine behaved similarly to levorphanol during analyte recoveries from urine and was also conveniently positioned in the electropherogram (Fig. 18).

Isolation and analysis of morphinan enantiomers and morphine from urine

The O-demethylated products dextrorphan and levorphanol are excreted as glucuronides. The parent hydroxymethorphans were obtained from the glucuronides either by acid or enzymic hydrolysis. Direct injection of the hydrolysed urine gave too many interfering peaks in the electropherogram to allow detection of the analytes, even at high concentrations. An isolation step based

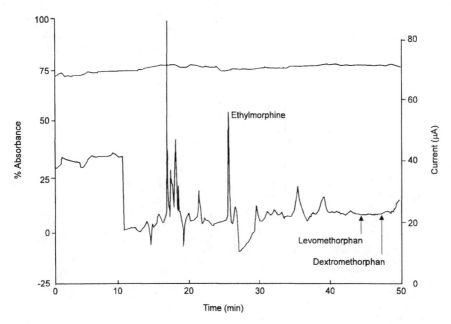

Figure 19. Electropherogram of urine spiked with 1 ppm ethylmorphine and taken through the extraction procedure.

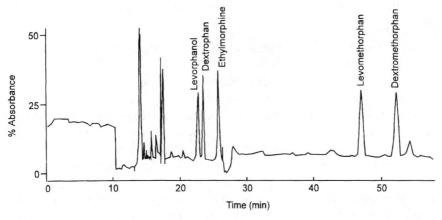

Figure 20. Electropherogram of urine spiked with 1 ppm levorphanol, dextrophan, levomethorphan and dextromethorphan and 0.5 ppm ethyl morphine.

on the use of a solid phase extraction cartridge was therefore incorporated into the work-up procedure for isolation of analytes.

Both ethylmorphine, dextromethorphan and levorphanol could be recovered in good yield from urine spiked with these compounds. Figure 19 shows a urine blank spiked with ethylmorphine and taken through the isolation procedure. Figure 20 shows the recoveries of drugs from urine spiked at 1.0 ppm. The concentration of analytes achieved during the work-up allowed the detection and quantitation of dextromethorphan and levorphanol spiked at 0.2 ppm in

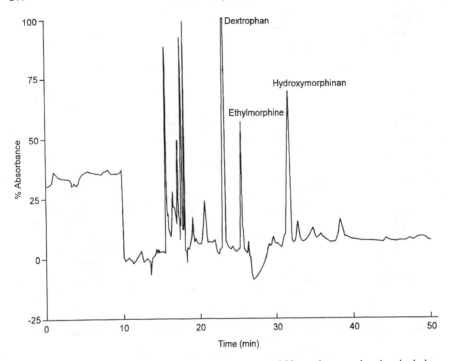

Figure 21. Electropherogram of urine 2 h after ingestion of 30 mg dextromethorphan hydrobromide.

original urine using ethylmorphine as the internal standard (added to the urine at the beginning of the analysis).

Morphine was readily detected under the same MEKC conditions used to separate the morphinans. Enzymic hydrolysis morphine 3-glucuronide in urine at 35 °C required 24 h and gave unsatisfactory results when subjected to the solid phase work-up procedure. Acid hydrolysis was therefore chosen for routine use because of its speed, efficiency and simplicity.

The validity of this procedure for the detection of dextrorphan and morphinan-3-ol *in vivo* was established by the analysis of incurred samples obtained from pharmacokinetic study. This study was conducted by the administration of a single 30 mg dose of dextromethorphan hydrobromide to a volunteer with urine samples collected over a 24-h period. Figure 21 shows the urinary metabolite profile 2 h after ingestion of the drug. Maximum metabolite concentrations in urine occurred between one and two hours and could still be readily detected at 24 h. MEKC results compared favourably with independent qualitative GC-MS measurements (Aumatell and Wells, 1993).

Opium alkaloids (Trenerry *et al.*, 1995b)

Morphine and codeine, the major alkaloids found in poppy straw and opium were reasonably well separated using the buffer system described for the separation and quantitation of illicit heroin seizures. Thebaine, narcotine, papaverine, narceine, cryptopine and salutaridine are also present in various amounts in poppy straw, crude morphine and opium derived from *Papaver somniferium* poppies (Dainis and Ronis, 1983). Oripavine is present in *P. somniferium* poppies that are grown in Australia and is found in various amounts in Australian grown poppy straw (Dainis *et al.*, 1984). Thebaine and oripavine are are not found in illicit heroin seizures as they rearrange when the crude morphine is converted to heroin. Initial attempts to separate morphine, codeine, thebaine, oripavine, narcotine, papaverine, narceine, cryptopine and salutaridine on a 50 μm uncoated fused silica column with the buffer used for the separation of the major components of illicit heroin were encouraging. Pholcodine was well separated from the other alkaloids and was therefore suitable for use as the internal standard. Optimum separation of the components was accomplished by increasing the amount of acetonitrile from 10 to 12.5% (Fig. 22). Morphine migrated before codeine and oripavine and thebaine were well separated. The separation of alkaloids was maintained over twenty repetitive injections, except for oripavine and cryptopine which coalesced (Fig. 23). Other organic modifiers were then tried in an attempt to achieve a stable separation of the alkaloids. Reasonable separation of the alkaloids was achieved with 27.5% methanol as the organic modifier. In contrast to the buffer containing acetonitrile, codeine migrated before morphine and the separation of oripavine and thebaine was not as pronounced. The run time for the separation also increased from 10–15 min with this buffer. Dioxan and tetrahydrofuran were also unsuitable as organic modifiers as morphine and codeine could not be completely separated. The buffer containing tetrahydrofuran was particularly unsuitable as the migration times of the alkaloids changed markedly over twenty repetitive injections. Dimethylformamide (DMF), dimethylsulphoxide and hexamethylphosphoramide have also been used as organic modifiers and also affected the separation of the alkaloids. 10% DMF was the organic modifier of choice as it gave the best separation of the ten alkaloids. The separation was also extremely stable over twenty repetitions (Fig. 24). An electropherogram showing the separation of the alkaloids with a buffer modified with 10% DMF is displayed in Fig. 25.

The effect of the different additives on the separation of the alkaloids is shown in Fig. 26. Very little separation was seen when both DMF and CTAB were absent from the buffer. The addition of 0.05 M CTAB increased the separation of the alkaloids, but complete separation was only achieved when the buffer contained both 0.05 M CTAB and 10% DMF. The repeatability data (% CV) for seven consecutive injections of a number of standards of different concentrations are displayed in Table 13.

A number of crude morphine preparations that were available from a previous study (Dainis *et al.*, 1984) were analysed by both MEKC and by HPLC (Nobuhara *et al.*, 1980). The levels of the various alkaloids in a number of crude morphine samples and the % CV data for seven repetitive injections for one sample of crude morphine are listed in Table 14. In general, there was a

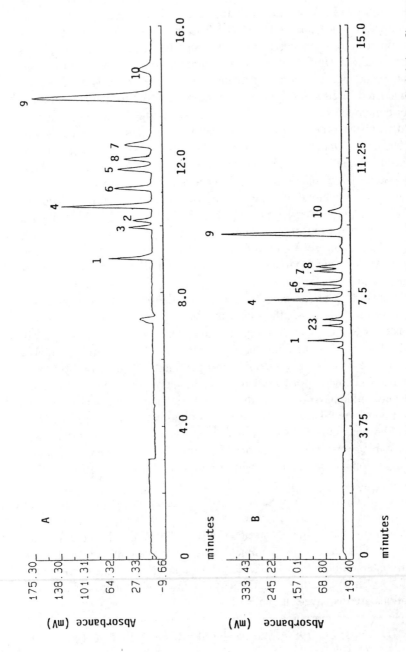

Figure 22. Separation of 1) pholcodine, 2) morphine, 3) codeine, 4) salutaridine, 5) oripavine, 6) cryptopine, 7) narceine, 8) thebaine, 9) papaverine and 10) narcotine using a mixture of A: 27.5% CH$_3$OH and 72.5% of a buffer consisting of 0.05 M CTAB and a 1:1 mixture of 0.01 M KH$_2$PO$_4$ and 0.01 M sodium tetraborate, pH 8.6 and B: 12.5% CH$_3$CN and 87.5% of a buffer consisting of 0.05 M CTAB and a 1:1 mixture of 0.01 M KH$_2$PO$_4$ and 0.01 M sodium tetraborate, pH 8.6. The x-axis gives migration time in minutes.

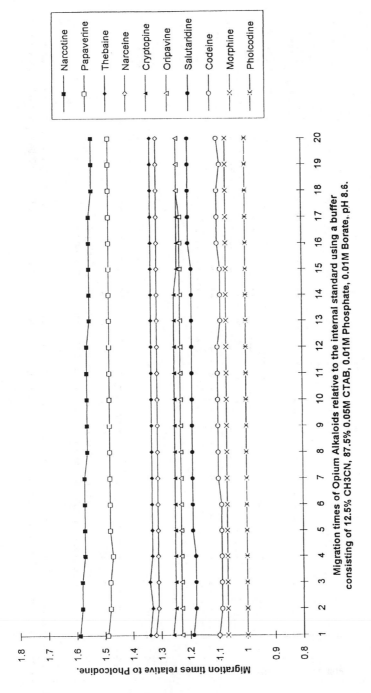

Figure 23. The retention times of morphine and related substances relative to the internal standard pholcodine on repetitive injection of a standard mixture 20 times using a mixture of 12.5% CH$_3$CN and 87.5% of a buffer consisting of 0.05 M CTAB and a 1:1 mixture of 0.01 M KH$_2$PO$_4$ and 0.01 M sodium tetraborate, pH 8.6.

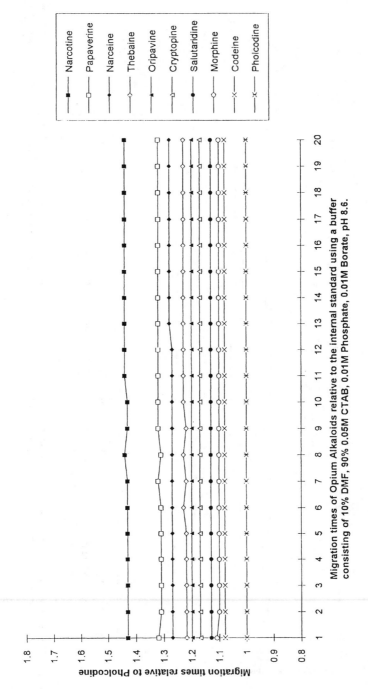

Figure 24. The retention times of morphine and related substances relative to the internal standard pholcodine on repetitive injection of a standard mixture 20 times using a mixture of 10% DMF and 90% of a buffer consisting of 0.05 M CTAB and a 1:1 mixture of 0.01 M KH_2PO_4 and 0.01 M sodium tetraborate, pH 8.6.

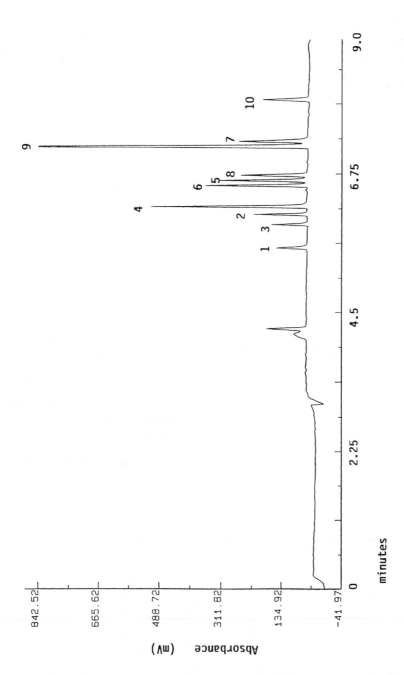

Figure 25. Separation of morphine and related substances using a mixture of 10% DMF and 90% of a buffer consisting of 0.05 M CTAB and a 1:1 mixture of 0.01 M KH$_2$PO$_4$ and 0.01 M sodium tetraborate, pH 8.6. Peak assignments as for Fig. 22. The x-axis gives migration time in minutes.

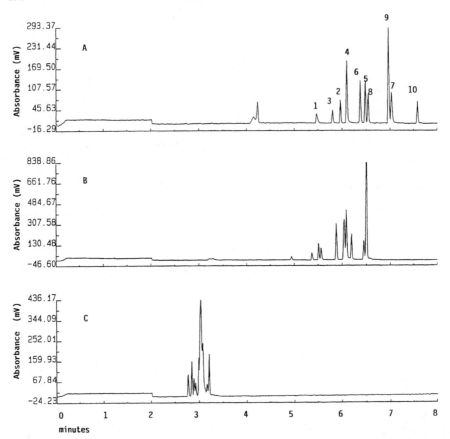

Figure 26. Separation of morphine and related substances using A: a mixture of 10% DMF and 90% of a buffer consisting of 0.05 M CTAB and a 1:1 mixture of 0.01 M KH₂PO₄ and 0.01 M sodium tetraborate, pH 8.6, B: a buffer consisting of 0.05 M CTAB and a 1:1 mixture of 0.01 M KH₂PO₄ and 0.01 M sodium tetraborate, pH 8.6 and C: a buffer consisting of a 1:1 mixture of 0.01 M KH₂PO₄ and 0.01 M sodium tetraborate, pH 8.6. Peak assignments as for Fig. 22. The *x*-axis gives migration time in minutes.

Table 13.
% CV for standard solutions by MEKC for seven repetitions

Conc. (mg/ml)	Codeine	Morphine	Oripavine	Thebaine	Papaverine	Narcotine
0.01	–	–	10	12	4.3	–
0.02	10	8.0	3.6	6.1	4.4	7.8
0.05	4.8	4.3	2.4	2.1	1.5	3.5
0.1	2.4	1.7	2.4	1.7	1.8	1.8
0.2	–	–	1.4	3.2	0.9	2.7
0.5	3.8	2.5	–	–	–	–
1.0	2.4	2.9	–	–	–	–

Table 14.

Comparison of quantitative results and the % CVs from MEKC and HPLC for a number of crude morphine preparations

Drug MEKC/HPLC	Codeine		Morphine		Oripavine		Thebaine	
Sample								
1-70	3.9	3.6	85.4	81.6	1.6	1.4	2.0	2.0
% CV[1]	9.0	0.6	4.3	0.3	4.1	0.8	6.4	0.9
% CV[2]	1.8	–	–	–	0.9	–	1.0	–
1-49	1.0	0.8	67.3	61.9	ND	ND	ND	ND
1-54E	1.5	1.4	95.5	93.2	0.4	0.4	ND	ND
% CV[3]	3.1	–	–	–	3.0	–		
1-44C	5.5	5.9	62.4	58.2	1.6	1.9	1.4	1.4

[1] 11.4 mg sample / 10 ml 0.01 M HCl.
[2] 51.6 mg sample / 10 ml 0.01 M HCl.
[3] 150.0 mg sample / 10 ml 0.1 M HCl.
– % CVs not determined.
ND – not detected.

Table 15.

Comparison of quantitative results from MEKC and HPLC for a sample of poppy straw, opium and opium dross

Drug MEKC/HPLC	Codeine		Morphine		Oripavine		Thebaine		Papaverine		Narcotine	
Sample												
Poppy straw[1]	0.1	0.1	0.9	0.9	0.01	0.01	0.05	0.05				
Poppy straw[2]	0.1	0.1	1.2	1.2	0.01	0.01	0.02	0.02				
Opium[3]	5.3	4.7	11.5	9.9			2.0	1.7	0.5	0.4	0.4	0.6
Dross[3]	1.7	1.6	2.9	2.9			0.2	0.2	0.6	0.6	0.9	0.9

[1] Soxhlet extraction.
[2] Lime-water extraction.
[3] Acetic acid extraction.

good agreement for the levels of the alkaloids in the samples, when determined by MEKC and HPLC. The levels of the minor components of crude morphine preparations could be more accurately determined by analysing more concentrated solutions. The % CV data for the minor alkaloids also improved when a more concentrated solution was analysed. The excellent separation of the alkaloids was maintained even when a solution containing 15 mg/ml of crude morphine was loaded onto the column. The % CVs for codeine (3.1%) and oripavine (3.0%) in sample 1-54E were quite satisfactory for the low levels in the sample. The partial electropherograms showing the separation of codeine, morphine and oripavine in sample 1-54E at sample concentrations of 1 mg/ml and 15 mg/ml are displayed in Fig. 27.

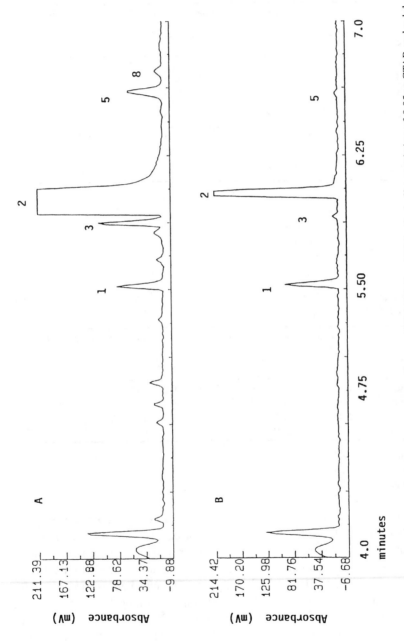

Figure 27. Separation of morphine and related substances using a mixture of 10% DMF and 90% of a buffer consisting of 0.05 M CTAB and a 1:1 mixture of 0.01 M KH$_2$PO$_4$ and 0.01 M sodium tetraborate, pH 8.6 for A) a solution containing 15 mg/ml of sample 1-54 and B) a solution containing 1 mg/ml of sample 1-54. Peak assignments as for Fig. 22. The *x*-axis gives migration time in minutes.

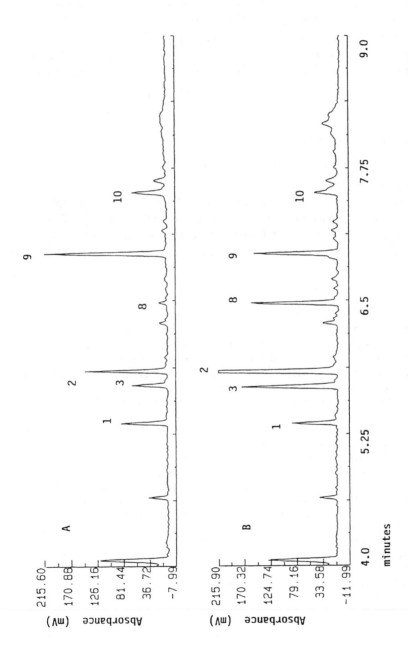

Figure 28. Separation of morphine and related substances using a mixture of 10% DMF and 90% of a buffer consisting of 0.05 M CTAB and a 1:1 mixture of 0.01 M KH$_2$PO$_4$ and 0.01 M sodium tetraborate, pH 8.6 for A) a sample of opium dross and B) a sample of opium. Peak assignments as for Fig. 22. The x-axis gives migration time in minutes.

The procedure was then extended to the analysis of a sample of poppy straw, opium and opium dross. One sample each of poppy straw, opium and opium dross was extracted with either lime-water, acetic acid or chloroform/ethanol and analysed by both MEKC and HPLC. The levels of the alkaloids are displayed in Table 15 and are in good agreement. The different levels of alkaloids extracted from the poppy straw with the lime-water and solvent was due to the incomplete extraction with the conditions used. Partial electropherograms of the opium and opium dross are displayed in Fig. 28.

ACKNOWLEDGEMENTS

We are grateful to Dr James Robertson of the Australian Federal Police for his continuing help and encouragement and to the General Manager of the Australian Government Analytical Laboratories for permissision to publish this work. Various diagrams and tables have been reproduced from original publications with the permission of the Editors. We thank the editors of *Electrophoresis*, *Journal of Chromatographic Science* and *Journal of Chromatography*.

REFERENCES

Aumatell, A. and Wells, R. J. (1993). *J. Chromatogr. Sci.* **31**, 502–509.
Bender, M. L. and Komiyama, M. (1978). *Cyclodextrin Chemistry*. Springer-Verlag, Berlin.
Benson, W. M., Stefko, P. L. and Randall, L. O. (1953). *J. Pharm. Ther.* **109**, 189.
Dainis, I. and Ronis, M. (1983). AGAL National Research Report Number 106.
Dainis, I., Ronis, M. and Trenerry, V. C. (1984). AGAL National Research Report Number 116.
Huizer, J., Logenberg, H. and Steenstro, A. J. (1977). Bulletin on Narcotics XXIX, 4, 65.
Kuhn, R. and Hoffstetter-Kuhn, S. (1993). *Capillary Electrophoresis: Principles and Practice*. Springer-Verlag, Berlin.
Li, S. F. Y. (1992). Capillary Electrophoresis: principles, practice and applications. *J. Chromatogr. Library* **52**, 259–270. Elsevier, Amsterdam.
Nobuhara, Y., Hirano, S., Namba, K. and Hashimoto, M. (1980). *J. Chromatogr.* **190**, 251.
Noggle, F. T. Jr. and Clark, C. R. (1982). *J. Assoc. Off. Anal. Chem.* **65**, 756.
Srivastava, V. K. and Maheshawari, M. L. (1985). *J. Assoc. Off. Anal. Chem.* **64** (4), 801–803.
Trenerry, V. C., Robertson, J. and Wells, R. J. (1994a). *J. Chromatogr. Sci.* **32**, 1.
Trenerry, V. C., Robertson, J. and Wells, R. J. (1994b). *Electrophoresis* **15**, 103.
Trenerry, V. C., Robertson, J. and Wells, R. J. (1995a). *J. Chromatogr.* **708**, 169–176.
Trenerry, V. C., Wells, R. J. and Robertson, J. (1995b). *J. Chromatogr.* **718**, 217–225.
Weinberger, R. (1993). *Practical Capillary Electrophoresis*. Academic Press, San Diego, CA, USA.
Weinberger, R. and Lurie, I. S. (1991). *Anal. Chem.* **63**, 823–827.

Progress in HPLC-HPCE, Vol. 5, pp. 255–275
H. Parvez *et al.* (Eds)
© VSP 1997.

Study of protein–drug binding by capillary zone electrophoresis: Prospects and problems

J. C. KRAAK, M. H. A. BUSCH and H. POPPE

Laboratory for Analytical Chemistry, Amsterdam Institute for Molecular Studies,
University of Amsterdam, Nieuwe Achtergracht 166, 1018 WV Amsterdam,
The Netherlands

INTRODUCTION

It is well recognized that the pharmacological activity of a drug is closely related to the free drug concentration in the blood. The free drug concentration is usually significantly smaller than the overall concentration because most drugs are, during their transport through the blood circulation, reversibly bound to blood constituents particularly the plasma proteins, albumin and α-acid glycoproteins (Jusko and Gretch, 1976). In order to be able to adjust the optimum therapeutic dose of a drug in man, it is therefore necessary to gather information about the extent of drug binding. Although the binding of a drug should preferably be studied with whole blood, valuable information can be abstracted from the interaction of drugs with a plasma protein such as albumin. Important binding parameters are the value of the binding constants and the maximum number of drug molecules bound to the protein. Also, it is desired to know the effects of displacement of the drug from the plasma proteins by simultaneously administered exogenous compounds. This is crucial with toxic drugs having a strong protein binding, since a small decrease in the binding may lead to serious side effects for the patient.

In protein–drug binding studies it is assumed that the drug can interact independently with a number of binding regions on the protein. The reversible binding of a drug to these regions can be described by the multiple equilibria theory (Klotz and Hunston, 1971).

$$r = \frac{[D]_b}{[P]} = \sum_{i=1}^{m} \frac{n_i K_i [D]_f}{1 + K_i [D]_f}, \tag{1}$$

where r is the mean of number of drug molecules adsorbed per protein molecule, $[D]_b$ is the concentration of bound drug and $[P]$ the total protein concentration, m is the number of identical regions of independent adsorption sites, n_i is the number of sites in a region i with an association (binding) constant of K_i, $[D]_f$ is the concentration of the unbound (free) drug. Usually it is assumed that only two binding regions on the protein and two binding constants dominate the binding of a drug to a protein.

Equation (1) describes the relationship between the bound drug concentration, $[D]_b$, and the free drug concentration, $[D]_f$, e.g. the distribution isotherm. By measuring the bound drug concentration as a function of the free drug concentration without disturbing the equilibrium, principally the association constants can be calculated by applying a proper mathematical treatment of the experimental data.

Various analytical techniques have been applied to determine either the unbound (free) drug or the bound drug concentration in solution (Sebille *et al.*, 1990). Of these methods, the techniques whereby the unbound drug and the protein–drug complex are separated, such as dialysis (Zini *et al.*, 1988), ultracentrifugation (Bowers *et al.*, 1984), chromatographic techniques and particularly size-exclusion chromatography (SEC) (Hummel and Dreyer, 1962; Wood and Cooper, 1970; Sebille and Thuaud, 1984) are nowadays frequently applied. Although the principle of the chromatographic techniques are similar, several variants have been developed and are known as the Hummel–Dreyer method (Hummel and Dreyer, 1962; Sebille *et al.*, 1978), the vacancy peak method (Sebille *et al.*, 1979), frontal analysis (Cooper and Wood, 1968) and zonal elution (Loo *et al.*, 1984).

An alternative separation technique to study protein–drug interactions is capillary zone electrophoresis (CZE). It is a very simple and extremely efficient separation technique capable of separating a drug and its protein–drug complex according to their net charge and size difference (Jorgenson and Lukacs, 1981). This means that in principle, the methods analogous to SEC can be adapted to CZE. Compared to the chromatographic techniques, CZE has favourable features such as being a one-phase separation system, and having a relatively low surface-to-volume ratio and a small sample requirement. This diminishes the risk of undesired interactions between the protein or drug and the surface of the capillary. The great potential of CZE to study protein binding is now recognized and the number of applications is increasing steadily.

The prospects and limitations of CZE as an analysis technique to determine drug-binding parameters will be discussed in this chapter.

PRINCIPLE OF CZE METHODS FOR PROTEIN–DRUG BINDING STUDIES

The CZE principle is based on the separation of the drug, the protein and the protein–drug complex on the basis of their difference in electrophoretic mobilities, which depend on the net charge and size of the compounds. The net charge of a protein depends on its pI value and on the pH of the background electrolyte.

The CZE methods described below can only be applied when the protein and protein–drug complex have very close electrophoretic mobilities, e.g. when the charge and size of the protein is not significantly changed by adsorbed drug molecules. In the case where the mobilities differ, the binding constants must be determined by use of migration times and this will be discussed in a later section.

In order to apply the CZE methods succesfully, it is necessary that the electrophoretic mobilities of the free drug and protein–drug complex differ, but also that the drug gives a response to the applied detection system. The following CZE methods can be applied to determine protein–drug binding:
– zonal migration method,
– Hummel–Dreyer method,
– vacancy peak method,
– frontal analysis,
– migration time method (Affinity Capillary Electrophoresis).
As the last method is entirely different from the first four, it is discussed separately in a later section.

The zonal migration method is suitable for binding studies in which the dissociation rate of the protein–drug complex is very small and the other three methods have to be used when the dissociation rate of the protein–drug complex is relatively large, which is usually the case.

Zonal migration method

In this method, a mixture of the drug and protein in buffer is incubated until equilibrium is attained. Then, a small amount of this mixture is injected into the capillary and the protein–drug complex (and protein) is electrophoretically separated from the unbound (free) drug. The free drug concentration can then be determined from a previously constructed calibration line. By keeping the protein concentration constant but varying the drug concentration in the mixture, a distribution isotherm (r versus $[D]_f$) can be constructed and the binding parameters can be calculated by a proper fitting procedure. The zonal migration method is only applicable when the protein–drug complex is kept intact during the separation, e.g. when the dissociation rate of the complex is small compared to the electrophoretic migration process. Since most protein–drug complexes have relatively large dissociation rates, methods are preferred in which the shift of the equilibrium during the measurements does not pose a problem. Suitable techniques for this purpose are the Hummel–Dreyer method, the vacancy peak method and the frontal analysis.

Hummel–Dreyer method

In this method, the capillary is filled with a buffer containing a given concentration of the drug. As the drug responds to the detector, its presence in the buffer causes a large background detector signal. Then a small sample plug, containing the same buffer and drug and in addition a known amount of protein, is injected into the capillary and the voltage is switched on. It may be noticed that, in principle, the protein can be dissolved in the buffer and the drug in the sample (Chu and Whitesides, 1992). The migration of the injection plug is illustrated schematically in Fig. 1. In this example, it is assumed that the electrophoretic mobility of the protein and protein–drug complex are equal but larger than that of the drug.

The total concentration of the drug in the sample plug is equal to the drug concentration in the buffer, but part of the drug is bound to the protein. The protein and protein–drug complex migrate faster from the sample plug then the drug, leaving a deficiency in drug concentration in the injection plug. Since the detector monitors the drug in the buffer, this deficiency creates a negative peak because the free drug concentration in the sample plug is less than that in the buffer. The negative peak migrates with the mobility of the drug. The area of the negative peak depends on the amount of drug bound to the protein. During its migration, the protein–drug complex is always in equilibrium with the free drug permanently present in the buffer. Therefore, the presence of the protein–drug complex will cause a positive peak. When the free protein has no response in the detector and the detector response of the drug and protein–drug complex are the same, then in principle the area of the positive and negative peak will be equal. However, in practice, these circumstances are almost never

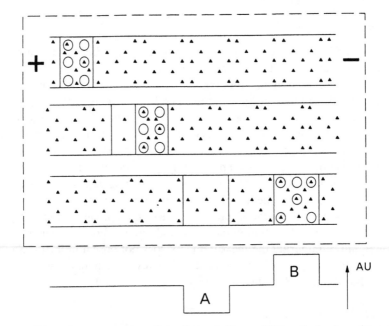

Figure 1. Schematic representation of the Hummel–Dreyer CZE method. ▲ = drug; ○ = protein; ⊛ = protein–drug complex. A: negative area reflecting the amount of drug bound. B: area reflecting the amount of complex formed.

met; moreover, in practice, the positive peak is often distorted, which impedes the proper determination of the area.

For simplicity, the peaks in Fig. 1 are represented as blocks but in practice, owing to dispersion, the blocks appear as more or less Gaussian peaks, as can be seen in Fig. 2.

It may be noticed that when the mobility of the drug is larger than that of the protein–drug complex, the negative peak appears before the positive peak.

The bound drug concentration, reflected in the negative peak, can be quantified in two ways: according to the original Hummel–Dreyer method, which involves the construction of a calibration line, and the modified Hummel–Dreyer method, which requires fewer data points. In the original Hummel–Dreyer method, a set of samples is prepared with incremental excess of drug in the sample. By plotting the absorbance response against the excess of drug added to the sample (see Fig. 3), the amount of drug required to fill the vacancy can then be determined by extrapolating to zero absorbance.

It will be clear that many sample injections have to be done to obtain one data point. In that respect, the modified Hummel–Dreyer method (Pinkerton and Koeplinger, 1990) is a good alternative. The bound drug concentration

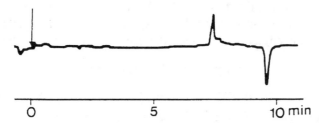

Figure 2. A typical electropherogram of BSA–Warfarin obtained with the Hummel–Dreyer method.

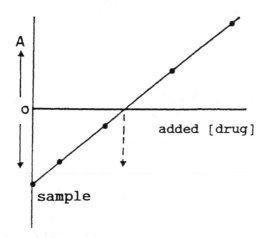

Figure 3. Illustration of the determination of the bound drug concentration with the original Hummel–Dreyer method.

can be calculated from the areas of only two injections, the protein sample and only buffer, according to:

$$[D]_b = \frac{A_s - A_e}{A_e} c_d, \tag{2}$$

where A_s is the area of the sample peak, A_e is the area of the blank buffer peak and c_d is the concentration of the drug in the buffer.

Vacancy peak method

In this method, the capillary is filled with a running buffer containing a known amount of protein and varying amounts of drug. The presence of drug and protein–drug complex causes a large background detector response. Then, a small amount of only buffer is injected into the capillary and the power supply is switched on. Two negative peaks will appear in the electropherogram. The appearance of two negative peaks is schematically illustrated in Fig. 4, again assuming that the electrophoretic mobilities of the protein and protein–drug complex are similar and larger than that of the drug.

At the front edge of the injection plug the free drug is migrating more slowly than the protein and thus stays behind. At the rear edge of the injection plug the free protein migrates faster then the drug. This continues until the two

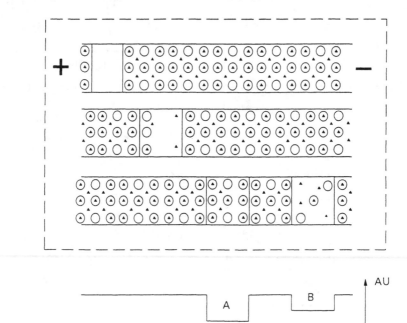

Figure 4. Schematic representation of the vacancy peak CZE method. Compounds: same notation as in Fig. 1. A: negative area reflecting the free drug concentration. B: negative area reflecting the deficiency in complex.

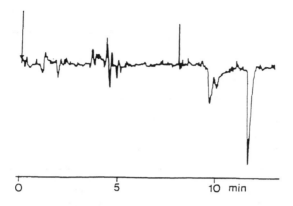

Figure 5. Typical electropherogram of BSA–Warfarin with the vacancy peak CZE method.

fronts reach each other. In that middle region, the protein again adsorbs drug molecules until the equilibrium is attained again. From that point on, a steady state is reached, resulting in two negative bands (peaks) in the electropherogram. The first peak reflects the bound drug and the second peak the free drug concentration. A typical electropherogram obtained with the vacancy method in practice is shown in Fig. 5.

When the electrophoretic mobility of the drug is larger than that of the protein and protein–drug complex, the first negative peak reflects the free drug concentration and the second peak reflects the bound drug concentration. Sebille *et al.* (1979) have provided evidence that the negative peaks do indeed reflect the bound and free drug in liquid chromatography and a similar reasoning can be applied to CZE.

In principle, both negative peaks can be used to construct the distribution isotherm. However, in practice, the peak representing the free drug has preference because the peak related to the bound drug is often distorted, which makes it difficult to determine properly the area or height. The determination of the free drug concentration from the second peak can be done by internal calibration, as shown before in Fig. 3. For that purpose, a series of samples with increasing drug concentrations are injected into the capillary and the added amount of drug is plotted versus the area or height. The free drug can then be determined by interpolating to zero area or height (the concentration to fill up the vacancy peak).

Frontal analysis

In the frontal analysis method, the capillary is filled with the buffer. Then, a very large sample plug, containing blank buffer + protein + drug, is injected into the capillary. Since the mobility of the protein–drug complex and protein is larger than that of the drug, the free drug leaks out of the plug at the rear edge (as occurs in the vacancy peak method) and a plateau is formed, as is schematically represented in Fig. 6.

J. C. Kraak et al.

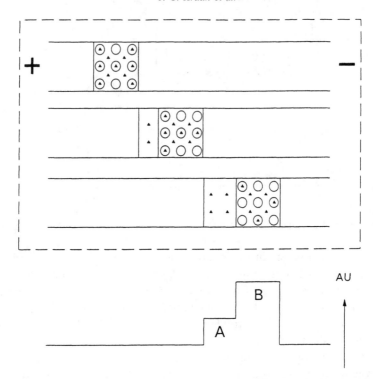

Figure 6. Schematic representation of the frontal analysis CZE method. Compounds: same notation as in Fig. 1. A: signal height reflecting the free drug concentration. B: signal height reflecting the complex concentration.

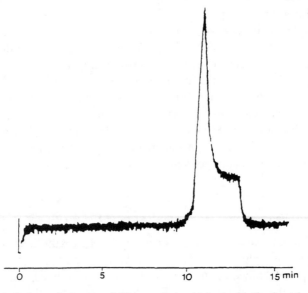

Figure 7. Typical electropherogram of BSA–warfarin obtained with the frontal analysis CZE method.

In the case that the protein and complex have the same electrophoretic mobility, the final elution profile consists of two parts: at the front edge, a plateau related to the free protein and protein–drug complex; and at the rear edge, a plateau reflecting the free drug concentration. By determining the free drug concentration in samples with constant protein amount but with varying drug concentrations, the distribution isotherm can be constructed. Figure 7 shows a typical electropherogram obtained with the frontal analysis.

The free drug concentration can be calculated from the height of the drug plateau in the sample and from the height of the drug plateau obtained when injecting a plug of buffer containing the same total amount of drug, c_d, as present in the sample solution according to:

$$[D]_f = \frac{\text{sample height}}{\text{reference height}} \, c_d. \tag{3}$$

From the calculated free drug concentration and the known total drug concentration in the sample, the bound drug concentration can then be calculated.

SELECTION OF EXPERIMENTAL CONDITIONS

Dimensions of the capillary

In order to preserve the large separation efficiency of CZE, excessive Joule heating in the capillary has to be avoided (Knox, 1988). Apart from preserving efficiency, proper temperature control is also required in protein–drug binding studies because the position of the equilibria depends on the temperature and thus influences the binding parameters.

The temperature increase due the Joule heating in the capillary depends on the field strength, the current and on the diameter of the capillary. The current is determined by the conductivity of the background electrolyte and also by the diameter of the capillary. In CZE, the field strength commonly ranges from 200–300 V/cm and the background electrolyte concentration is kept low to avoid Joule heating. In binding studies, the electrolyte concentrations can be relatively high and thus one has to be aware of too large a Joule heating effect. For instance, in the Hummel–Dreyer and vacancy peak methods, varying amounts of drug and/or protein are added to the buffer. When these compounds are charged, the conductivity of the solution can increase considerably which produces joule heating. Therefore, it is very important to adjust the diameter of the capillary correctly to effectively dissipate the heat produced. In practice, a diameter in the range of 50–75 μm i.d. looks a good compromise. However, when larger electrolyte or solute concentrations are used, it is recommended to make a plot of voltage *versus* current. When the curve deviates from linearity at larger voltages, there is a significant temperature rise and a smaller diameter capillary has to be used. However, a decrease of the diameter of the capillary affects the sensitivity in the case of on-column detection and therefore often a compromise has to be made.

Pretreatment of the capillary

As a result of the high temperatures used during their fabrication, the surface of fused silica capillaries contain, apart from silanol groups, also siloxane groups. In aqueous solutions, these siloxane groups are slowly transformed into silanol groups and thus change the nature of the surface. This change results in a gradual increase of the electroosmotic flow and consequently affects the migration times. To avoid a slow change in migration times during a set of measurements, it is recommended to flush the original capillary respectively for 3 h with 1 mol/l KOH, 15 min with distilled water, 30 min with 0.03 mol/l HCl and finally with the buffer.

Adsorption of protein or drug on the surface

It is well known that proteins can interact rather strongly with the silica surface of the capillary (Lauer and McManigill, 1986). The interaction is mainly attributed to electrostatic interactions between the protein and ionized silanol groups on the silica surface. Also, drug molecules can adsorb on the silica surface. The adsorption of protein (or drug) on the silica surface has to be avoided for the following reasons: i) it can result in less sharp or even distorted peak-shapes; ii) it will change the nature of the surface and thus the electroosmotic flow; iii) adsorption of protein (or drug) may introduce systematic errors in the calculated binding parameters. This last effect can be particularly disastrous with the Hummel–Dreyer method because only a small amount of protein (or drug) is injected. The occurrence of adsorption of protein or drug on the surface can be recognized from the peak shapes when injecting the solutes separately in buffer. When the peaks are not symmetrical and sharp, significant adsorption is likely and the interaction with the silica surface has to be diminished. Of the several approaches available to avoid adsorption of the protein (or drug) to the silica surface (Bruin *et al.*, 1989), the ones in which additives are added to the buffer are undesirable because this may affect protein–drug binding. Also, when protein–drug binding is studied with uncharged drugs, surface modifications in which the electroosmotic flow is largely reduced, cannot be applied. So far, no attention has been paid to surface modification of the capillary in protein–drug binding studies, one reason being that some coatings are not stable at pH > 7.

  Up till now the effects of adsorption of protein or drug on the surface can partly be circumvented by flushing the capillary with an alkaline solution and running buffer after each analysis. However, it is worthwhile to investigate the effects of surface modification in binding studies in more detail.

Injection mode

Two ways of sample introduction are frequently used in capillary electrophoresis: hydrodynamic injection and electrokinetic injection. With hydrodynamic injection the sample is introduced into the capillary by applying a pressure on the sample vial or by sucking the sample into the capillary by applying a vacuum at the other end of the capillary. In this case, the sample is homogeneously injected.

With the electrokinetic injection the sample is introduced into the capillary by a combined action of the electroosmotic flow and the electrophoretic mobility of the solutes. Neutral solutes are transported into the capillary solely by the electroosmotic flow, while charged solutes are introduced with a different velocity because their apparent migration is the result of their own mobility and electroosmotic flow. Due to this discrimination a non-homogeneous sample is injected and this may introduce a systematical error in the determination of the binding parameters with the Hummel–Dreyer, vacancy and frontal analysis. Because of this discrimination of solutes by electrokinetic injection, the application of hydrodynamic injection is preferred in binding studies.

DETECTION

Because of the extremely small dimensions, detection is still the weak link in capillary electrophoresis. The applied detection systems for CE can be distinguished as on-column or off-column. The on-column detectors are similar to the UV-visible and fluorescence detectors applied in HPLC, with that difference that a small cross-section of the capillary is used as detection cell.

The most popular on-column detection is UV-visible and it can be applied to a large number of compounds. Unfortunately, the sensitivity is relatively poor because of the small optical path length equal to the internal diameter of the capillary. A high sensitivity is required to determine properly the first points of the distribution isotherm (i.e. small r values). The sensitivity can be improved by enlarging the internal diameter of the capillary and thus the optical path length. However, in practice, increasing the inner diameter may led to excessive heat production. An alternative way of enlarging the optical path length without introducing extra joule heating is the use of the so-called U-shaped or Z-shaped cells (Chervet *et al.*, 1991) or the bubble cell (Hewlett–Packard). In the frontal analysis, UV-visible detection is direct while in the Hummel–Dreyer and vacancy peak method the indirect mode (Bruin *et al.*, 1992) is used. The indirect detection mode is usually less sensitive than the direct mode. This is due to the higher noise level caused by the high back ground signal of the detector.

For drugs with a fluorophore, fluorescence detection is preferred over UV-visible detection, because it has a higher sensitivity and selectivity. Particularly, the application of laser induced fluorescence (LIF) detection looks very attractive in binding studies.

Mass spectrometry (MS) and electrochemical detection (ED) are possible candidates as off-line detection systems for binding studies.

Mass spectrometric detection is very sensitive but requires an electrospray interface to couple the CE with the MS (Kostiainien et al., 1993) and this somewhat complicates its application. However, it can be expected that in the near future MS detection will be important in binding studies.

For electrochemically active drugs, electrochemical detection in the direct or indirect mode looks attractive because of its extremely high sensitivity and selectivity of detection. However, in practice, the coupling of the electrochemical detector to the CE is not easy and requires decoupling of the high electrical field. An elegant method is to use a palladium low dead volume connector as ground electrode (Kok and Sahin, 1993). Apart from the complexity of the coupling, the fouling of the measuring electrodes due to adsorption of solutes or surface reactions is one of the most serious problems in electrochemical detection. Particularly, proteins appear to adsorb irreversibly on the surface of the measuring electrodes. This can partly be overcome by applying electrodes which are coated with a permselective layer (Tüdös et al., 1991).

CONSTRUCTION OF THE DISTRIBUTION ISOTHERM

The binding parameters can be calculated from the constructed distribution isotherm. It is of paramount importance, first, that sufficient data points (r values) are gathered and, second, that the data points are spread out over the entire r-range. In order to obtain reliable values of the binding parameters, 15–20 data points are desired. For the modified Hummel–Dreyer method and the frontal analysis technique, two measurements are needed to calculate one data point. This means that 30–40 measurements have to be performed to construct a single distribution isotherm. With the original Hummel–Dreyer method and the vacancy peak method four measurements are necessary to calculate one data point; thus, 60–80 measurements have to be performed to construct a distribution isotherm. If we assume an analysis time of about 10 min, the whole measurement of the distribution isotherm will last 5–7 h for the Hummel–Dreyer and frontal analysis methods, and 10–14 h for the original Hummel–Dreyer and the vacancy peak methods. It will be obvious that an automated CE system is desirable to perform this large number of measurements. Moreover, an automated CE system has, compared to manual manipulation, the additional advantage that the whole cycle of washing and measuring is performed according to a stringent protocol. This last facility appears to be needed to obtain reproducible results. A typical distribution isotherm of BSA–warfarin obtained with an automated CE system is given in Fig. 8.

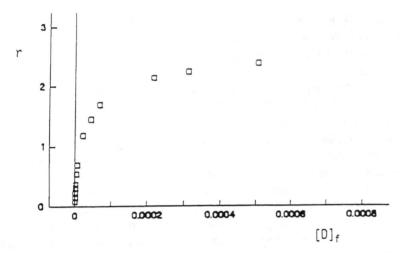

Figure 8. A typical BSA–warfarin distribution isotherm measured with frontal analysis on an automated CE system (ABI).

DISPLACEMENT STUDIES

The extent of binding of a drug can be significantly altered by displacement of the drug from the protein by simultaneously administered exogenous compounds. The study of the displacement of the parent drug by another compound is, in principle, possible with the CZE methods (Chu *et al.*, 1993). This requires that the parent and competitive drug have different mobilities or the competitive drug shows no detector response. A difference in mobility occurs when one of the drugs is neutral and the other charged or both drugs have a different charge. In these situations the Hummel–Dreyer and the vacancy peak methods can be applied. When the competitive drug shows a response to the detector, three peaks will appear in the electropherograms: two negative peaks and one positive peak in the case of the Hummel–Dreyer and three negative peaks with the vacancy peak method. Only two peaks will be found with both methods when the competitive drug shows no detector response.

In principle, the frontal analysis method can also be applied, for instance, when the mobility of the protein–complex is in between the mobilities of both drugs or the competitive drug shows no detector response. In the first case, the plateaus appearing at the front and rear edge of the injection plug represent the free drug concentration of both drugs. In the second case, only the plateau of one free drug is visible.

LIMITATIONS OF THE CZE METHODS

Effect of the mobility of protein and protein–drug complex

As indicated before, the described CZE methods to study protein–drug binding, e.g. the Hummel–Dreyer, the vacancy peak and frontal analysis methods can only be applied when the mobilities of the protein and that of the protein–complex are very close.

The migration behaviour of the various species with the three CZE methods can be simulated and the results are compiled in Figs 9–11. These figures reflect the concentration profiles of the three species at certain positions in the capillary when the mobilities of the protein and protein–drug complex are equal (Figs 9A–11A) or different (Figs 9B–11B).

With the Hummel–Dreyer method (Fig. 9), the drug is present in the buffer at a certain concentration and a small amount of protein is injected. It is reasonable to assume that the equilibrium in the injection zone is attained after a small distance. The concentration of the free drug in that injection zone is smaller because part of the drug is bound to the protein. Figure 9 shows the migration profiles of the species recorded at two positions in the capillary. It shows that when the mobilities of the protein and the protein–drug complex are similar (Fig. 9A), the concentration of the free drug at the position of the protein and protein–drug complex is not altered and thus indicates that the equilibrium between the species in that zone is maintained during the migration. However, when the mobilities of the protein and the protein–drug complex differ (Fig. 9B), the concentration of the free drug is increased at the position of the protein–drug complex. This indicates that the equilibrium is shifted during the migration. Since the negative peak reflects the amount of drug bound to the protein, the excess of free drug at the position of the protein–drug complex leads, according to the mass balance, to a larger negative peak. This causes a systematic error in the calculation of the r-value.

With the vacancy peak method the capillary is filled with buffer containing protein + drug and a small amount of buffer is injected. The concentration profiles of the species at various distances are given in Fig. 10.

In the case where the mobilities of the protein and protein–drug complex are the same (Fig. 10A), the introduced vacancy of the free drug is maintained during the migration. However, the concentration profiles of the protein and protein–drug complex indicate the appearance of an extra peak which migrates at the same velocity as the free drug. The extra peak is positive for the protein and negative for the protein–drug complex. The areas of these peaks are equal and thus have no influence on the negative free drug peak. When the mobilities of the protein and complex differ, the concentration profiles are different, as can be seen in Fig. 10B. Under these circumstances, again, an extra peak originates for the protein and complex but the areas are different. Moreover, a second peak appears in the drug profile. This peak migrates with the velocity of the protein and complex. The extra peaks in the profile influence the area of the negative peaks and thus introduce a systematic error in the calculation of the binding parameters.

With the frontal analysis method, a large plug of the sample is injected, containing the buffer + protein + drug. The migration profiles of the individual species at two positions in the capillary are given in Fig. 11.

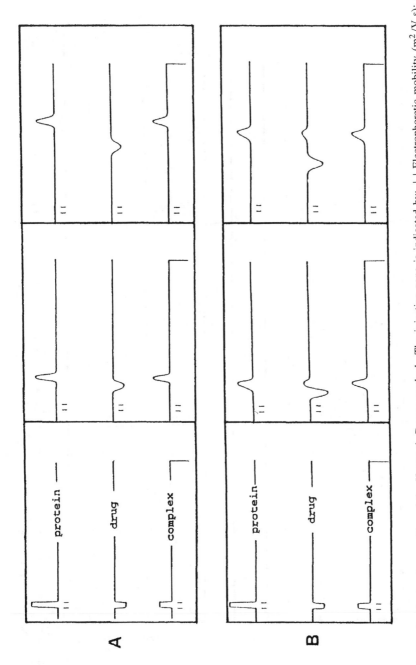

Figure 9. Simulated migration profiles of the Hummel–Dreyer method. The injection zone is indicated by: | | Electrophoretic mobility (m^2/V s): $\mu_{drug} =$ 35 × 10^{-9}, A: $\mu_{protein} = \mu_{complex} = 50 × 10^{-9}$, B: $\mu_{protein} = 50 × 10^{-9}$; $\mu_{complex} = 70 × 10^{-9}$.

270 J. C. *Kraak* et al.

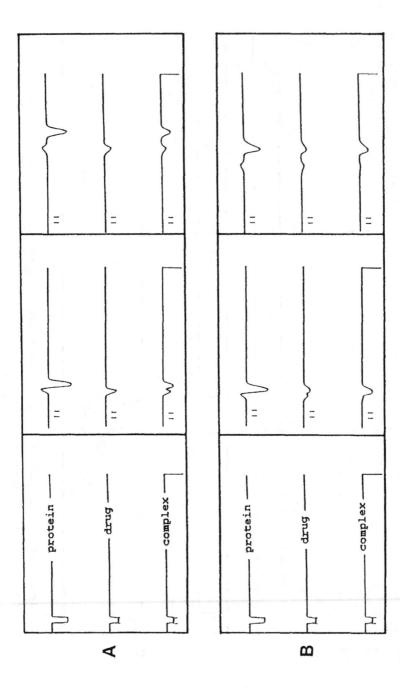

Figure 10. Simulated migration profiles of the vacancy peak method. The injection zone is indicated by: || Electrophoretic mobility ($m^2/V\,s$): $\mu_{drug} = 35 \times 10^{-9}$, A: $\mu_{protein} = \mu_{complex} = 50 \times 10^{-9}$, B: $\mu_{protein} = 50 \times 10^{-9}$; $\mu_{complex} = 70 \times 10^{-9}$.

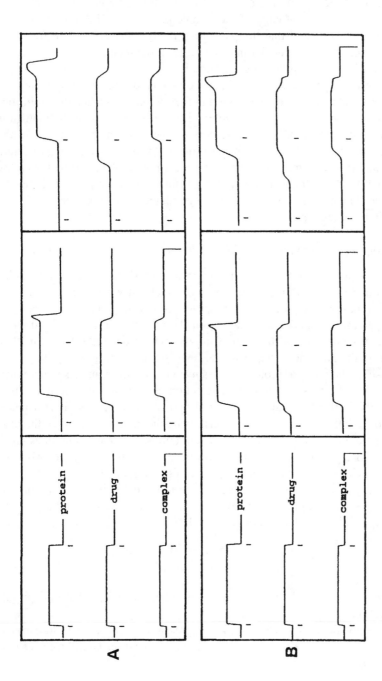

Figure 11. Simulated migration profiles of the frontal elution method. The injection zone is indicated by: | | Electrophoretic mobility (m^2/V s): $\mu_{drug} = 35 \times 10^{-9}$, A: $\mu_{protein} = \mu_{complex} = 50 \times 10^{-9}$, B: $\mu_{protein} = 50 \times 10^{-9}$; $\mu_{complex} = 70 \times 10^{-9}$.

From Fig. 11A it can be seen that the free drug concentration is not altered when the mobilities of the protein and protein–drug complex are equal. This indicates that, at the front edge of the sample plug, the equilibrium is attained during the migration and that the plateau correctly reflects the right free drug concentration. It can be noticed that the larger protein concentration is due to the dissociation of the protein–drug complex at the rear edge of the sample plug. From Fig. 11B it can be seen that the plateau of the free drug is smaller than the original free drug concentration in the sample plug when the mobilities of the protein and protein–drug complex differ. The decrease of the plateau is due to the fact that the protein–drug complex migrates away faster from the front edge of the sample plug than the protein. In order to attain equilibrium in that zone, free drug molecules are adsorbed on the protein and will decrease the original free drug concentration at the front edge of the plug. This results in a lower plateau and thus introduces a systematic error in the calculation of the binding parameters.

The use of migration times for the determination of binding parameters

On the basis of the simulations of the migration behaviour it can be concluded that the three CZE methods cannot be recommended when the mobilities of the protein and protein–drug complex differ significantly. For such cases, the binding parameters can be estimated from the migration times of the drug or protein as has been demonstrated by Chu and Whitesides (1992). The method is similar to the Hummel–Dreyer method but is named 'affinity capillary electrophoresis'. The drug (or protein) is dissolved in the buffer and a small amount of protein (or drug) is injected. The selection as to which compound is best added to the buffer is usually determined by economic factors. For instance, when choosing the protein as buffer constituent, the migration times of the drug are measured with a range of protein concentrations. The migration times range between two extremes: t_d, the migration time of the free drug, i.e. when no protein is present in the buffer and t_{max}, the migration time of the protein–drug complex, i.e. when the protein concentration is large. At intermediate protein concentrations, the migration time of the drug is t_i (between t_d and t_{max}). The value of t_i is related to the fraction, f, of drug that is complexed according to:

$$f = \frac{[P-D]}{[P]_{tot}} = \frac{(t_i - t_d)}{(t_{max} - t_d)} = \frac{\Delta t}{\Delta t_{max}}, \qquad (4)$$

where $[P-D]$ is the bound drug concentration and $[P]_{tot}$ the total protein concentration.

For a monovalent complex, a simple expression can be derived to estimate the binding constant, K_b, by combining the association equilibrium, $K_b = [P-D]/[D][P]$, with (4):

$$\frac{\Delta t}{[P]} = K_b(\Delta t_{max} - \Delta t). \qquad (5)$$

K_b is then given by the slope of $\Delta t/[P]$ versus $(\Delta t_{max} - \Delta t)$, the Scatchard plot (Scatchard, 1949; Gomez *et al.*, 1994).

Equation (5) is valid only when the electroosmotic flow is not changed over the investigated concentration range. In the case that the electroosmotic flow changes, the migration times have to be corrected, using a neutral compound as marker of the electroosmotic flow (Gomez *et al.*, 1994).

RECENT APPLICATIONS OF CZE FOR BINDING STUDIES

Whitesides and co-workers (Chu and Whitesides, 1992; Gomez *et al.*, 1994) have extensively studied the migration time method for the determination of the binding constants of synthetic peptides to vancymocin and of benzenesulfon-amides to carbonic anhydrase B. They also describe an interesting procedure, based on competitive binding, to identify in a peptide library the peptide that binds most tightly to vancomycin. These same authors demonstrated the applicability of the migration time method for simultaneous measurements of binding constants (Chu *et al.*, 1993).

Honda *et al.* (1992) have applied CZE to determine the association constant of protein–sugar interactions. They used β-galactose-specific lectins and lactobionic acid as protein and sugar respectively. Since the mobility of the protein differs from that of the protein–sugar complex, the shift in migration time of the complex with increasing sugar content in the buffer could be used to calculate the association constants.

Kraak *et al.* (1992) have investigated the suitability of the Hummel–Dreyer, the vacancy peak and frontal analysis CZE methods for binding studies using warfarin and BSA as test compounds. The frontal analysis appeared to be the preferred method.

The zonal migration method has been applied by Heegaard and Robey (1992) to study the binding of anionic carbohydrates to synthetic peptides. The authors stated that the method can be applied to any ligand as long as the charge/mass ratio between the free and complexed peptide differs sufficiently. The same authors (Heegaard and Robey, 1993) used the zonal migration method to study the interactions between oligonucleotides and synthetic peptides derived from amyloid P.

Thomas *et al.* (1993) have used the zonal migration method to determine the irreversible binding of procainamide metabolites to hemoglobin and histone proteins. The technique can be applied to mixtures of proteins and a common binding ligand. Adsorption of proteins to the wall still limits the full exploitation of the technique.

The binding of vancomycin to cytoplasmic peptidoglycan precursors have been studied by Jinping Lui *et al.* (1994) using the migration time CZE method. The authors recommend the use of biological buffers in binding studies.

SUMMARY

CZE is an attractive one-phase micro-separation technique to study interactions between species in solution. It can be applied to determine the binding between proteins and drugs but also between smaller compounds (Chu and Whitesides, 1992). A great advantage of CZE is its simplicity and the small quantity of compound which is needed to measure the binding constants.

Various CZE methods can be applied, viz. the zonal migration, the Hummel–Dreyer, the vacancy peak, the frontal analysis and the migration time method. The selection of one of these methods depends on the dissociation rate of the protein–drug complex and on the mobilities of the protein and protein–drug complex.

In the case where the dissociation rate of the protein–drug complex is very small, the zonal migration method can be applied. For larger dissociation rates one of the other methods has to be used. The Hummel–Dreyer, the vacancy peak and the frontal analysis methods can be used when the mobilities of the protein and protein–drug complex are very close. The migration time method has to be used when the mobilities of the protein and protein–drug complex differ significantly.

The Hummel–Dreyer, the vacancy peak and the frontal analysis can be applied to study the bivalent binding of a drug to a protein. Of these three methods the frontal analysis method has preference. So far, the migration time method has been used only to determine the binding constant of monovalent complexes.

REFERENCES

Bowers, W. F., Fulton, S. and Thompson, J. (1984). *Clin. Pharmcokinet.* **9** (Suppl.), 49.

Bruin, G. J. M., Huisden, R., Kraak, J. C. and Poppe, H. (1989). *J. Chromatogr.* **480**, 339.

Bruin, G. J. M., van Asten, A. C., Xu, X. and Poppe, H. (1992). *J. Chromatogr.* **608**, 97.

Chervet, J. P., van Soest, N., Ursem, M. and Salzman, J. P. (1991). *J. Chromatogr.* **543**, 439.

Chu, Y.-H. and Whitesides, G. M. (1992). *J. Org. Chem.* **57**, 3524.

Chu, Y.-H., Avila, L. Z., Biebuyck, H. A. and Whitesides, G. M. (1993). *J. Org. Chem.* **58**, 648.

Cooper, P. F. and Wood, G. C. (1968). *J. Pharm. Pharmacol.* **20**, 1503.

Gomez, F. A., Avila, L. Z., Chu, Y.-H. and Whitesides, G. M. (1994). *Anal. Chem.* **66**, 1785.

Heegaard, N. H. H. and Robey, F. A. (1992). *Anal. Chem.* **64**, 2479.

Heegaard, N. H. H. and Robey, F. A. (1993). *J. Liquid Chromatogr.* **16**, 1923.

Hewlett–Packard (1992). *High Performance Capillary Electrophoresis*, Heiger, D. N. (Ed.). Waldbron, Germany, publ. nr 12-5091-6199E.

Honda, S., Taga, A., Suzuki, K., Suzuki, S. and Kakehi, K. (1992). *J. Chromatogr.* **597**, 377.

Hummel, J. P. and Dreyer, W. J. (1962). *Biochim. Biophys. Acta* **63**, 530.

Jorgenson, J. W. and Lukacs, K. D. (1981). *Anal. Chem.* **53**, 1298.

Jusko, W. J. and Gretch, M. (1976). *Drug Metab. Rev.* **5**, 43.

Klotz, I. M. and Hunston, D. L. (1971). *Biochemistry* **16** (10), 3065.

Knox, J. H. (1988). *Chromatographia* **26**, 329.

Kok, W. Th. and Sahin, Y. (1993). *Anal. Chem.* **65**, 2497.

Kostiainen, R., Franssen, E. J. F. and Bruins, A. P. (1993). *J. Chromatogr.* **647**, 361.

Kraak, J. C., Busch, M. H. A. and Poppe, H. (1992). *J. Chromatogr.* **608**, 257.

Lauer, H. H. and McManigill, D. (1986). *Anal. Chem.* **58**, 166.

Liu, J., Volk, K. J., Lee, M. S., Pucci, M. and Handwerger, S. (1994). *Anal. Chem.* **66**, 2412.

Loo, J. C. K., Jordan, N. and Ngoc, A. H. (1984). *J. Chromatogr.* **305**, 194.

Pinkerton, T. C. and Koeplinger, K. A. (1990). *Anal. Chem.* **62**, 2114.

Scatchard, G. F. (1949). *Ann. N.Y. Acad. Sci.* **51**, 660.

Sebille, B. and Thuaud, N. (1984).In: *Handbook of HPLC for the Separation of Amino Acids, Peptides and Proteins*, Vol. II, Hancock, W. S. (Ed.). CRC Press, Boca Raton, FL, pp. 379–391.

Sebille, B., Thuaud, N. and Tillement, J. P. (1978). *J. Chromatogr.* **167**, 159.

Sebille, B., Thuaud, N. and Tillement, J. P. (1979). *J. Chromatogr.* **180**, 103.

Sebille, B., Zini, R., Madjar, C. V., Thuaud, N. and Tillement, J. P. (1990). *J. Chromatogr.* **531**, 51.

Thomas, C. V., Cater, A. C. and Wheeler, J. J. (1993). *J. Liquid Chromatogr.* **16**, 1903.

Tüdös, A. J., Ozinga, W. J. J. and Kok, W. Th. (1991). *J. Chromatogr.* **547**, 1.

Wood, G. C. and Cooper, P. F. (1970). *Chromatogr. Rev.* **12**, 88.

Zini, R., Morin, D., Jouenne, P. and Tillement, J. P. (1988). *Life Sci.* **43**, 2103.

Progress in HPLC-HPCE, Vol. 5, pp. 277–291
H. Parvez *et al.* (Eds)
© VSP 1997.

HPCE of catecholamines, bioactive molecules and other transmitters

TAKASHI KANETA[1] and SHUNITZ TANAKA[2]

[1]*Department of Chemical Science and Technology, Faculty of Engineering, Kyushu University, Japan*
[2]*Division of Material Science, Graduate School of Environmental Earth Science, Hokkaido University, Japan*

INTRODUCTION

This chapter describes the separation and determination of some neurotransmitters, represented by catecholamines, and some bioactive molecules, such as tranquilizers and antidepressants, using HPCE. The separation of catecholamines has been achieved in various migrating buffer systems of CZE or MEKC modes. Particularly, the use of a complexation reaction with borate, because of its selectivity, facilitates excellent separation of catecholamines and also the separation of catecholamines and serotonin, which have a similar electrophoretic mobility. The host-guest interaction of cyclodextrin is useful for the improvement of the resolution of catecholamines and for the chiral separation of some bioactive molecules. These migrating buffer systems are discussed and summarized. As the detection method, UV spectrometry, an electrochemical method and laser-induced fluorometry are currently used. The methodology and the characteristics of these methods are discussed. The analysis of catecholamines by HPCE is widely applied to many actual biological samples. The usefulness of HPCE in its application to biological samples with very small volume is proved by the success in monitoring catecholamines in a single nerve cell by HPCE. The separation of other transmitters and some bioactive molecules is also discussed.

Information in the human brain is transmitted by the passage of chemical substances called neurotransmitters from the end of one neuron to an adjacent neuron (Snyder, 1984). Neurotransmitters are stored in the terminal region of a neuron. The arrival of an action at the terminal leads to the release of the neurotransmitter to the synaptic cleft; the neurotransmitter then diffuses across the

synaptic cleft and alters the ion permeability of the adjacent neuron by binding to receptors on its membrane. There are many kinds of neurotransmitters
in the human brain, e.g. catecholamines, γ-aminobutyric acid, glutamic acid,
acetylcholine, serotonin and histamine. A number of electroanalytical techniques have been developed to investigate the role of these neurotransmitters
in the brain. Electroanalytical methods are useful in studying brain functions
because of their high sensitivity (Adams, 1976). However, by electroanalytical
techniques, it is difficult to determine selectively a neurotransmitter in complex matrices of biological fluid and tissues. Thus, these methods are used,
combined with some separation technique, such as high performance liquid
chromatography (HPLC) (Wightman et al., 1988).

Currently, high performance capillary electrophoresis (HPCE) is a rapid
and excellent separation technique (Kuhr, 1990; Kuhr and Monnig, 1992).
HPCE has several advantages over the other conventional separation techniques:
great efficiency, small sample volume, and rapid separation. Therefore, HPCE
has advanced rapidly in various fields, especially in analysis of biochemical
molecules. This chapter will focus on the analytical methodology of HPCE
which includes the separation, detection and injection techniques for catecholamines, some bioactive molecules and the other neurotransmitters, both
in vivo and in vitro.

CATECHOLAMINES

Separation

Catecholamines are one of the most important neurotransmitters in the central nervous system. They act through their binding with a receptor. Catecholamines possess a primary or a secondary amino group which is protonated
at neutral pH and thus exists as a positively charged species at such pH. However, conventional HPCE is not preferable to the separation of cationic species
because of poor resolution and adsorption of the analytes on the capillary wall.
Wallingford and Ewing (1987) reported the first application of HPCE to the
separation of catecholamines in which they achieved the separation of the three
catecholamines, dopamine, epinephrine, and norepinephrine by using 0.05 M
N-[tris-(hydroxymethyl)-methyl]-2-aminoethane sulfonic acid buffer solution
(pH 7.42) as the migrating buffer. However, pronounced tailing was observed
in the peak of the catecholamines as well as in that of proteins (Jorgenson and
Lukacs, 1983). This tailing of the peak was due to the electrostatic interactions
between positively charged catecholamines and the negatively charged capillary
wall, which is caused by the proton dissociation of the silanol group on the
wall.

One idea to prevent this adsorption is to use the complexation between boric
acid and catecholamines. The borate-catechol complex is formed through a
reversible reaction with a strong pH dependency as follows:

When catecholamines complex with boric acid, their cationic species are converted into zwitterions and nonionic species into anions. At neutral pH, most of the borate complexes with catecholamines exist as zwitterions (zero net charge) since the dissociation constants of amino groups in catecholamines are above 10^{-9} mol dm^{-3}. Electrically neutral samples cannot be separated by conventional HPCE. However, micellar electrokinetic chromatography (MEKC) is an excellent alternative technique for electrically neutral compounds (Terabe *et al.*, 1984, 1985).

There have been a few reports on MEKC retention behavior of ionic and non-ionic catechols (Wallingford and Ewing, 1988a, b; Ong *et al.*, 1991a). As shown in Fig. 1 (Wallingford and Ewing, 1988b), the selectivity for catecholamines obtained with phosphate buffer is completely different from that with phosphate-borate buffer. In Fig. 1A, the retention mechanism of catecholamines probably includes five equilibria: (1) ion-pairing formation with the surfactant monomer in solution; (2) penetration of the ion-pair into the micelle interior; (3) solubilization of the ion-pair by insertion of the monomer end into the micelle; (4) solubilization of the ion-pair with the catechol moiety penetrating into the micelle; (5) solubilization of the free cation at the anionic surface of the micelle. A theoretical study on the retention behavior of catecholamines was reported by Strasters and Khaledi (1991). The ion-association constants estimated by them were 380 mol^{-1} dm^3 for norepinephrine and 2000 mol^{-1} dm^3 for epinephrine, while the distribution constants were 110 for norepinephrine and 180 for epinephrine. The values of distribution coefficients estimated took no account of the ion-pairing formation constant. However, the ion association equilibrium is considered to be a more important factor for the explanation of the change in selectivity.

The retention mechanism of catecholamines as borate-complexes appeared to be more complicated. Catecholamines form net-neutral borate complexes at neutral pH, but still maintain a cationic moiety which can form an ion-pair with surfactant monomers or interact with the micelle exterior. The above five interactions listed in the phosphate buffer system may occur even in separating catecholamines as borate complexes. The retention times of catecholamines in Fig. 1B is however shorter than those in Fig. 1A. This phenomenon is explained as follows: the anionic charge of the complexes appears to reduce the distribution coefficient of the complexes into negatively charged micelles resulting in short retention. The data in Fig. 1B indicate greater selectivity and shorter analysis time for the separation of catecholamines than those in Fig. 1A.

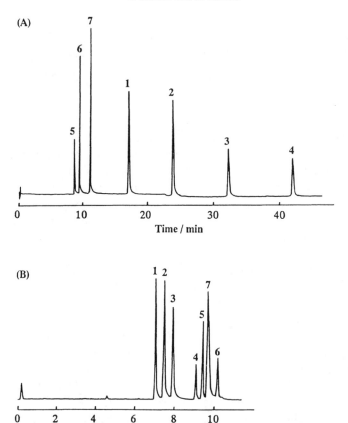

Figure 1. A: Electrokinetic separation of non-ionic and cationic catechols. 5 mM dibasic sodium phosphate-5 mM monobasic sodium phosphate at pH with 20 mM SDS; separation capillary length, 66.5 cm; detection capillary length, 1.6 cm; separation potential, 20 kV (5 μA); injection, 2 s at 20 kV. B: Electrokinetic separation of catechols as borate complexes; 10 mM dibasic sodium phosphate-6 mM sodium borate at pH 7 with 10 mM SDS; separation capillary, 64.3 cm; detection capillary, 1.7 cm; separation potential, 20 kV (7 μA); injection, 4 s at 20 kV. 1. Norepinephrine; 2. Ephinephirine; 3. 3,4-Dihydroxybenzylamine; 4. Dopamine; 5. L-DOPA; 6. Catechol; 7. 4-Methylcatechol. (Adapted from Wallingford and Ewing (1988b) with permission of Elsevier Science Publishers.)

On the other hand, we have also attempted the separation of catecholamines as the anionic borate complexes without any surfactant (Tanaka *et al.*, 1990; Kaneta *et al.*, 1991). When the migration buffer containing 100 mM potassium hydroxide – 200 mM boric acid (pH 9.1) was used, ten components involving catecholamine metabolites could be separated. Catecholamines appeared to migrate as the anionic borate complexes, but a somewhat lesser degree of tailing was observed. Furthermore, we attempted to improve the resolution by controlling the electroosmotic velocity with cationic surfactant. This challenge led to a significant improvement of the resolution, although the analysis time became very long (Kaneta *et al.*, 1991).

One of the other separation techniques is based on host-guest complexation with cyclodextrin, which was reported by Fanali (1989). He has achieved enantiomeric separation of catecholamines (epinephrine, norepinephrine, and isoproterenol) by using heptakis(2,6-di-O-methyl-β-cyclodextrin) (di-OMe-β-CD) as an additive. It was noteworthy that no peak tailing was observed under this experimental condition. Using the migration buffer solution (pH 2.4) containing 18 mM di-OMe-β-CD, symmetrically sharp peaks for six enantiomeric catecholamines could be observed. The peak tailing in HPCE is caused by the adsorption of cationic samples on the negatively charged capillary wall by dissociated silanol groups. The dissociation constant for silanol groups is reported as *ca.* $10^{-5.3}$ mol dm^{-3} (Schwer and Kenndler, 1991); thus, the adsorption would be prevented by using a low pH buffer system. When considering Fanali's contribution, it is expected that acidic conditions will be preferable for the separation of catecholamines.

In Table 1, optimum conditions in the separation techniques mentioned above are summarized in detail. Most catecholamines can be separated from each other by using the migration buffer systems listed in Table 1. However, readers must optimize separation conditions when they analyze catecholamines in real biological samples, e.g. urine, serum, and cell preparations. The applications to real samples will be described later.

Detection

An on-column UV-visible absorbance detector is currently the most popular in HPCE. In most works on the separation of catecholamines, a UV detector has been employed and the detection limit is about 10^{-5} M level. Laser-induced fluorometric detection (LIF) is probably the most sensitive detection technique (Gozel *et al.*, 1987; Cheng and Dovichi, 1988; Nickerson and Jorgenson, 1988), but there have been no reports on LIF of catecholamines yet. The other detection technique used is amperometry, which is potentially one of the most sensitive detection techniques for HPCE. An amperometric detector was first introduced by Wallingford and Ewing (1987) and there have been several reports on further improvements and applications since. The advantages of amperometric detection are that the method offers both good sensitivity and selectivity. Not only catecholamines but also other biological analytes usually exist in complex matrices. Therefore, for the detection of biological samples, the selectivity offered by an amperometric detector is required. Here, we summarize recent developments in using this technique. For additional details, readers should refer to other publications.

Problems in amperometric detection are the requirements of placing the electrochemical detector at the end of the column in a buffer reservoir, and also of performing electrochemical measurement in a high-voltage electric field in HPCE (Wallingford and Ewing, 1987). These problems could be resolved by separating the capillary for the electrochemical detector from the migrating capillary in high voltage. The two capillaries are connected with a porous

Table 1.
Electrolyte systems and experimental conditions for separation of catecholamines and their metabolites

Electrolyte (pH)	Migration velocity	Dimensions of capillary		Migration voltage (current)
		i.d.	length	
20 mM MES buffer (6.05)	DA > NE > E > CA	26 μm	879 mm	25 kV (2 μA)
10 mM phosphate buffer with 10 mM SDS (6.98)	CA > 4-MC > 4-EC > NE	52 μm	651 mm	13 kV (8 μA)
5 mM NaH$_2$PO$_4$ and 5 mM Na$_2$HPO$_4$ with 20 mM SDS (7)	L-DOPA > CA > 4-MC > NE > E > DBA > DA	26 μm	665 mm	20 kV (5 μA)
10 mM NaH$_2$PO$_4$ and 6 mM NaB(OH)$_4$ with 10 mM SDS	NE > E > DBA > DA > DOPA > CA > 4-MC	26 μm	643 mm	20 kV (7 μA)
25–30 mM MES with 20% 2-propanol (5.5–5.55)	5-HT > DA > NE > E	14 μm	690 mm	30 kV
10 mM Tris and H$_3$PO$_4$ with 18 mM di-OMe-β-CD (2.4)	(−)NEP > (+)NEP > (−)EP > (+)EP or (−)E > (+)E > (−)NE > (+)NE > (−)IP > (+)IP	25 μm	200 mm	8 kV (6.8 μA)

Table 1.
(continued)

100 mM KOH and 200 mM B(OH)₃ (9.1)	MN > NM > DEP > DA > DBA > IP > E > NE > DOPA > VMA	100 μm	700 mm	10 kV (100 μA)
10 mM NaH₂PO₄ and 6 mM NaB(OH)₄ with 10 mM SDS (7)	NE > E > DA > CA > 5-HT	26 μm	616.5 mm	20 kV
100 mM borate and 50 mM phosphate buffer with 80 mM SDS (7)	NE > TRP > DBA > DPG > HIA > DPA > NEP > EP	50 μm	450 mm	15 kV
200 mM acetate buffer (4.10)	HMG > MIA = HIA > HMBA > VMA > HGA > HVA	75 μm	500 mm	25 kV (115 μA)

Symbols: DA – dopamine; NE – norepinephrine; E – epinephrine; CA – catechol; DOPA – 3,4-dihydroxyphenylalanine; 4-MC – 4-methylcatechol; 4-EC – 4-ethylcatechol; DBA – 3,4-dihydroxybenzylamine; NEP – norephedrine; EP – ephedrine; IP – isoproterenol; MN – metanephrine; NM – normetanephrine; DEP – deoxyepinephrine; VMA – vanillylmandelic acid; 5-HT – serotonin; TRP – 5-hydroxytriptophan; DPG – 3,4-dihydroxyphenylglycol; HIA – 5-hydroxyindole-3-acetic acid; DPA – 3,4-dihydroxyphenylacetic acid; HMG – 4-hydroxy-3-methoxyglycol; MIA – 5-methoxyindole-3-acetic acid; HMBA – 4-hydroxy-3-methoxybenzoic acid; HGA – homogentisic acid; HVA – homovanillic acid.

joint to keep electrical conductivity. In this system, the applied voltage was dropped across the separation capillary prior to the porous joint and the resulting electroosmotic flow acts as a 'pump' to push buffer and analyte bands through the detection capillary behind the joint. Since amperometric measurement was performed by a micro-carbon fiber electrode inserted into the end of the detection capillary, the effect of the high-voltage electric field was removed. The detection limits obtained in such a system were 0.2–0.4 fmol for catechol and catecholamines (the sample volume injected is approximately 0.23 nl) (Wallingford and Ewing, 1988a).

Another advantage of amperometric detection compared to conventional spectrophotometric detection is in the ability to use a small-bore capillary. The small-bore capillary could improve the mass detection limit due to the small volume injected but not necessarily the concentration detection limit. For example, the mass detection limits were 6 amol in 12.7 μm capillary and 0.7 amol in 9 μm capillary but the concentration detection limits were 8.5×10^{-9} mol in a 12.7 μm capillary and 2.4×10^{-8} mol in a 9 μm capillary (Wallingford and Ewing, 1988c, 1989). Thus, the smaller-bore capillary is more useful for combining with a microinjector rather than for improving the sensitivity and it leads to the direct determination of neurotransmitters in very small biological tissues such as nerves cell and cytoplasma. So far, use of the smallest capillary of 2 μm was reported in application to cytoplasmic analysis (Olefirowicz and Ewing, 1990a).

Improved amperometric detection without a porous glass joint (end-column detection) has been reported by Haung and co-workers (Haung et al., 1991). The detection system consisted only of a fused-silica capillary and a carbon fiber microelectrode. In the detection system, it is not necessary to isolate the sensing electrode from the high-voltage electric field because the internal diameter of the separation capillary is so small (5 μm) that very little current passes through it. The microelectrode (internal diameter, 10 μm) was aligned with the bore of the capillary and positioned up against but not into the capillary. The primary advantage of this detector is its simple design and construction while the limitation is that small internal diameters must be used.

A problem in end-column detection is the difficulty of aligning the electrode at the end of the capillary and maintaining the alignment throughout an electrophoretic run. This was improved by etching the inside of the detector end of the capillary (Sloss and Ewing, 1993). The detection end of the capillary was etched with hydrofluoric acid, resulting in a conical-shaped etched bore at the tip of the capillary. The dimensions of the conical area in a 2 μm i.d. capillary were 300 μm length with a 20 μm diameter at the outlet of the capillary. The optimized end-column detection with etched capillaries easily maintained the alignment of the capillary bore and the electrode, and allowed the routine use of small-bore capillaries. Amperometric detection with microelectrodes has been shown to be one of the most sensitive techniques. However, only a few groups have experimented with an amperometric detector in HPCE since an amperometric detector is not commercially available yet.

Other detection methods for small-bore HPCE have been attempted. However, the use of a small-bore capillary requires very sensitive and low-volume detectors because the volume involved in detection is very small. An LIF detector is probably the most sensitive in HPCE and provides enough sensitivity for a small-bore capillary. However, LIF detection often requires pre- or post-column derivatization of the sample and application of this technique has not yet been reported for catecholamines, although LIF may be easy to apply for catecholamines as well as for amino acids by using a labeling agent such as fluorescein isothiocyanate (FITC) (Cheng and Dovichi, 1988).

Consequently, either a commercially available amperometric detector or the development of other conventional sensitive detection techniques are expected for trace analysis of catecholamines.

Application

Issaq *et al.* (1992) have carried out a model experiment to determine catecholamine metabolites in infant urine by HPCE with spectrophotometric detector. They accomplished the separation of homovanillic acid (HVA) and vanillylmandelic acid (VMA) from other possible catecholamine metabolites by using 200 mM acetate buffer of pH 4.10 (applied voltage, 25 kV). Under this experimental condition, HVA and VMA spiked in infant urine could be detected without any interference, but the detection limit at 214 nm was 36 mg/l for VMA and 64 mg/l for HVA. Thus, a concentration step was necessary before samples could be analyzed by HPCE with a UV detector since the concentration of VMA and HVA in the urine of normal infants is less than 5 mg/l.

Ewing and co-workers have extensively studied the analysis of catecholamines in a single cell (Olefirowicz and Ewing, 1990a, b, 1991; Chien *et al.*, 1990). The HPCE system consisted of a small-bore capillary with an internal diameter of 5 μm, an amperometric detector, and a microinjector which were prepared by etching the injection end of the capillary. In this system, the detection end was housed in a Faraday cage in order to minimize the effect of external noise sources. The injector had a length of approximately 300 μm and an external diameter of 8–10 μm (Olefirowicz and Ewing, 1990a). The tip of the microinjector was inserted directly into a cell body immersed in a saline solution, and then cytoplasmic sampling was carried out by electromigration. After the cytoplasm was drawn into the capillary, a normal electrophoretic separation was accomplished by placing the microinjector in the migration buffer reservoir and applying the separation voltage.

This microinjector was applied to the analysis of single large nerve cells in the brain of the pond snail, *Planorbis corneus*, which provides an excellent model system to study neurotransmitters in single cells because the cell is relatively large (75–200 μm diameter) and is known to contain neurotransmitters. A schematic illustration of the system is shown in Fig. 2. The concentration of dopamine in the cell body of the giant neuron was estimated as 2.2 μM by

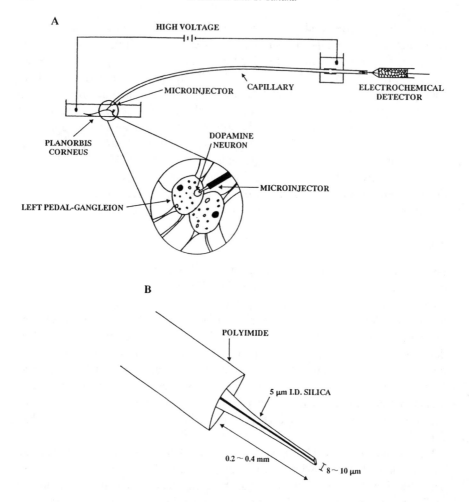

Figure 2. A: Schematic diagram of the capillary electrophoresis system used for the injection, separation and detection of cytoplasmic samples. Also shown is an exploded view of the left and right pedal ganglia of *Planorbis corneus* with the microinjector inserted directly into the dopamine neuron. B: A schematic of an etched microinjector. In order to construct the microinjector, approximately 1 cm of the polyimide coating is removed from the high voltage end of the electrophoresis capillary. The exposed fused silica is then placed into a 40% solution of hydrofluoric acid for 35 min to etch the outer diameter to the appropriate size. (Adapted from Olefirowicz and Ewing (1990b) with permission of Elsevier Science Publishers.)

using the microinjector of 5 μm (Olefirowicz and Ewing, 1990b). Both the distribution of neurotransmitter in each cellular compartment and the whole single cell concentration could be measured by using the microinjectors of 5 μm and 25 μm i.d., respectively (Olefirowicz and Ewing, 1991). This technique need not be limited only to the study of neurotransmitter concentrations, but could be applicable to the investigation of many biological substances in cytoplasmic samples.

BIOACTIVE MOLECULES AND OTHER TRANSMITTERS

In this section, we will describe the separation and determination by HPCE of other neurotransmitters besides catecholamines, and several bioactive molecules such as antidepressants, tranquilizers and hypnotics, which function through the central nervous system.

Serotonin, a compound having an indole group, is biosynthesized from tryptophan and acts as a neurotransmitter through binding with a receptor and release to a synaptic cleft. The concentration of serotonin in brain is similar to that of catecholamines and these two transmitters have similar electrophoretic mobilities. Therefore, to monitor serotonin in brain, complete separation of serotonin from catecholamines is required. Improvement in the resolution of serotonin from catecholamines was obtained by the addition of 2-propanol to the migrating MES buffer. Complete separation was achieved by MEKC in a phosphate/borate buffer with SDS. In this buffer system, both the borate complex formation and the distribution of it into micelles contributed to the separation of serotonin from catechols. Catecholamines form the catecholborate complex which has a negative charge that hinders solubilization into the negatively charged SDS micelles: this results in early elution. Conversely, serotonin does not complex with borate and is therefore distributed into micelles to be more highly retained (Wallingford and Ewing, 1989). HPCE with an electrochemical detection system was applied to the determination of serotonin in a cytoplasmic sample obtained from a serotonin neuron of *Planorbis corneus*. As described previously, the positive end of the electrophoretic capillary etched with hydrofluoric acid was inserted directly into a single nerve cell and the electrophoretic separation was carried out after sampling electrophoretically. The neurotransmitters in the nerve cell, serotonin and dopamine, could be detected sensitively by an off-column electrochemical detector (Olefirowicz and Ewing, 1990a, b).

Histamine is another biogenic amine which is distributed widely in mammalian tissues. Histamine plays an important role in allergy and inflammation. These functions of histamine occur through its binding with receptors. However, there are few reports of the separation of histamine and related substances by HPCE. The determination of histamine in red wine was performed by using a post-capillary fluorescence detector that measured the fluorescence of amines derivatized by the reaction with *o*-phthaldialdehyde as the fluorescent tagging reagent in the post-column (Rose and Jorgenson, 1988). The use of an ion-selective microelectrode as a detector for HPCE was tested, and protonated histamine and dopamine as well as alkali metals could be detected selectively (Nann *et al.*, 1993). Cimetidine, a histamine H2 receptor antagonist, was determined in serum after preconcentration by an electrochromatographic solid-phase extraction (the combination of a traditional solid-phase extraction with electrically driven elution) (Soini *et al.*, 1991). Chiral separation of promethazine, a histamine H1 receptor antagonist, was achieved by using a glycoprotein orosomucoid as a chiral complexing agents. In this case, the addition of organic modifiers such as 1-propanol was essential to obtain complete chiral separation (Buscu *et al.*, 1993). Nine antihistamine drugs (pheniramine, doxylamine, methapyrilene, thonylamine, triprolidine, dimenhydrinate, cyclizine, promethazine and chlorpheniramine) were separated by MEKC with β-cyclodextrin and tetrabutylammonium (Ong *et al.*, 1991b).

It is known that some amino acids such as glutamic acid and γ-aminobutyric acid (GABA) act as neurotransmitters in the mammalian central nervous system. The separation behavior of these amino acids by HPCE has been investigated with many other amino acids. In this section, only the detection of glutamic acid in brain by HPCE is described. Glutamic acid, an excitatory neurotransmitter in the corpus striatum, could be monitored *in vivo* in freely moving rats by combining brain microdialysis with HPCE. In this method, glutamic acid and aspartic acid in 23 pl of brain dialysate were detected by laser-induced fluorescence detector after derivatization with naphthalene-2,3-dicarboxaldehyde (Hernandez *et al.*, 1993a, b).

Adrenergic β-receptor blocking agents, which are commonly known as β-blockers, are widely used to treat angina pectoris and depression. In order to separate some β-blockers, Lukkari investigated the effects of pH of phosphate buffer and organic modifier on the resolution of β-blockers in MEKC using a cationic surfactant such as cetyltrimethylammonium bromide (Lukkari *et al.*, 1993a, b). The resolution of these compounds is greatly affected by pH. The method was applied to the determination of ten β-blockers in human urine. The best resolution was achieved by the addition of 2-propanol into migrating buffer of MEKC (Lukkari *et al.*, 1993c). Chiral separations of β-blockers were attempted by several researchers. Fanali studied the resolution of propranolol and terbutaline enantiomers in phosphate buffer containing β-cyclodextrin as chiral selector and urea (Fanali, 1991). Separation of enantiomers of β-blockers was also achieved by using the chiral recognition property of an enzyme such as cellulase (Valtcheva *et al.*, 1993).

Benzodiazepines are representative of antidepressant and anticonvulsant drugs. These drugs bind with the binding site of the benzodiazepine/GABA receptor complex to stimulate GABA receptor functions. The separation and detection of benzodiazepines were investigated by HPCE with an atmospheric pressure ionization mass spectrometer (Johansson *et al.*, 1991). The utility of the technique was demonstrated for the detection of flurazepam metabolites in human urine. The major metabolite of flurazepam in man, *N*-1-hydroxyethylflurazepam, could be monitored in human urine extracts after a single 30 mg oral dose of flurazepam dihydrochloride.

Seven tricyclic antidepressants — protriptyline, desipramine, nortriptyline, nordoxepin, imipramine, amitriptyline and doxepin — were separated by HPCE. Full resolution was achieved by the addition of methanol to the buffer, which decreased both the electroosmotic flow and the electrophoretic mobilities of the sample (Salomon *et al.*, 1991). Lee also demonstrated the determination of tricyclic antidepressants in human plasma by MEKC using dodecyltrimethyl ammonium bromide and urea (Lee *et al.*, 1993). Separation of some antide-pressants (clomipramine, imipramine and maprotiline) was also achieved by capillary isotachophoresis (Buzinkaiova *et al.*, 1993).

Barbiturates, which act on the GABA receptor/Cl ion channel complex to open the ion channel, are widely used for sedation and hypnosis. Barbitu-rates in human serum and urine were analyzed by HPCE with on-column fast-scanning multi-wavelength detection. The results of the spectrum obtained by the detector allow the identification of barbiturates in the sample (Thormann *et al.*, 1991). Meier investigated the determination of the barbiturate, thiopental, after its extraction from human serum and plasma with pentane, and seven bar-biturates containing thiopental were separated by MEKC. The HPCE data of thiopental from 66 patient samples were compared with that of HPLC (Meier and Thormann, 1991). Tellez demonstrated the combined use of *in vivo* micro-dialysis and HPCE (Tellez *et al.*, 1992). After an intraperitoneal injection of phenobarbital, dialysis was performed by a microdialysis probe inserted in the brain and the jugular vein of a freely moving rat. The dialysates were collected and analyzed by MEKC. It was confirmed that the method is useful for phar-macokinetic studies of bioactive compounds. Moreover, the separation of four barbiturates — hexobarbital, secobarbital, phenobarbital and barbital — was reported using anionic borate complexes of N-D-gluco-N-methylalkanamide (MEGA) surfactant as a new micellar phase (Smith *et al.*, 1994). Chiral sepa-ration of several barbitals was investigated with cyclodextrin-modified micellar CE. The addition of a chiral selector such as sodium d-camphor-10-sulphonate into the migrating buffer containing γ-CD-SDS enhanced the enantioselectivity for the separation of barbiturates (Nishi *et al.*, 1991).

REFERENCES

Adams, R. N. (1976). Probing brain chemistry with electroanalytical techniques. *Anal. Chem.* **48**, 1128A–1137A.

Busch, S., Kraak, J. C. and Poppe, H. (1993). Chiral separations by complexation with proteins in capillary zone electrophoresis. *J. Chromatogr.* **635**, 119–126.

Buzinkaiova, T., Sadecka, J., Polonsky, J., Vlasicova, E. and Korinkova, V. (1993). Isotachophoretic analysis of some antidepressants. *J. Chromatogr.* **638**, 231–234.

Cheng, Y. and Dovichi, N. J. (1988). Subattomole amino acid analysis by capillary zone elec-trophoresis and laser-induced fluorescence. *Science* **242**, 562–564.

Chien, J. B., Wallingford, T. M. and Ewing, A. G. (1990). Estimation of free dopamine in the cytoplasm of the giant dopamine cell of *Planorbis corneus* by voltammetry and capillary electrophoresis. *J. Neurochem.* **54**, 633–638.

Fanali, S. (1989). Separation of optical isomers by capillary zone electrophoresis based on host-guest complexation with cyclodextrins. *J. Chromatogr.* **474**, 441–446.

Fanali, S. (1991). Use of cyclodextrins in capillary zone electrophoresis. Resolution of terbutaline and propranolol enantiomers. *J. Chromatogr.* **545**, 437–477.

Gozel, P., Gassmann, E., Michelsen, H. and Zare, R. (1988). Eleetrokinetic resolution of amino acids enantiomers with copper(II)-aspartame support electrolyte. *Anal. Chem.* **59**, 44–49.

Haung, X., Zare, R. N., Sloss, S. and Ewing, A. G. (1991). End-column detection for capillary zone electrophoresis. *Anal. Chem.* **63**, 189–192.

Hernandez, L., Tucci, S., Guzman, N. and Paez, X. (1993a). *In vivo* monitoring of glutamate in the brain by microdialysis and capillary electrophoresis with laser-induced fluorescence detection. *J. Chromatogr.* **652**, 393–398.

Hernandez, L., Joshi, N., Murzi, E., Verdeguer, P., Mifsud, C. and Guzman, N. (1993b). Collinear laser-induced fluorescence detector for capillary electrophoresis. Analysis of glutamic acid in brain dialysates. *J. Chromatogr.* **652**, 399–405.

Issaq, H. J., Delviks, K., Janini, G. M. and Muschik, G. M. (1992). Capillary zone electrophoretic separation of homovanillic and vanillylmandelic acids. *J. Liq. Chromatogr.* **15**, 3193–3201.

Johansson, I. M., Pavelka, R. and Henion, J. D. (1991). Determination of small drug molecules by capillary electrophoresis–atmospheric pressure ionization mass spectrometry. *J. Chromatogr.* **559**, 515–528.

Jorgenson, J. W. and Lukacs, K. D. (1983). Capillary zone electrophoresis. *Science* **222**, 266–272.

Kaneta, T., Tanaka, S. and Yoshida, H. (1991). Improvement of resolution in the capillary electrophoretic separation of catecholamines by complex formation with boric acid and control of electroosmosis with a cationic surfactant. *J. Chromatogr.* **538**, 385–391.

Kuhr, W. G. (1990). Capillary electrophoresis. *Anal. Chem.* **62**, 403R–414R.

Kuhr, W. G. and Monnig, C. A. (1992). Capillary electrophoresis. *Anal. Chem.* **64**, 389R–407R.

Lee, K. J., Lee, J. J. and Moon, D. C. (1993). Determination of tricyclic antidepressants in human plasma by micellar electrokinetic capillary chromatography. *J. Chromatogr.* **616**, 135–143.

Lukkari, P., Vuorela, H. and Riekkola, M. (1993a). Effect of buffer solution pH on the elution and separation of β-blockers by micellar electrokinetic capillary chromatography. *J. Chromatogr.* **652**, 451–457.

Lukkari, P., Vuorela, H. and Riekkola, M. (1993b). Effects of organic mobile phase modifiers on elution and separation of β-blockers in micellar electrokinetic capillary chromatography. *J. Chromatogr.* **652**, 317–324.

Lukkari, P., Siren, H. and Riekkola, M. (1993c). Determination of ten β-blockers in urine by micellar electrokinetic capillary chromatography. *J. Chromatogr.* **632**, 143–148.

Meier, P. and Thormann, W. (1991). Determination of thiopental in human serum and plasma by high-performance capillary electrophoresis-micellar electrokinetic chromatography. *J. Chromatogr.* **559**, 505–513.

Nann, A., Silvestri, I. and Simon, W. (1993). Quantitative analysis in capillary zone electrophoresis using ion-selective microelectrodes as on-column detectors. *Anal. Chem.* **65**, 1662–1667.

Nickerson, B. and Jorgenson, J. W. (1988). High sensitive laser induced fluorescence detection in capillary zone electrophoresis. *J. High Res. Chromatogr. Chromatogr. Commun.* **11**, 878–881.

Nishi, H., Fukuyama, T. and Terabe, S. (1991). Chiral separation by cyclodextrin-modified micellar electrokinetic chromatography. *J. Chromatogr.* **553**, 503–516.

Olefirowicz, T. M. and Ewing, A. G. (1990a). Capillary electrophoresis in 2 and 5 μm diameter capillaries: application to cytoplasmic analysis. *Anal. Chem.* **62**, 1872–1876.

Olefirowicz, T. M. and Ewing, A. G. (1990b). Dopamine concentration in the cytoplasmic compartment of single neurons determined by capillary electrophoresis. *J. Neurosci. Meth.* **34**, 11–15.

Olefirowicz, T. M. and Ewing, A. G. (1991). Capillary electrophoresis for sampling single nerve cells. *Chimia* **45**, 106–108.

Ong, C. P., Pang, S. F., Low, S. P., Lee, H. K. and Li, S. F. Y. (1991a). Migration behavior of catechols and catecholamines in capillary electrophoresis. *J. Chromatogr.* **559**, 529–536.

Ong, C. P., Ng, C. L., Lee, H. K. and Li, S. F. Y. (1991b). Determination of antihistamines in pharmaceuticals by capillary electrophoresis. *J. Chromatogr.* **588**, 335–339.

Rose, Jr. D. J. and Jorgenson, J. W. (1988). Post-capillary fluorescence detection in capillary zone electrophoresis using *o*-phthaldialdehyde. *J. Chromatogr.* **447**, 117–131.

Salomon, K., Burgi, D. S. and Helmer, J. C. (1991). Separation of seven tricyclic antidepressants using capillary electrophoresis. *J. Chromatogr.* **549**, 375–385.

Schwer, C. and Kenndler, E. (1991). Electrophoresis in fused-silica capillaries: The influence of organic solvents on the electroosmotic velocity and the ζ potential. *Anal. Chem.* **63**, 1801–1807.

Sloss, S. and Ewing, A. G. (1993). Improved method for end-column amperometric detection for capillary electrophoresis. *Anal. Chem.* **65**, 577–581.

Smith, J. T., Nashabeh, W. and Rassi, Z. E. (1994). Micellar electrokinetic capillary chromatography with *in situ* charged micelles. 1. Evaluation of N-D-gluco-N-methylalkanamide surfactants as anodic borate complexes. *Anal. Chem.* **66**, 1119–1133.

Snyder, S. H. (1984). Drug and neurotransmitter receptors in the brain. *Science* **224**, 22–31.

Soini, H., Tsuda, T. and Novotny, M. V. (1991). Electrochromatographic solid-phase extraction for determination of cimetidine in serum by micellar electrokinetic capillary chromatography. *J. Chromatogr.* **559**, 547–588.

Strasters, J. K. and Khaledi, M. G. (1991). Migration behavior of cationic solutes in micellar electrokinetic capillary chromatography. *Anal. Chem.* **63**, 2503–2508.

Tanaka, S., Kaneta, T. and Yoshida, H. (1990). Separation of catecholamines by capillary zone electrophoresis using complexation with boric acid. *Anal. Sci.* **6**, 467–468.

Tellez, S., Forges, N. and Roussin, A. (1992). Coupling of microdialysis with capillary electrophoresis: A new approach to the study of drug transfer between two compartments of the body in freely moving rats. *J. Chromatogr.* **581**, 257–266.

Terabe, S., Ohtsuka, K., Ichikawa, K., Tsuchiya, A. and Ando, T. (1984). Electrokinetic separations with micellar solutions and open-tubular capillaries. *Anal. Chem.* **56**, 111–113.

Terabe, S., Ohtsuka, K. and Ando, T. (1985). Electrokinetic chromatography with micellar solution and open-tubular capillary. *Anal. Chem.* **57**, 834–841.

Thormann, W., Meier, P., Marcolli, C. and Binder, F. (1991). Analysis of barbiturates in human serum and urine by high-performance capillary electrophoresis-micellar electrokinetic capillary chromatography with on-column multi-wavelength detection. *J. Chromatogr.* **545**, 445–460.

Valtcheva, L., Mohammad, J., Pettersson, G. and Hjerten, S. (1993). Chiral separation of β-blockers by high-performance capillary electrophoresis based on non-immobilized cellulase as enantioselective protein. *J. Chromatogr.* **638**, 263–267.

Wallingford, R. A. and Ewing, A. G. (1987). Capillary zone electrophoresis with electrochemical detection. *Anal. Chem.* **59**, 1762–1766.

Wallingford, R. A. and Ewing, A. G. (1988a). Amperometric detection of catechols in capillary zone electrophoresis with normal and micellar solutions. *Anal. Chem.* **60**, 258–263.

Wallingford, R. A. and Ewing, A. G. (1988b). Retention of ionic and non-ionic catechols in capillary zone electrophoresis with micellar solutions. *J. Chromatogr.* **441**, 299–309.

Wallingford, R. A. and Ewing, A. G. (1988c). Capillary zone electrophoresis with electrochemical detection in 12.7 μm diameter columns. *Anal. Chem.* **60**, 1972–1975.

Wallingford, R. A. and Ewing, A. G. (1989). Separation of serotonin from catechols by capillary zone electrophoresis with electrochemical detection. *Anal. Chem.* **61**, 98–100.

Wightman, R. M., May, L. J. and Michael, A. C. (1988). Detection of dopamine dynamics in the brain. *Anal. Chem.* **60**, 769A–779A.

Progress in HPLC-HPCE, Vol. 5, pp. 293–317
H. Parvez *et al.* (Eds)
© VSP 1997.

Capillary gel electrophoresis and antisense therapeutics: The analysis of DNA analogs

LAWRENCE A. DeDIONISIO and DAVID H. LLOYD

Lynx Therapeutics Incorporated, 3832 Bay Center Place, Hayward, California 94545, USA

INTRODUCTION

Capillary Gel Electrophoresis (CGE) has become an effective tool for the analysis of antisense oligonucleotides (oligos). As these compounds begin to show promise in the pharmaceutical field, CGE is often used to determine the quality of chemically synthesized DNA analogs, which are presently being studied as potential antisense therapeutics. The demand for gel capillaries to possess high resolving power and provide statistically meaningful data has indirectly provided a better understanding of what is required to denature single-stranded oligos. For CGE to be useful for the analysis of oligos in general, an internal standard is often employed; however, apart from being a strictly quantitative tool, CGE has the capability to be useful in a wide range of applications within the field of antisense therapeutics. CGE can be used in conjunction with HPLC to determine an effective method for the purification of crude oligo solutions. It has also proven useful in determining whether or not a DNA analog can promote the ribonuclease H (RNase H) mediated hydrolysis of RNA. An understanding of the interactions between antisense oligos and nucleases in general is critical for determining how antisense oligos function within a biological system.

CGE has been demonstrated to be effective in the separation of phosphodiester deoxyoligonucleotides (ODNs) (Cohen *et al.*, 1988; Paulas and Ohms, 1990; Demorest and Dubrow, 1991). A polyacrylamide solution polymerized in buffer containing 7–8 M urea is the gel matrix most commonly used for these separations. Entangled polymer matrices such as Micro-Gel™ 100 have also been employed in the separation of ODNs (Dubrow, 1991). However, applying the above techniques to the separation of DNA analogs (ODNs that have been synthetically modified) can be challenging (DeDionisio, 1993). Various DNA analogs are presently being studied for pharmaceutical uses, and CGE has been envisaged as an analytical tool for purity determination after their chemical synthesis. Initial experiments into the separation of DNA analogs by CGE led to the evaluation of certain electrophoretic parameters such as pH, buffer concentration, and organic additives (DeDionisio, 1993, 1994).

Varying these parameters in a capillary gel gave significant insight into what effects the electrophoretic separation of oligos in general. It has been demonstrated that varying the pH has a substantial effect on the mobilities of homo-oligodeoxyribonucleotides (homooligos) (Guttman et al., 1992a, b). It is also known that with polyacrylamide gel electrophoresis (PAGE), homooligos of the same length migrate with different mobilities, which do not correspond to their size, i.e. molecular weight (Frank and Köster, 1979). This phenomenon, observed with CGE as well, calls into question whether or not a particular electrophoretic method can be described as 'denaturing'. Consequently, certain single-stranded ODNs can form secondary structures that 7–8 M urea in a 20% gel cannot disrupt (Sanger et al., 1977; Bowling et al., 1991). Standard PAGE analysis is also ineffective at denaturing secondary structures formed by certain single-stranded RNA oligos (Kramer and Mills, 1978).

Discovering ideal conditions for the separation of oligos has recently accelerated due to the inherent advantages of CGE. A capillary gel can be reliably re-used for at least 40 runs (DeDionisio, 1993). It is unnecessary to prepare a new gel, as must be done with PAGE, for each set of samples. Due to the durability of CGE, a capillary gel can therefore handle a larger number of samples than can a typical slab gel (PAGE) which contains about 12 wells per gel. The consequence of this feature has led to a better understanding of the conditions necessary to denature single-stranded oligos.

CGE and the analysis of DNA analogs

DNA analogs can be defined as ODNs that have been chemically modified in a specific manner. This modification can be placed anywhere within the nucleotide subunit and is often repeated throughout the length of the oligo. DNA analogs have shown promise as possible antisense therapeutics in the treatment of viral infections and certain cancers (Matsukura et al., 1989; Agrawal et al., 1989; Perlaky et al., 1993). Antisense DNA can be defined as a strand of DNA that hybridizes in a complimentary manner to a target strand of RNA or DNA. The therapeutic utility of antisense compounds were first investigated with unmodified DNA (Zamecnik and Stephenson, 1978; Wickstrom et al., 1988). However, DNA nucleases within cells degrade unmodified DNA; thus, in order to insure nuclease resistance, it is necessary to chemically modify an antisense oligo thereby producing a DNA analog.

At present, most antisense studies involving DNA analogs have utilized phosphorothioate DNA (SODN) (Fig. 1). As yet, this class of DNA analog has not shown efficacy in a clinical setting. CGE is used to determine the purity of chemically synthesized SODNs, and there are instances where CGE is used routinely within quality control departments (Srivatsa et al., 1994). As in the

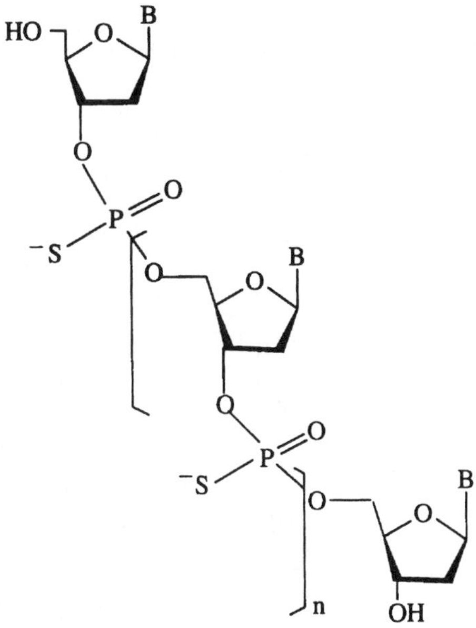

Figure 1. Molecular sructure of phosphorothioate DNA (SODN). Phosphodiester deoxyoligonu-cleotides (ODNs) possess an O^- in place of the S^- atom found in SODNs. The length of the oligo $= n + 2$. 'B' represents one of the four nucleotide bases, adenine (A), guanine (G), cytidine (C) and thymine (T) found in naturally occurring DNA.

analysis of ODNs, polyacrylamide based gels are most commonly used for the analysis of SODNs. However, we have found that utilizing an entangled polymer gel matrix such as Micro-Gel™ is sufficient for the analysis of synthetic SODNs (DeDionisio, 1993, 1994). We have also utilized these capillary gels for DNA/enzyme digestion assays and for monitoring HPLC purification processes.

CGE and the quantitative analysis of oligos

As a quantitative tool for oligo analysis, CGE should be used with an internal standard (Demorest and Dubrow, 1991; Guttman *et al.*, 1992a). This is neces-sary for two reasons. First, the electrokinetic injection, which is the method of sample loading for CGE, inherently causes variations in sample size. Second, migration times generally increase with the age of the capillary gel. Choos-ing an appropriate internal standard allows for the correction of sample load variation and inconsistent migration time.

The detection system used most often with CGE is on-line UV (260 nm) detection. In most automated CE instruments, the detector is positioned at some point along the length of the capillary. Analytes move past the detector window at different rates based on their charge to mass ratios (q/m) (Demorest and Dubrow, 1991). In general, a compound with a relatively low q/m resides in the detector window for a longer period of time than a compound with a

L. A. DeDionisio and D. H. Lloyd

Table 1.

Molecular weight (M_w), charge to mass ratio (q/m), and the change in q/m ($\Delta q/m$) for poly(dA)$_{2-20}$. Mass (m) = M_w

Oligonucleotide	M_w	q/m	$\Delta q/m$
(dA)$_2$	565	0.001770	
			0.000511
(dA)$_3$	877	0.002281	
			0.000242
(dA)$_4$	1189	0.002523	
			0.000142
(dA)$_5$	1501	0.002665	
			0.000093
(dA)$_6$	1813	0.002758	
			0.000066
(dA)$_7$	2125	0.002824	
			0.000048
(dA)$_8$	2437	0.002872	
			0.000038
(dA)$_9$	2749	0.002910	
			0.000030
(dA)$_{10}$	3061	0.002940	
			0.000025
(dA)$_{11}$	3373	0.002965	
			0.000020
(dA)$_{12}$	3685	0.002985	
			0.000017
(dA)$_{13}$	3997	0.003002	
			0.000015
(dA)$_{14}$	4309	0.003017	
			0.000013
(dA)$_{15}$	4621	0.003030	
			0.000011
(dA)$_{16}$	4933	0.003041	
			0.000010
(dA)$_{17}$	5245	0.003051	
			0.000008
(dA)$_{18}$	5557	0.003059	
			0.000008
(dA)$_{19}$	5869	0.003067	
			0.000007
(dA)$_{20}$	6181	0.003074	

relatively high q/m. The peak resulting from the slower moving compound will be broad, and thus will integrate inaccurately. Table 1 lists the molecular weights (M_ws), q/m, and the change in q/m ($\Delta q/m$) for 20 homooligos (poly(dA)$_{2-20}$). These relationships are graphically represented in Fig. 2. Figure 2A indicates that the (dA)$_4$-mer will move at a slightly slower rate than the (dA)$_{15}$-mer, and thus the (dA)$_4$-mer resides at the detector for a longer period of time. In general, longer oligos have larger q/ms, i.e. they migrate

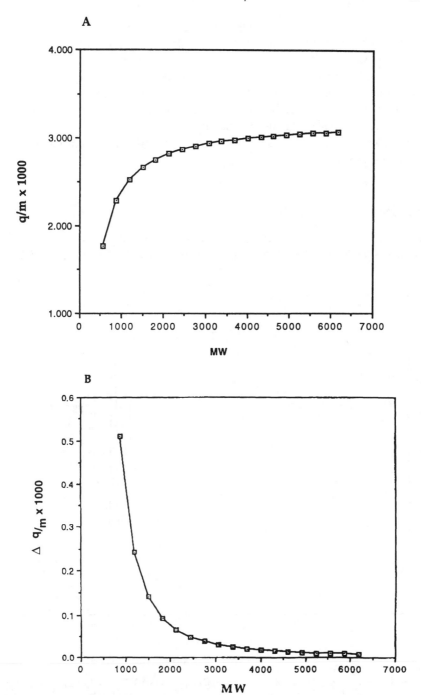

Figure 2. The relationship between the charge to mass ratio (q/m) of an oligo and its molecular weight (M_w). M_ws for homooligos of adenylic acid $(dA)_{2-20}$ (See Table 1) were plotted against their: (A) q/m and (B) $\Delta q/m$.

faster (Fig. 2A). However, the longer the oligo, the smaller the $\Delta q/m$ becomes between two oligos that differ in length by one nucleotide. For example, the $\Delta q/m$ between a 4-mer and a 5-mer is greater than the $\Delta q/m$ between an 18-mer and a 19-mer (Fig. 2B). Dividing the peak area by migration time is often done in an attempt to 'correct' for relative differences in the residence time at the detector caused by variations in the q/m of a sample's constituents (Srivatsa et al., 1994). Studies into the impact of this type of data manipulation have shown, that in some cases, these corrections provide more accurate peak integration (Altria, 1993).

CGE AND HPLC

Prior to CGE, HPLC was used as an analytical tool to determine the purity of synthetic ODNs and DNA analogs (specifically SODNs). Ion exchange HPLC

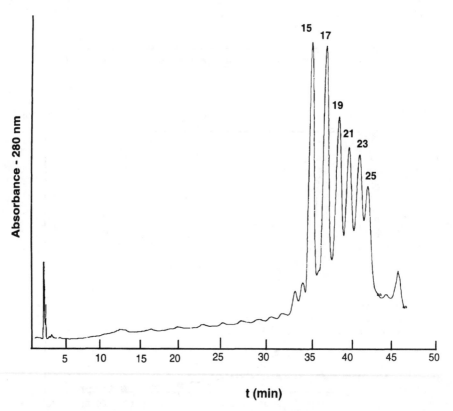

Figure 3. Ion exchange HPLC of a mixture containing six SODNs (15–25 mers) differing in length by two nucleotides. Chromatographic conditions were as follows: Buffer A was 50 mM Na_3PO_4 (pH 8.1), buffer B was 1.0 M NaSCN, buffer C was CH_3CN; the flow rate was 1 ml/min, and a linear gradient of 0.8%/min buffer B was applied over 50 min. Buffer C was kept constant at 30%. A dionex (Sunnyvale, CA, USA) nucleopac PA-100 (4 × 250 mm) column was used, and the detector was set at 280 nm.

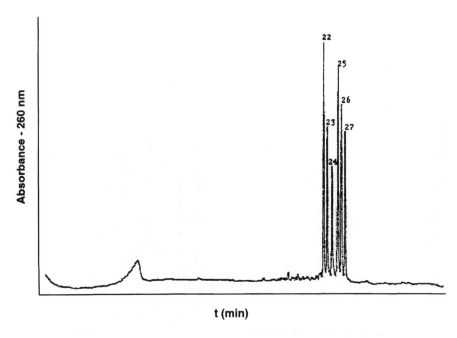

Figure 4. Electropherogram of a mixture of six ODNs (22–27 mers) differing in length by one nucleotide. Sample concentration was 0.2 absorbance units (AU)/ml. Electrophoresis was conducted with an electrokinetic injection at −5 kV for 5 s, and a constant running voltage of −10 kV was used. Gel matrix and running buffer consisted of 10% Micro-Gel^{TM} in 75 mM Tris phosphate and 10% methanol (pH 7.5) with 50 cm × 100 µm i.d. capillaries.

was one of the first methods utilized in an attempt to separate SODNs that differ in length by one nucleotide base. Figure 3 illustrates an ion exchange chromatogram of a series of SODNs that differ in length by two nucleotides. While ion exchange HPLC has shown itself useful in the separation of ODNs and some DNA analogs, its utility for analyzing SODNs, as demonstrated in Fig. 3, is inadequate. HPLC is presently incapable of baseline resolving two SODNs that differ in length by one nucleotide base (Cohen *et al.*, 1993).

On the other hand, CGE has proven itself effective in the separation of both ODNs and SODNs (DeDionisio, 1994). Figure 4 is an electropherogram of six ODNs that differ in length by one base. The system utilized for the separation illustrated in Fig. 4 was modified to achieve a similar separation of SODNs in Fig. 5. As in Fig. 4, the SODNs separated in Fig. 5 differ in length by one base; thus, CGE is capable of providing baseline resolution for the analysis of SODNs. The average resolution between peaks in the electropherogram represented in Fig. 5 is 1.95. Resolution (R_s) is defined here as $2(t_y - t_x)/(W_y + W_x)$, where y and x refer to the peaks of interest, t is retention time in decimal min, and W is the peak width in decimal min at the base.

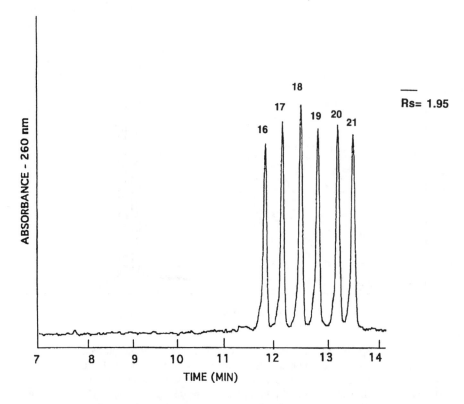

Figure 5. Electropherogram of a mixture of six SODNs (16–21 mers) differing in length by one nucleotide. Sample concentration was 0.5 (AU)/ml. Electrophoresis was conducted with an electrokinetic injection at −8 kV for 5 s, and a constant running voltage of −22 kV was used. Gel matrix and running buffer consisted of 10% Micro-Gel™ in 35 mM tris, 5.6 mM H_3BO_3 and 15% ethylene glycol (pH 9.0) with 50 cm × 100 μm i.d. capillaries. The average resolution (R_s) between the 6 major peaks is 1.95.

CGE used in conjunction with preparative HPLC

Preparative HPLC is often used to purify both ODNs and DNA analogs. In fact, HPLC is critical for the purification of SODNs (Zon and Geiser, 1991). CGE can be used to determine the effectiveness of a particular HPLC method chosen for the purification of synthetic oligos. Figure 6 is a chromatogram that illustrates an HPLC purification of a DNA analog. At 31.5 min, fractions (six in total) were collected, and the purity of each fraction was determined by CGE. Figure 7 illustrates six different electropherograms corresponding to the fractions collected from the HPLC purification represented by the chromatogram in Fig. 6. The CGE analysis of this purification revealed that fractions 2 and 3 were the most pure of those collected. This evaluation process has become routine in our laboratory, for it has provided a means to determine the most efficient HPLC parameters necessary for the purification of a given oligo.

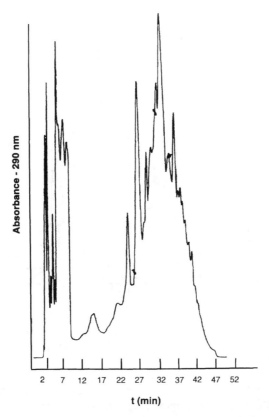

Figure 6. Ion exchange preperative HPLC of a crude DNA analog solution. Chromatographic conditions were as follows: buffer A was 10 mM NaOH, buffer B was 1.5 M NaCl, 10 mM NaOH (pH 12.0); the flow rate was 2 ml/min, and a linear gradient from 15% to 45% buffer B was applied over 40 min (0.75%/min). A Pharmacia (Piscataway, NJ, USA) mono-Q HR 10/10 (10 × 100 mm) column was used, and the detector was set at 290 nm.

THE PREPARATION AND TESTING OF GEL-FILLED CAPILLARIES

The gel matrix used in the preparation of our gel-filled capillaries consists of 10% Micro-Gel™ hydrated in 35 mM Tris, 5.6 mM H_3BO_3, and 15% ethylene glycol (EG), pH 9.0. Micro-Gel™ is an entangled polymer that was originally developed by Applied Biosystems Inc. (ABI) (Dubrow, 1991). Presently, Perkin-Elmer Corp. (PE), AB division in Foster City, CA, sells Micro-Gel™ 100. Micro-Gel™ 100 consists of a 50 μm inner diameter (i.d.) capillary filled with 8.5% Micro-Gel™ hydrated in 75 mM tris-phosphate (TP) and 10% methanol, pH 7.5. The buffer system used for Fig. 4 is the same as that used for Micro-Gel™ 100 except that the capillary gel used for Fig. 4 had a polymer concentration of 10%, and a 100 μm i.d. capillary was used instead of a 50 μm. We have found that 100 μm i.d. capillaries result in larger sample injections, which enhances detection. However, there are limitations on sample injections due to zone broadening effects (Paulus and Ohms, 1990;

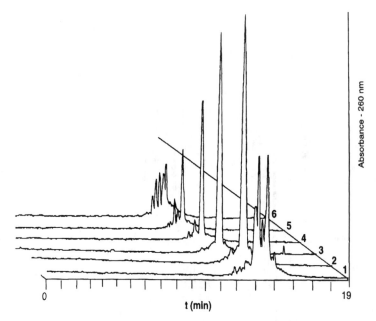

Figure 7. Electropherograms of six fractions collected at approximately 31 min from the ion exchange preparative HPLC illustrated in Fig. 6. Electrophoretic conditions were the same as those used for Fig. 5.

Macek *et al.*, 1991). In other words, as the sample size increases, resolution can decrease due to a general increase in the peak width.

Native buffers versus denaturing buffers

There is another important difference between Micro-Gel™ 100 capillaries and those that we prepare, besides the difference in the capillary i.d. Each system utilizes different hydrating and running buffers. The buffer system (75 mM TP; 10% methanol; pH 7.5) chosen by PE–AB division could be described as a native buffer, for these buffer conditions will promote the formation of secondary structure (DeDionisio, 1994). The ability for oligos to form secondary structures or compression artifacts is a function of the oligo's base sequence (Broido and Kearns, 1982; Bowling *et al.*, 1991; Satow *et al.*, 1993). Compression artifacts can be specifically defined as hairpin loops or generally defined as any occurrences of hydrogen bonding between adjacent nucleobases (Fig. 8A and 8B). Cations and a buffer pH between 7.0 and 7.5 gives rise to compression artifacts.

To counteract the tendency for single-stranded oligos to form compression artifacts, we have raised the pH of our buffer to 9.0, and we routinely perform electrophoresis at 55°C. These conditions disrupt most hydrogen bonding; however, we have discovered that oligos high in guanine (G) content require pre-heating at 95°C for 5 min. G-rich oligos are capable of forming both intramolecular hydrogen bonds as well as intermolecular hydrogen bonds in

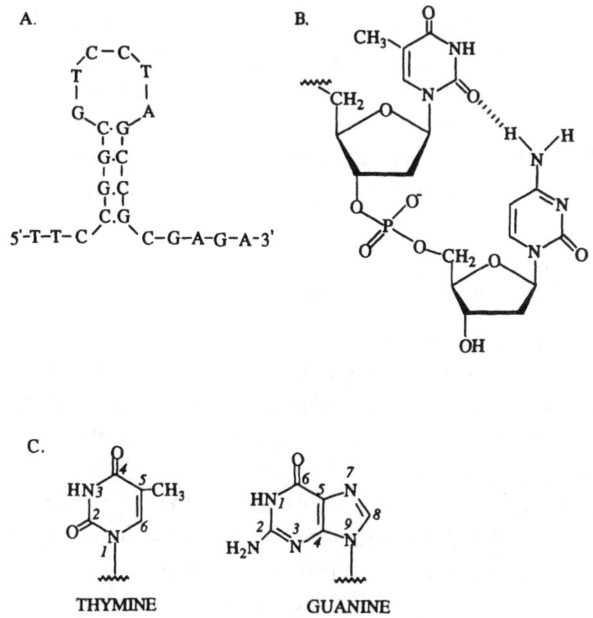

Figure 8. Possible secondary structures. A: Hairpin loop in a single-stranded oligo, B: Hydrogen bonding between 2 adjacent nucleotides (C and T) on the 3′ end, C: the molecular structure of thymine and guanine. At pH 10.0 the NH at position 3 on thymine and position 1 on guanine loses a proton which confers a negative charge on each of these molecules.

which up to four strands can orient themselves either parallel or anti-parallel to one another (Lee, 1990; Sen and Gilbert, 1991; Wang and Patel, 1992). The hydrogen bonds for these higher order intermolecular secondary structures occur between Gs that are adjacent to one another, and the presence of particular cations, namely, K^+, Na^+, NH_4^+, Ca^{2+}, Mg^{2+}, stabilizes them. We have observed intermolecular hydrogen bonding between G-rich oligos that have as few as 4 Gs in sequence. For example, the oligo with the following base sequence:

$$5'\text{-AAC GTT GA}\boxed{\text{G GGG}}\text{ CAT-3}'$$

gave two distinct peaks on electropherograms and two bands on denaturing PAGE (data not shown). The faster migrating peak/band corresponded to a single-stranded 15-mer while the slower moving peak/band corresponded to a 30-mer. After preheating the sample containing the above 15-mer to 95 °C, the peak/band corresponding to the 30-mer disappeared indicating that this peak/band was actually a duplex stabilized by hydrogen bonds between Gs. Thus, neither a pH 9.0 buffer utilized with CGE nor a 20% PAGE in 7 M urea were able to disrupt hydrogen bonds that occurred in the above G-rich oligo when preheating of the sample solution was omitted.

As yet, an electrophoretic system does not exist that will universally denature all oligos. A buffer that is stable at pH 10.0 or greater would disrupt all

L. A. DeDionisio and D. H. Lloyd

secondary structures caused by hydrogen bonds; however, the pK_as of T and G are approximately 10.0 (Fig. 8C). These bases become ionized at pH 10.0 or greater, and consequently, the overall charge and q/m of oligos rich in T or G would increase. Any effects on the migration properties of oligos caused by this potential increase in charge remains to be seen, for we have not found a suitable buffer of pH 10.0 or higher in conjunction with CGE and the analysis of oligos.

Testing the precision of gel-filled capillaries

Our CGE system (10% Micro-Gel™; 35 mM Tris; 5.6 mM H_3BO_3; 15% EG; pH 9.0 in 100 μm capillaries) is routinely tested for its consistency to resolve oligos like the separation illustrated in Fig. 5. Figure 9 represents an electropherogram of a standard solution that we utilize to test batch-to-batch consistency. The four main peaks in Fig. 9 represent four different homooligos. Peak 1 is a $(dT)_{10}$ ODN and is the reference peak. Relative mobilities (*RM*) of the three additional peaks are calculated by dividing the migration time for the peak of interest by the migration time of the reference peak, $(dT)_{10}$. For example, peak 2 in Fig. 9 is a $(dC)_{16}$ SODN and its *RM* is 10.58 min/8.22 min = 1.29. Similarly, peak 3, a $(dT)_{18}$, SODN, had a migration time of 11.13 min. Its *RM* is 11.13 min/8.22 min = 1.35, and the *RM* of peak 4, a $(dA)_{18}$ SODN is 11.52 min/8.22 min = 1.40. Table 2 represents *RM* data accumulated over a six month period from electropherograms of this

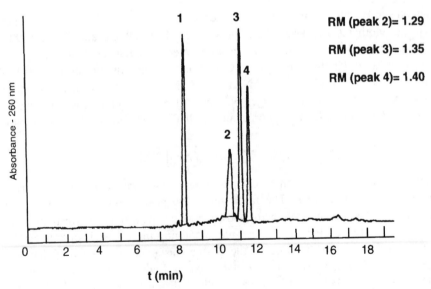

Figure 9. Electropherogram of a standard used to check batch-to-batch consistency. Peak 1 is the internal standard $(dT)_{10}$, peak 2 is a $(dC)_{16}$ SODN, peak 3 is $(dT)_{18}$ SODN, and peak 4 is a $(dA)_{18}$ SODN. Relative mobility (*RM*) for peaks 2, 3 and 4 was calculated by dividing their migration times by the migration time of the internal standard. Electrophoretic conditions were the same as those used for Fig. 5.

Table 2.
Accumulated relative mobility (*RM*) data for CGE standard gathered over 6 months on several different gel-filled capillaries

	RM (dC)$_{16}$	RM (dT)$_{18}$	RM (dA)$_{18}$
	1.3072	1.3824	1.4394
	1.3176	1.3889	1.4468
	1.3036	1.3738	1.4296
	1.2986	1.3616	1.4177
	1.3059	1.3796	1.4335
	1.2972	1.3616	1.4177
	1.2793	1.3400	1.3924
	1.2729	1.3301	1.3823
	1.2805	1.3374	1.3912
	1.3029	1.3717	1.4266
	1.2876	1.3514	1.4047
	1.2574	1.3122	1.3609
	1.2642	1.3189	1.3656
	1.2598	1.3128	1.3628
	1.3053	1.3768	1.4343
	1.2805	1.3433	1.3962
	1.2610	1.3166	1.3667
	1.2546	1.3084	1.3576
	1.2954	1.3625	1.4165
	1.2929	1.3599	1.4152
	1.2749	1.3355	1.3870
	1.2889	1.3490	1.4071
	1.3392	1.4153	1.4783
	1.2871	1.3534	1.4015
	1.2758	1.3416	1.3883
	1.3216	1.3955	1.4572
$\overline{X}$	1.2889	1.3531	1.4068
SD	0.02118	0.02788	0.03149
% RSD	1.64	2.06	2.24

RM for each homooligo [(dC)$_{16}$, (dT)$_{18}$ and (dA)$_{18}$] is calculated by dividing its migration time by the migration time of the internal standard [(dT)$_{10}$] (see Fig. 9).

standard mixture. This data was collected from several different batches of gel-filled capillaries. The statistical data from this table (% RSD < 2.5%) indicates that with a reference compound, serving as an internal standard, this system exhibits the precision necessary for quantitative analysis.

GENERATING A STANDARD CURVE FOR THE ANALYSIS OF ANTISENSE OLIGOS

The fact that homooligos of the same length migrate with different mobilities (Frank and Köster, 1979; Guttman *et al.*, 1992a, b) gives rise to difficulties

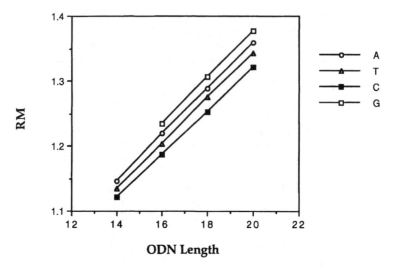

ODN Length

Figure 10. *RM* data for a series of homooligo ODNs. A linear correlation exists between the length of the homooligo and its *RM*. *RM*s were calculated by dividing the migration time of each ODN by the migration time of the internal standard, (dT)$_{10}$.

in determining the length or size of a mixed base oligo. In the electrophero-gram represented in Fig. 9, peaks 3 and 4 are two different SODNs, [(dA)$_{18}$ and (dT)$_{18}$] of the same length. Even though they are the same length, they clearly migrate with different mobilities and in fact, are baseline resolved. It is generally recognized that if electrophoresis is utilized as an analytical tool for the analysis of oligos, a series of homooligos of appropriate lengths must be synthesized and their electrophoretic mobilities established (Frank and Köster, 1979; Bauman *et al.*, 1989; Guttman *et al.*, 1992a). After the mobilities of the homooligos are determined, a series of equations can be applied to predict the *RM*s of mixed base oligos. Figure 10 illustrates a linear relationship between the *RM*s and the lengths of each of the homooligos. The following line equa-tion describes this relationship: $t'(Y_n) = sn + a$, where n = the number of nucleotides present in the oligo, Y represents one of the 4 nucleotide bases, A, G, C or T, $t'(Y) = RM$, s = the slope of the line; and a = the y-intercept. The values for s and a can be obtained from the graph in Fig. 10. Thus, the *RM* for a specific homooligo of length n can be described as follows. For:

$$Y = A, \qquad t'(A_n) = 0.035370(n) + 0.65216, \qquad (1a)$$

$$Y = G, \qquad t'(G_n) = 0.035600(n) + 0.66507, \qquad (1b)$$

$$Y = C, \qquad t'(C_n) = 0.033185(n) + 0.65653, \qquad (1c)$$

$$Y = T, \qquad t'(T_n) = 0.034865(n) + 0.64602. \qquad (1d)$$

The graph in Fig. 10 is reliable for the determination of *RM*s of homooligos in the size range 10–25 mers.

To determine the *RM* of a mixed based oligo, the following formula (Frank and Köster, 1979) is applied:

$$t'(T_t A_a C_c G_g) = t/nt'(T_n) + a/nt'(A_n) + c/nt'(C_n) + g/nt'(G_n), \quad (2)$$

where $t'(T_t A_a C_c G_g) = RM$, $t+a+c+g = n =$ the total number of nucleotides in the full-length oligo. Figure 11 illustrates two analyses of mixed base oligos. The base composition for the sequence analyzed in Fig. 11A is $A_{10}T_{10}C_4$. Applying equation (2), the *RM* for this sequence can be calculated as follows. First, after applying equations (1a), (1c) and (1d) respectively, with $n = 24$, the *RM*s of the corresponding homooligos are:

$$t'(A_{24}) = 1.5010, \quad t'(C_{24}) = 1.4530, \quad t'(T_{24}) = 1.4828,$$

$$t'(T_{10}A_{10}C_4G_0) = 10/24\,(1.4828) + 10/24\,(1.5010) + 4/24\,(1.4530) = 1.4893.$$

Since this calculation is based on the *RM* of the 3 corresponding homooligos, $(dA)_{24}$, $(dT)_{24}$, and $(dC)_{24}$, 1.4893 is a theoretical value for the *RM* of the mixed base 24-mer analyzed in Fig. 11A. The experimental *RM* for this analysis was 1.5047 (Fig. 11A), a 1.03% difference from the theoretical. Figure 11B is an analysis of a mixed base 11-mer. The base composition for this oligo is AT_6C_4, and the calculated *RM* is 1.0277. As indicated in Fig. 11B, the experimental value for the *RM* of this oligo is 1.0418, 1.4% higher than the theoretical *RM*.

There existed a drawback in the above approach to predicting experimental *RM*s for mixed base oligos. The mobilities of the dG homooligos were indirectly obtained. Oligos consisting solely of dG nucleotides form highly ordered aggregates (Ralph *et al.*, 1964), and these aggregates produce two obstacles. First, the aggregates consist of higher ordered secondary structures that, due to their higher M_w, will migrate at a slower rate than a single-stranded oligo. Secondly, the optical density at 260 nm decreases significantly when aggregate formation occurs. Since most automated CE units utilize UV detection for oligo analysis, dG homooligos become difficult to detect. To prevent these aggregates from forming, electrophoresis must be performed at temperatures in excess of 60 °C. We are unable to conduct CGE at these temperatures, and so three oligos, $(dAdG)_8$, $(dAdG)_9$, and $(dAdG)_{10}$ were synthesized and used to indirectly determine the *RM*s of the homooligos $(dG)_{16}$, $(dG)_{18}$ and $(dG)_{20}$, and these theoretical *RM* values were used to generate the standard curve for G in Fig. 10.

Figure 10 illustrates a unique aspect of these gel-filled capillaries that distinguishes them from acrylamide based capillary gels. Each set of homooligos is separated from the others in a manner that is directly proportional to their M_w or size. For the size range of homooligos utilized here (14–20 mers) the order of migration was poly: dG > dA > dT > dC, which equals the order of their

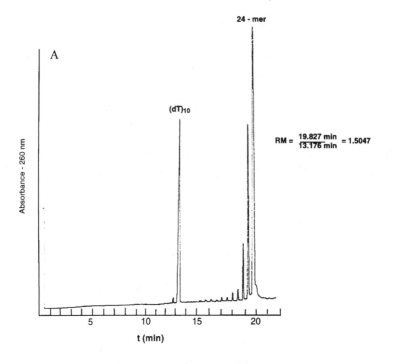

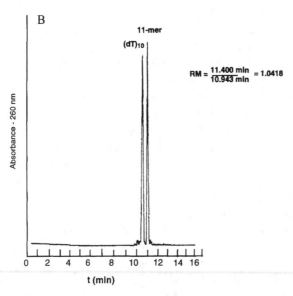

Figure 11. Electropherogram illustrating the use of an internal standard, (dT)$_{10}$. The *RM* for A: a mixed base 24-mer was calculated (equation (2)) to be 1.4893, a 1.03% difference from the *RM* (1.5047) taken from the electropherogram. The *RM* for B: a mixed base 11-mer was calculated (equation (2)) to be 1.0277, a 1.37% difference from the *RM* (1.0418) taken from the electropherogram. Electrophoretic conditions were the same as those used for Fig. 5. Experimental *RM*s were determined like those in Fig. 9.

Table 3.

The M_w and RM of 15 ODN homooligos used to generate a standard curve (see Figs 10 and 12)

Homooligo	M_w	RM
$(dA)_{14}$	3959	1.1215
$(dT)_{14}$	4169	1.1342
$(dA)_{14}$	4295	1.1466
$(dC)_{16}$	4533	1.1874
$(dT)_{16}$	4773	1.2031
$(dA)_{16}$	4917	1.2192
$(dC)_{18}$	5107	1.2529
$(dG)_{16}$	5173	1.2346
$(dT)_{18}$	5377	1.2749
$(dA)_{18}$	5539	1.2888
$(dC)_{20}$	5681	1.3209
$(dG)_{18}$	5827	1.3060
$(dT)_{20}$	5981	1.3427
$(dA)_{20}$	6161	1.3592
$(dG)_{20}$	6481	1.3770

The RM for each homooligo was calculated as described in Table 2.

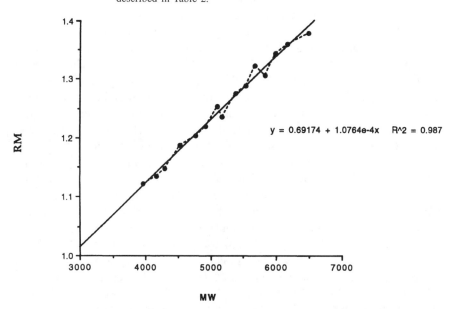

Figure 12. The linear relationship between the M_w of 15 different homooligos and their RMs. See Table 3 for the identity of each oligo. The square of the linear correlation coefficient (R) = 0.987 for the regression line generated by the data.

corresponding M_ws. Table 3 lists the 15 homooligos represented graphically in Fig. 10 and in Fig. 12. Figure 12 plots the RM values against the M_ws of each of the homooligos. The linear correlation coefficient (R) for this plot $= 0.993$ which suggests that with further studies and improvements to our running condition, that is, conducting the electrophoresis at temperature of 60°C or higher, our CGE system may be able to predict the M_w of any given oligo. It must be noted that the RMs which were indirectly determined for the dG homooligos are probably erroneously low. The M_w of (dG)$_{16}$ is larger than the M_w of (dC)$_{18}$ (Table 3); however, the calculated RM for (dG)$_{16}$ is smaller than the experimentally obtained RM for (dC)$_{18}$. This discrepency likewise occurs between (dG)$_{18}$ and (dC)$_{20}$ as indicated in Table 3 and Fig. 12.

Others have conducted similar studies with acrylamide-based electrophoresis systems. For polyacrylamide gel-filled capillaries the order of migration for the homooligos was dT > dG > dC > dA with poly dT having the longest migration or slowest mobility, and poly dA having the fastest mobility and shortest migration time (Guttman et al., 1992a). For PAGE analysis, the order of migration is dG > dT > dA > dC. Poly dG moves the slowest on slab gel, while poly dC moves the fastest (Frank and Köster, 1979).

Micro-Gel™ and 7 M urea

We have aslo filled capillaries with 10% Micro-Gel™ hydrated in a 7 M urea, tris-borate buffer (pH 9.0). The result of adding urea to the buffer is illustrated in Fig. 13. The SODN mixture analyzed in this figure is the same as the mixture analyzed in Fig. 5. Comparing the average resolution in the electropherogram of Fig. 13 to that of Fig. 5, there was an increase in resolution from 1.95 for the capillary gels without urea to 2.61 for gels containing urea. This increase in resolution resulted in shoulders on the primary peaks resolving into separate peaks (Fig. 13). Although the addition of urea resulted in higher resolving gel capillaries that allowed for the detection of small impurities between the major peaks in the SODN mixture, the durability of these gel-filled capillaries was unsuitable for practical use. They last for approximately 20 runs.

The addition of urea to the gel matrix and the running buffer causes the buffer conductivity to increase, which leads to higher running currents. Higher current produces higher temperatures from Joule heating (Macek et al., 1991), which leads to the formation of air bubbles or gaps and eventual gel failure. To counter this, the running voltage must be decreased in order to keep the current below 10 μA. For gel-filled capillaries without urea, the running voltage is kept constant at −22 kV which results in a current of approximately 5 μA. Decreasing the running voltage consequently leads to longer run times.

The longest SODN in the mixture analyzed in Figs 5 and 13 was a 21-mer. The migration time for this SODN doubled from approximately 15 min in Fig. 5 to 30 min in Fig. 13. Thus, the gel-filled capillaries that contained urea lasted half as long as the capillaries filled with gel matrix without urea which are dependable for at least 40 runs. Consequently, we feel that the resilience of gel-filled capillaries without urea outweigh the benefits gained in resolution from gel capillaries that contain 7 M urea. However, if a particular analysis requires a higher degree of resolution, then gel-filled capillaries with urea are used.

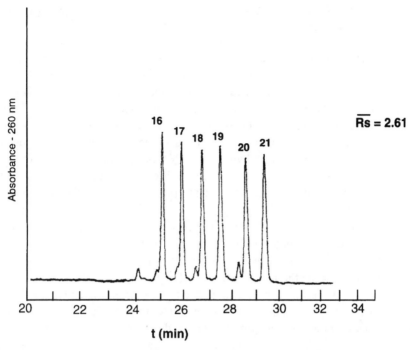

Figure 13. Electroherogram illustrating increased resolution after the addition of 7 M urea to the gel matrix. Sample mixture is the same as that used in Fig. 5. Electrophoresis was conducted with an electrokinetic injection at −8 kV for 5 s, and a constant running voltage of −16 kV was used. Gel matrix and running buffer consisted of 10% Micro-Gel^{TM} in 35 mM tris, 5.6 mM H_3BO_3, 15% ethylene glycol, and 7 M urea. The R_s between the 6 major peaks was 2.61. For capillary gels that do not contain urea, the $R_s = 1.95$ for this sample (see Fig. 5).

Establishing the purity of an antisense oligo with CGE

Presently, we do not use CGE to quantitatively determine the purity of our synthetic antisense oligos. However, others who are involved in developing antisense therapeutics routinely use CGE as a quantitative tool for establishing an oligo's purity (Srivatsa *et al.*, 1994). If used in conjunction with a suitable internal standard, CGE can provide the precision and accuracy necessary for determining the amount of full-length compound present.

The most commonly occuring impurity present in a typical oligo sample is termed the '$n - 1$' mer. Synthetic antisense oligos are normally manufactured with an automated synthesizer that builds the DNA chain one nucleotide at a time (Zon and Geiser, 1991). This process yields small quantities of 'failure sequences' or shorter pieces of DNA. The full-length oligo is referred to as the n-mer and failure sequences are categorized as $n - 1$, $n - 2$, etc. with the n-1 mer present as the largest impurity. While most quantitative CGE has been performed with acrylamide based gels, other gel matrices such as the one described here (with and without urea) should prove equally effective for determining the purity of synthetic antisense oligos.

CGE AND THE ANALYSIS OF ENZYME DIGESTION ASSAYS INVOLVING ANTISENSE OLIGOS

Synthetic DNA analogs are designed to be resistant to certain enzymes, called nucleases, which are present in cells for the purpose of hydrolyzing DNA and RNA. Resistance of an antisense oligo to nuclease activity is critical for any therapeutic activity to occur. One measure of antisense activity occurs when the antisense compound hybridizes to RNA in a complementary orientation. This short double stranded nucleic acid can potentially provide a substrate for RNase H, an enzyme that hydrolizes RNA only if the RNA strand exists in a duplex form with DNA or a DNA analog. CGE has proven useful in determining if an antisense oligo can elicit the RNase H mediated hydrolysis of RNA.

Several studies involving RNase H, RNA, and various types of antisense oligos have been conducted (Furdon *et al.*, 1989; Hoke *et al.*, 1991; Giles and Tidd, 1992), but these investigations utilized either PAGE or agarose gel electrophoresis as methods of analysis. Conducting a CGE analysis of an enzyme digest is viewed as problematic due to restrictions on the sample buffer. Customarily, samples are dissolved in pure water to minimize zone broadening that can occur with UV detection (Paulus and Ohms, 1990). However, if the sample is dissolved in a solvent that has a lower ionic strength than the running buffer, and if the injection parameters are optimized, zone broadening is readily controlled (Macek *et al.*, 1991).

Figure 14 illustrates the analysis of an RNase H digestion assay executed with an ODN and its complementary strand of RNA. The digestion buffer consisted of 10 mM Tris–HCl (pH 7.2) and 10 mM $MgCl_2$ in the presence of 1.1 unit of RNase H purchased from Pharmacia (Piscataway, NJ, USA). The diminishing RNA peak over time indicated that the RNA strand was enzymatically cleaved due to its hybridation to the complementary ODN. This was the expected result. A similar ODN/RNA reaction without the presence of RNase H was also incubated in parallel for 3 h. There was no indication of RNA hydrolysis for this control (data not shown). The injection voltage and injection time for this experiment were increased from 8 kV to 15 kV (negative polarity) and from 5.0 s to 60.0 s, respectively. This served to counter any zone broadening that may have resulted from salts present in the sample buffer.

Figure 15 represents the results of a similar experiment conducted with a DNA analog, that we are presently studying, in place of the ODN used in Fig. 14. The RNA peak for this experiment was identified in a separate analysis by its *RM* relative to an internal standard, $(dT)_{10}$ (data not shown). The experimental conditions for this digestion assay were the same as those used for the ODN/RNA assay of Fig. 14. A control sample consisting of the DNA analog/RNA duplex without RNase H was also run in parallel. This particular antisense DNA analog did not evoke an RNase H mediated hydrolysis of RNA.

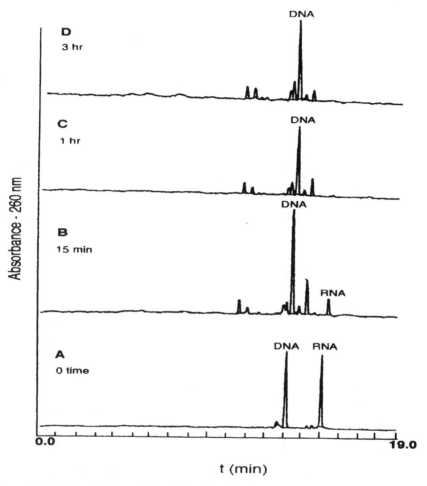

Figure 14. CGE analysis for an RNase H digestion of an ODN/RNA duplex. Sample (1.0 AU/ml) was buffered in 10 mM Tris–HCl (pH 7.2) and 10 mM MgCl$_2$. Electrophoresis was conducted with an electrokinetic injection of −15 kV for 60 s, and a constant running voltage of −22 kV was used. The gel matrix and running buffer were the same as that used for Fig. 5. Sample was incubated at room temperature (RT) with 1.1 units of RNase H. Electropherograms were obtained at A: 0 min, B: 15 min, C: 1 h and D: 3 h.

After 24 h (Fig. 15B), the RNA peak was relatively unchanged. In contrast, the RNA peak completely disappeared for the ODN/RNA digestion assay after only 1 h (Fig. 14C).

One advantage that CGE has over conventional slab gel electrophoresis for the analysis of an enzyme digestion like the one described here is that samples can be injected directly from the reaction mixture. No sample work-up was necessary since the sample buffer (10 mM Tris–HCl; 10 mM MgCl$_2$) was lower in ionic strength than the running buffer (35 mM Tris; 5.6 vM H$_3$BO$_3$; 15% EG). Injecting directly from the reaction vial allows continuous monitoring of the enzymatic assay without the inconvenience of aliquoting samples at various

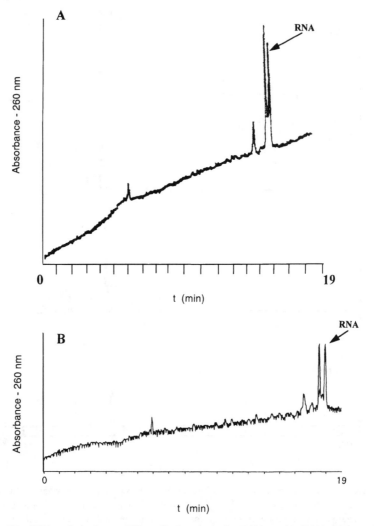

Figure 15. CGE analysis for an RNase H digestion of a DNA analog/RNA duplex. Buffer conditions and electrophoresis were the same as those used for Fig. 14. Electropherograms were obtained at A: 0 min at RT and B: after sample was incubated at 37°C for 24 h.

time intervals for a later analysis by slab gel electrophoresis. Conventional PAGE or agarose gels require samples to be dissolved in formamide or some other suitable denaturant; thus, the digestion buffer must first be evaporated. This requirement introduces the possibility of sample contamination. Due to the ubiquitous presence of ribonucleases in general, RNA can be and so excessive sample handling, such as aliquoting from the reaction mixture into separate eppendorf tubes, could result in the undesirable hydrolysis of the RNA strand. This CGE analysis, designed to determine if a particular DNA analog could activate an RNase H mediated hydrolysis of RNA, could be extended to other enzyme digestion experiments.

CONCLUSIONS

Although the primary goal for the CGE method described here is to determine the purity of antisense oligos after their chemical synthesis, this separation technique has potential for a range of applications within the field of antisense technology. The advantage of CGE for the analysis of certain DNA analogs such as SODNs over HPLC lie in its superior resolving power. CGE is able to separate two SODNs that differ in length by one nucleotide whereas HPLC presently cannot. The advantages of CGE over conventional slab gel electrophoresis stem from its on-line detection system which does not require an indirect post-electrophoretic chemical reaction such as staining. Also, the ability for some samples to be injected directly into the capillary gel without prior sample preparation provides another advantage over slab gel electrophoresis.

The electrokinetic injection required for CGE analysis has, in the past, been viewed as an encumbrance (Paulus and Ohms, 1990; Macek *et al.*, 1991). We feel that this unavoidable method for sample introduction into the gel can be viewed as an advantage because of the flexibility that the electrokinetic injection provides. The electrokinetic injection has two parameters (voltage and time), that can be adjusted to ensure reproducible peak shape and to discourage the propensity for zone broadening that can occur when samples are not dissolved in pure water. The latitude of the electrokinetic injection permits us to sample crude solutions of synthetic oligos before they are HPLC purified, and it allows us to inject directly *in vitro* enzyme digestion reaction mixtures.

While others have developed similar CGE systems for the analysis of DNA analogs (Srivatsa *et al.*, 1994), our unique system possesses the potential for more than just providing the purity of synthetic oligos. Micro-Gel™ is an entangled polymer that does not require additional polymerization after it has been hydrated and pumped into the capillary. It is stable above pH 8.3 which makes it compatable with a wider range of buffers than polyacrylamide. We have shown that buffers above pH 9.0 are sufficient for the denaturation of single-stranded oligos that do not possess the tendency to form aggregates (caused by stretches of G within the sequence), and that urea is not necessary to provide baseline resolution for oligos that differ in length by one nucleotide. Our preliminary data indicate that this system could be used to verify the M_w of oligos. The linear correlation between an oligo's RM and its M_w is evidence that our system has a greater capacity to denature single-stranded oligos than other electrophoretic systems which do not enjoy this attribute. Results like this give insight into the potential that CGE possesses as the field of antisense therapeutics pushes separation technologies beyond their present limits.

REFERENCES

Agrawal, S., Ikeuchi, T., Sun, D., Sarin, P. S., Konopka, A., Maizel, J. and Zamecnik, C. (1989). Inhibition of human immunodeficiency virus in early infected and chronically infected cells by antisense oligodeoxynucleotides and their phosphorothioate analogues. *Proc. Natl. Acad. Sci. USA* **86**, 7790–7794.

Altria, K. D. (1993). Essential peak area normalisation for quantitative impurity content determination by capillary electrophoresis. *Chromatographia* **35**, 177–182.

Baumann, U., Frank, R. and Blöcker, H. (1989). Probing hairpin structures of small DNAs by nondenaturing polyacrylamide gel electrophoresis. *Anal. Biochem.* **183**, 152–158.

Bowling, J. M., Bruner, K. L., Cmarik, L. and Tibbetts, C. (1991). Neighboring nucleotide interactions during DNA sequencing gel electrophoresis. *Nucl. Acids Res.* **19**, 3089–3097.

Broido, M. S. and Kearns, D. R. (1982). ^{1}H NMR evidence for a left-handed helical structure of poly(ribocytidylic acid) in neutral solution. *J. Am. Chem. Soc.* **104**, 5207–5216.

Cohen, A. S., Najarian, D. R., Paulus, A., Guttman, A., Smith, J. A. and Karger, B. L. (1988). Rapid separation and purification of oligonucleotides by high-performance capillary gel electrophoresis. *Proc. Natl. Acad. Sci. USA* **85**, 9660–9663.

Cohen, A. S., Vilenchik, M., Dudley, J. L., Gemborys, M. W. and Bourque, A. J. (1993). High performance liquid chromatography and capillary gel electrophoresis as applied to antisense DNA. *J. Chromatogr.* **638**, 293–301.

DeDionisio, L. (1993). Capillary gel electrophoresis and the analysis of DNA phosphorothioates. *J. Chromatogr.* **652**, 101–108.

DeDionisio, L. (1994). Capillary gel electrophoresis and the analysis of phosphodiester and phosphorothioate DNA. *Am. Lab.* July, 30–34.

Demorest, D. and Dubrow, R. (1991). Factors influencing the resolution and quantitation of oligonucleotides separated by capillary electrophoresis on a gel-filled capillary. *J. Chromatogr.* **559**, 43–56.

Dubrow, R. (1991). Analysis of synthetic oligonucleotide purity by capillary gel electrophoresis. *Am. Lab.* March, 64–67.

Frank, R. and Köster, H. (1979). DNA chain length markers and the influence of base composition of electrophoretic mobility of oligodeoxyribonucleotides in polyacrylamide-gels. *Nucl. Acids Res.* **6**, 2069–2087.

Furdon, P. J., Dominski, Z. and Kole, R. (1989). RNase H cleavage of RNA hybridized to oligonucleotides containing methylphosphonate, phosphorothioate and phosphodiester bonds. *Nucl. Acids Res.* **17**, 9193–9204.

Giles, R. V. and Tidd, D. M. (1992). Enhanced RNase H activity with methylphosphonodiester/phosphodiester chimeric antisense oligodeoxynucleotides. *Anti-Cancer Drug Des.* **7**, 37–48.

Guttman, A., Nelson, R. J. and Cooke, N. (1992a). Prediction of migration behavior of oligonucleotides in capillary gel electrophoresis. *J. Chromatogr.* **593**, 297–303.

Guttman, A., Arai, A. and Magyar, K. (1992b). Influence of pH on the migration properties of oligonucleotides in capillary gel electrophoresis. *J. Chromatogr.* **608**, 175–179.

Hoke, G. D., Draper, K., Freier, S. M., Gonzalez, C., Driver, V. B., Zounes, M. C. and Ecker, D. J. (1991). Effects of phosphorothioate capping on antisense oligonucleotide stability, hybridization and antiviral efficacy versus *Herpes simplex* virus infection. *Nucl. Acids Res.* **19**, 5743–5748.

Kramer, F. R. and Mills, D. R. (1978). RNA sequencing with radioactive chain-terminating ribonucleotides. *Proc. Natl. Acad. Sci. USA* **75**, 5334–5338.

Lee, J. S. (1990). The stability of polypurine tetraplexes in the presence of mono- and divalent cations. *Nucl. Acids Res.* **18**, 6057–6060.

Macek, J., Tjaden, U. R. and van der Greef, J. (1991). Resolution and concentration detection limit in capillary gel electrophoresis. *J. Chromatogr.* **545**, 177–182.

Matsukura, M., Zon, G., Shinozuka, K., Robert-Guruff, M., Shimada, T., Stein, C. A., Mitsuya, H., Wong-Staal, F., Cohen, J. S. and Broder, S. (1989). Regulation of viral expression of human immunodeficiency virus *in vitro* by an antisense phosphorothioate oligodeoxynucleotide against *Rev* (*Art/Trs*) in chronically infected cells. *Proc. Natl. Acad. Sci. USA* **86**, 4244–4248.

Paulus, A. and Ohms, J. I. (1990). Analysis of oligonucleotides by capillary gel electrophoresis. *J. Chromatogr.* **507**, 113–123.

Perlaky, L., Saijo, Y., Busch, R. K., Bennett, C. F., Mirabelli, C. K., Crooke, S. T. and Busch, H. (1993). Growth inhibition of human tumor cell lines by antisense oligonucleotides designed to inhibit p120 expression. *Anti-Cancer Drug Des.* **8**, 3–14.

Ralph, R. K., Connors, W. J. and Khorana, H. G. (1962). Secondary structure and aggregation in deoxyguanosine oligonucleotides. *J. Am. Chem. Soc.* **84**, 2265–2266.

Sanger, F., Nicklen, S. and Coulson, A. R. (1977). DNA sequencing with chain-terminating inhibitors. *Proc. Natl. Acad. Sci. USA* **74**, 5463–5467.

Satow, T., Akiyama, T., Machida, A., Utagawa, Y. and Kobayashi, H. (1993). Simultaneous determination of the migration coefficient of each base in heterogeneous oligo DNA by gel filled capillary electrophoresis. *J. Chromatogr.* **652**, 23–30.

Sen, D. and Gilbert, W. (1991). The structure of telomeric DNA: DNA quadriplex formation. *Current Opinion in Structural Biology* **1**, 435–438.

Srivatsa, G. S., Batt, M., Schuette, J., Carlson, R. H., Fitchett, J., Lee, C. and Cole, D. L. (1994). Quantitative capillary gel electrophoresis assay of phosphorothioate oligonucleotides in pharmaceutical formulations. *J. Chromatogr.* **680**, 469–477.

Wang, Y. and Patel, D. J. (1992). Guanine residues in $d(T_2AG_3)$ and $d(T_2G_4)$ form parallel-stranded potassium cation stabilized G-quadruplexes with anti glycosidic torsion angles in solution. *Biochemistry* **31**, 8112–8119.

Wickstrom, E. L., Bacon, T. A., Gonzalez, A., Freeman, D. L., Lyman, G. H. and Wickstrom, E. (1988). Human promyelocytic leukemia HL-60 cell proliferation and C-*myc* protein expression are inhibited by an antisense pentadecadeoxynucleotide targeted against C-*myc* mRNA. *Proc. Natl. Acad. Sci. USA* **85**, 1028–1032.

Zamecnik, P. C. and Stephenson, M. L. (1978). Inhibition of Rous Sarcoma virus replication and cell transformation by a specific oligodeoxynucleotide. *Proc. Natl. Acad. Sci. USA* **75**, 280–284.

Zon, G. and Geiser, T. G. (1991). Phosphorothioate oligonucleotides: Chemistry, purification, analysis, scale-up and future directions. *Anti-Cancer Drug Des.* **6**, 539–568.

HPCE OF CARBOHYDRATES,
CHIRAL AND ENANTIOMERS

Progress in HPLC-HPCE, Vol. 5, pp. 321–353
H. Parvez *et al.* (Eds)
© VSP 1997.

Capillary electrophoresis of heparin-derived carbohydrate chains

JAN B. L. DAMM

Department of Analytical Chemistry, N.V. Organon, Akzo Nobel Pharma Group, PO Box 20, NL-5340 BH Oss, The Netherlands

INTRODUCTION

The separation of compounds by capillary electrophoresis (CE) is based on charge, charge distribution and molecular mass of the analyte. This chapter describes the application of CE for the analysis of natural and synthetic low molecular weight heparin fragments, as well as of heparin-like oligosaccharides. Compared to High Performance Anion Exchange Chromatography (HPAEC), CE affords a similar or better resolution for various heparin fragments and it is concluded that, at least for this type of molecule, CE forms an attractive alternative to HPAEC. The high resolution and mass sensitivity render CE an attractive analytical tool for the determination of the purity of natural and synthetic heparin fragments. Moreover, the method is suitable for the analysis of other types of glycosaminoglycans and it can be used, for instance, to determine structural differences between heparin and heparan sulphate by analysis of their disaccharide components after enzymatic breakdown of the parent molecules. Likewise, it is possible to distinguish between heparin preparations derived from different species on the basis of subtle differences in the disaccharide composition. In combination with indirect UV detection, CE provides a qualitative and quantitative analysis of synthetic low molecular weight heparin fragments, allowing an accurate determination of the purity of a heparin preparation. In contrast to direct UV detection, indirect UV detection of various synthetic heparin pentasaccharides yields a signal that is nearly independent of the structure of the heparin molecule. Moreover, the sensitivity of indirect UV detection is at least one order of magnitude higher than that of direct UV detection, taking the limit of detection to about 5 fmol pentasaccharide. The limit of quantitation for synthetic pentasaccharides in the indirect detection mode is about 25 fmol. The method shows excellent repeatability and is linear in the femtomole–picomole range. The technique is exemplified by the analysis of several preparations of synthetic heparin pentasaccharides.

Heparin is a proteoglycan consisting of a small protein core to which multiple large glycosaminoglycan (GAG) side chains are attached. The glycosamino-glycan chains vary in length and are initially biosynthesized as a sequence of repeating disaccharides consisting of uronic acid and glucosamine. Follow-ing the initial synthesis of the carbohydrate chains, a series of consecutive modifications may occur, involving N-deacetylation and N-sulphation of the glucosamine residues, C5-epimerization and 2- and/or 6-O-sulphation of the uronic acid residues, leading to sulphated iduronic acid (Casu, 1989). Because these modifications occur to varying degrees, very heterogeneous polymers are formed that vary not only in molecular mass but also in structure. A final cause for heterogeneity is the 3-O-sulphation of a particular N-sulphated glucosamine residue in the unique D-glucosamine-N,6-disulphate (α1-4) L-iduronic acid-2-sulphate (β1-4) D-glucosamine-N,3,6-trisulphate (α1-4) D-glucuronic acid (β1-4) D-glucosamine-N,6-disulphate pentasaccharide sequence, that is present in approximately one-third of the heparin molecules. This pentasaccharide con-fers the well-documented anticoagulant activity to heparin through increased inhibition of factor II_a-, X_a- and XII_a-activity (Gallus and Hirsh, 1976; Lin-dahl et al., 1979; Thunberg et al., 1982; Choay et al., 1981, 1983; Thomas and Merton, 1982; Folkman and Ingber, 1989; Grootenhuis and van Boeckel, 1991).

Despite its apparent merits in the field of anti-thrombosis, the vast structural heterogeneity of heparin renders the biological activity undefined, both qual-itatively and quantitatively, and presents a serious limitation of its value as a pharmaceutical drug. This concern grows larger as it is becoming increasingly clear that unfractionated heparin displays a wide variety of biological effects not related to anticoagulant or antithrombotic activity (Nakajima et al., 1988; Ruoslathi, 1989; Wright et al., 1989; Bashkin et al., 1989; Lider et al., 1990; Ruoslathi and Yamaguchi, 1991; Yayon et al., 1991; Gorsky et al., 1991). Therefore, during the past decade, attempts have been made to produce better defined heparin preparations. This has been achieved in part by generation of low molecular weight heparin fragments of more or less uniform mass (Pan-grazzi et al., 1985; Linhardt et al., 1986; Guo and Conrad, 1988) since it has been demonstrated that low molecular weight heparin fragments of molecu-lar mass 4000–5000 Da can be applied effectively as prophylactic drugs after surgery (Albada et al., 1989; Holmer, 1989). A particularly attractive way to produce a well-defined heparin fragment with a high specific anticoagulant ac-tivity was found in chemical synthesis of the unique pentasaccharide sequence and various derivatives thereof (Choay et al., 1981; Sinay et al., 1984; van Boeckel et al., 1985, 1987, 1988, 1989; Beetz and van Boeckel, 1986; Petitou et al., 1987, 1988). These preparations have shown to be safe and effective an-ticoagulant drugs (Thunberg et al., 1982; Walenga et al., 1986) and at present clinical studies are in progress.

An important aspect in the production of natural and synthetic heparin fragments for pharmaceutical use is the availability of analytical procedures for characterization of intermediates and final products. Since the heparin fragments have an inherently high potential for microheterogeneity, the analytical procedures must meet stringent criteria. Traditional methods for separation of mixtures of heparin fragments rely on high performance size exclusion (de Vries, 1989), anion exchange (Blake and McLean, 1990) and reversed phase ion-pairing chromatography (Guo and Conrad, 1988; Linhardt *et al.*, 1989) and polyacrylamide gel electrophoresis (Turnbull and Gallagher, 1988). Recent years have witnessed the application of CE as a sensitive and high resolution method for the determination of the disaccharide composition of several proteoglycans (Carney and Osborne, 1991; Al-Hakim and Linhardt, 1991; Damm *et al.*, 1993). In these studies, the disaccharides were fractionated either as borate complexes at relatively high pH (Carney and Osborne, 1991; Al-Hakim and Linhardt, 1991) or in their native form at low pH (Damm *et al.*, 1993). Furthermore, CE has been used to analyze complex mixtures of heparin fragments (Damm *et al.*, 1992; Damm and Overklift, 1994).

For a reliable quantitative analysis, for example, as needed in the quality control of synthetic heparin fragments, the glycans should produce an equal detection response. In principle, there are several options for non-selective detection of heparin fragments, namely, detection after chromophoric, fluorescence or radioactive labelling, refractive index detection, or detection by mass spectrometry. Furthermore, conductivity detection of uniformly charged analytes and amperometric detection may serve as (pseudo) non-selective detection modes for CE. It should be kept in mind however, that specific, unimolar attachment of a chromophore or fluorescence label still is not a straightforward procedure or might even be impossible (in the case that the reducing terminus of the carbohydrate chain is blocked).

CE in combination with conductivity or (indirect) amperometric detection has been reported for the quantitative analysis of carboxylic acids (Huang *et al.*, 1989), amino acids (Olefirowicz and Ewing, 1990) and carbohydrates (Colón *et al.*, 1993; O'Shea *et al.*, 1993). However, at present, these detectors are not yet commercially available for CE. On-column laser-based refractive index detection for CE of carbohydrates has been described by Bruno *et al.* (1991), but this application also is not yet commercially available. The feasibility of fast atom bombardment (Moseley *et al.*, 1989; Caprioli *et al.*, 1989), ion-spray and electron-spray ionization (Smith *et al.*, 1988; Hallen *et al.*, 1989) mass spectrometric detection and multi-channel Raman spectroscopic detection (Chen and Morris, 1991) for the on-line detection and characterization of GAGs by CE still needs to be substantiated. Nonetheless, the micro-flow characteristics render CE ideally suited for 'hyphenation' with, for example, electro-spray or ion-spray mass spectrometry and therefore CE-MS techniques are regarded as potentially powerful analytical tools for GAG characterization.

Garner and Yeung (1990) reported indirect fluorescence detection as a universal detection method for CE of charged carbohydrates. Indirect UV detection has been applied successfully to the analysis by CE of organic (Foret *et al.*, 1989) and inorganic (Jandik and Jones, 1991) anions as well as of various monosaccharides (Foret *et al.*, 1989) and heparin fragments (Damm and Overklift, 1994). In the latter two studies, 6 mM sorbic acid, pH 12.1 and 5 mM 5-sulphosalicylic acid, pH 2–3.5 were applied as electrophoresis buffer and chromophore. Ionization of neutral sugars by high pH is a prerequisite as indirect UV detection is dependent on charge displacement (Vorndran *et al.*, 1992). For the analysis of synthetic heparin fragments by CE/indirect UV detection, high pH-induced ionization is not necessary since these compounds contain multiple carboxylic acid and sulphate groups. Wang and Hartwick (1992) recently documented the use of binary buffers in CE/indirect UV detection, allowing for a wide range of pH of the electrophoresis buffers and mobility of the analyte ions. A key question then is whether the background electrolyte that is required in the indirect detection mode interferes with an efficient resolution. Furthermore, it is essential that the background electrolyte has a high molar extinction coefficient at the selected detection wavelength to warrant sufficient detection sensitivity, and an effective mobility similar to that of the analyte ions in order to prevent fronting or tailing of analyte peaks.

In this chapter, some examples are given of the application of CE in combination with direct or indirect UV detection as an analytical method for the qualitative and quantitative analysis of low molecular weight heparin fragments. Furthermore, the determination of structural differences between heparin and other GAGs, namely, heparan sulphate, by CE on the basis of differences in the respective disaccharide compositions is shown.

CE OF HEPARIN-LIKE GLYCANS USING DIRECT UV DETECTION

The capillary electrophoresis separations described in this section are carried out using a Beckman PACE 2100 or 2210 CE system (Beckman Instruments, Palo Alto, CA, USA) equipped with a UV absorbance detector and a fused silica capillary tube (75 or 50 μm i.d. × 57 cm, detector at 50 cm). System operation and data handling are fully controlled and integrated via the System Gold software (version 6 or 711, Beckman) running on an IBM 55 SX personal computer. The apparatus is operated in the reversed polarity mode, i.e. the sample is introduced at the cathodic side of the capillary. Samples are dissolved in Milli Q water to a concentration of 0.1–5 mg/ml and are loaded by applying pressurized N_2 for 10 s resulting in injection of 25 nl sample solution. Before introduction of the first sample, the capillary is rinsed for 15 min each with 0.1 M H_3PO_4, 0.5 M NaOH, 0.1 M NaOH (all from J.T. Baker B.V., Deventer, The Netherlands), Milli Q water and running buffer, respectively. Between runs, only the last three wash steps are applied. Separations are carried out using 200 mM NaH_2PO_4 (J.T. Baker), adjusted to pH 2–3.5 with concentrated H_3PO_4, as running buffer at a potential of 7.5 kV (131.5 V/cm) and 40°C. The CE experiments are carried out using low pH ($\leqslant 3.5$) electrophoresis buffers to prevent dissociation of the silanol groups of the capillary inner wall, resulting in arheic or nearly-arheic separation conditions, without the need for anticonvective gel filling or coating of the capillary inner wall (Damm *et al.*, 1992; Chiesa and Horváth, 1993). On-capillary detection is performed by UV absorbance at 214 or 230 nm.

Heparin disaccharides

Commercially obtainable, well-defined heparin disaccharides are attractive model compounds that allow the establishment of optimal conditions for the separation of sulphated GAGs by CE. In Fig. 1 the CE electropherograms obtained at pH 3 for nine heparin disaccharides (Grampian Enzymes, Aberdeen, UK) are compiled. The abbreviations and structures (denoted **1–9**) are indicated in the figure. As expected, the disaccharides are separated primarily according to their charge, giving rise to four groups of compounds having **4**, **3**, **2** and **1** negative charged groups, respectively. Because the apparatus is operated in the reversed polarity mode, the migration time is inversely correlated with the negative charge of the disaccharide. The operation at low pH ensures that electroosmotic flow is nearly eliminated and is smaller than the electrophoretic flow, affording migration of the analytes towards the detector. As is also clear from Fig. 1, the molecules of uniform charge, for example, δUA2S-GlcNS and δUA-GlcNS6S, which differ only in the position of the sulphate group occurring either on C2 of uronic acid or on C6 of glucosamine, are separated. This illustrates that the separation is not only based on the amount of (negative) charge, but also on the charge distribution.

The CE conditions established for the heparin disaccharides are a suitable starting point for the development of conditions for the separation of larger heparin-derived oligosaccharides.

Natural heparin oligosaccharides

In the following example, the application of CE for the separation of larger heparin fragments is discussed. To obtain discrete, oligomeric heparin fragments, heparin was treated with heparinase I, that cleaves the glycosidic bonds between N-sulphated glucosamine and iduronic acid-2-sulphate residues. The procedure affords a heterogeneous mixture of fragments, differing not only in degree of polymerization, but also in the sequence of residues. The pool of heparin fragments was fractionated by gel permeation chromatography to obtain mixtures of heparin fragments of more or less uniform mass (Damm *et al.*, 1992). In Fig. 2, the subfractionation by CE of the tetramer to decamer fractions is shown. The separation is carried out at pH 2 which nearly eliminates the electroosmotic flow and results in relatively short retention times. It is clear that the structural heterogeneity increases in the order tetra-, hexa-, octa-, decamers. The migration times of all heparin preparations are virtually in the same range, namely, from 16 to 24 min. Since with increasing molecular

J. B. L. Damm

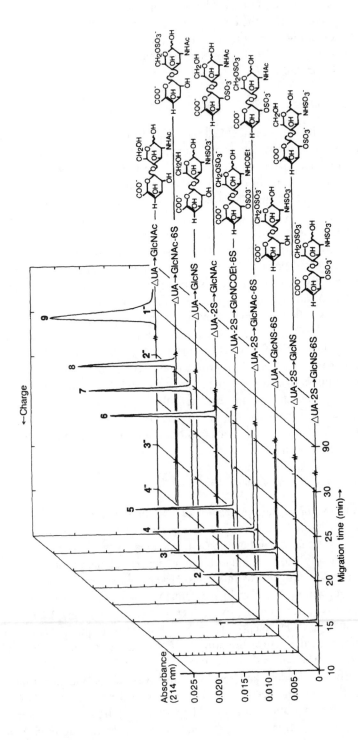

Figure 1. CE of nine heparin disaccharides. The electropherograms are compiled in a 3D plot. Electrophoresis was carried out in 200 mM phosphate buffer, pH 3 at 40°C and 7.5 kV in the reversed polarity mode. Column: 57 cm (50 cm effective length) × 75 μm i.d. On-capillary detection was at 214 nm. Injection by pressurized N_2: 25 nl from a solution containing approximately 0.1 mg/ml of each disaccharide. The structures of disaccharides **1–9** are indicated in the figure. One part of the absorbance scale corresponds to 0.005 AUFS.

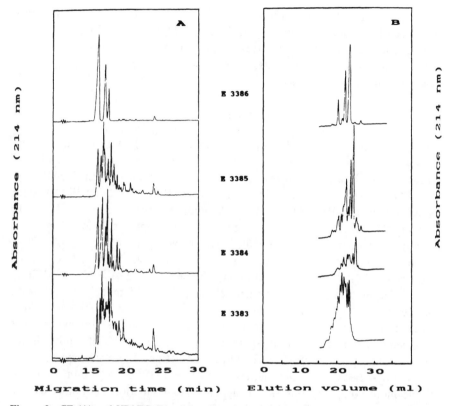

Figure 2. CE (A) and HPAEC (B) of four heparin oligosaccharide pools. The oligosaccharide pools were obtained as follows: Heparin was partially degraded by heparinase and batch-adsorbed to an anion-exchanger. The material that desorbed from the anion exchange column between 5 and 10% (w/v) NaCl (Merck) was fractionated by a gel permeation chromatography system, consisting of two columns in tandem, namely a TSK Fractogel HW 40 S column (70 × 2.6 cm, Merck) and a BioGel P-2 column (100 × 2.5 cm, 400 mesh, BioRad). The elution was carried out with 0.5 M ammonium acetate (Sigma), pH 5.0 at a flow rate of 0.7 ml/min. Four fractions of increasing molecular mass were collected and fractionated by CE and HPAEC. CE conditions as in Fig. 1, except that the total concentration of the oligosaccharides in the sample solutions was 4 mg/ml and the phosphate buffer for CE was pH 2. HPAEC was carried out on a Mono Q HR 5/5 column using a linear concentration gradient of 0–2 M NaCl in 30 ml Milli Q water at a flow rate of 1 ml/min. Detection at 214 nm. Injection volume, 25 µl from a sample solution containing 4 mg/ml of the oligosaccharides.

mass, the negative charge increases accordingly, this means that the separation is based not only on charge and charge distribution, as already discussed, but also on molecular mass, which apparently exerts an opposite effect on the migration time.

This result is in agreement with the Debye–Hückel–Henry theory on electrophoretic mobility. According to this theory, the electrophoretic mobility is, to a good approximation, proportional to the charge-to-mass (radius) ratio of the analyte (Mosher *et al.*, 1989). The relationship between the molecular mass and charge of carbohydrate molecules and their electrophoretic mobility was recently studied in detail by Chiesa and Horváth (1993) using malto-oligosaccharides derivatized with 8-aminonaphthalene-1,3,6-trisulphonic acid as model compounds. In the latter study it was shown that the electrophoretic mobility can be expressed as

$$\mu_{ep} = Cq(M_r)^{-2/3}, \tag{1}$$

where μ_{ep} represents the electrophoretic mobility, C is a constant, q is the electrical charge of the analyte and M_r is the molecular mass of the analyte.

Figure 2B shows the corresponding patterns obtained with HPAEC on Mono Q. HPAEC is carried out on a Hewlett Packard 1050 HPLC system (Hewlett Packard, Palo Alto, CA, USA) equipped with a HR 5/5 Mono Q column (Pharmacia, Uppsala, Sweden). 100 µl from a solution containing 1 mg heparin fragments per ml Milli Q water (Millipore, Milford, MA, USA) is brought onto the column and elution is carried out with a flow rate of 1 ml/min using a linear concentration gradient of 0–2 M NaCl in Milli Q water. The elution profiles are recorded at 214 or 215 nm. The increasing complexity of the tetra- to decamer fractions is confirmed by HPAEC. When the electropherograms are compared with the chromatograms, it is evident that the general pattern remains, although as expected, the order of elution is reversed, as is especially clear for the tetramer fraction. The resolution obtained for the low molecular weight samples with CE and HPAEC are comparable but for the more complex samples, CE appears to be superior. It should also be noted that, for analysis by CE, only 0.1 µg carbohydrate material is injected compared with 0.1 mg for HPAEC.

Synthetic heparin pentasaccharides

As mentioned earlier, the determination of the purity of synthetic heparin (penta)saccharides is of crucial importance in the quality control of these preparations. Due to its high resolution and mass sensitivity, CE offers excellent opportunities in this field. Figure 3 shows the CE electropherograms obtained for three synthetic pentasaccharides, denoted Org 31213, Org 31540 and Org 31550. The pentasaccharides are either (for Org 31540) identical to, or (in the cases of Org 31550 and Org 31213) are derivatives of, the unique sequence responsible for the anticoagulant activity of heparin and were synthesized using a multistep (> 50) procedure. The three compounds have an identical pentasaccharide backbone (structures in Fig. 3) and differ from each other only with respect to the number of sulphate substituents, which can be 7 (Org 31213), 8 (Org 31540) or 9 (Org 31550). The purity of the preparations has been determined previously by 360- and 500-MHz ^{1}H-NMR spectroscopy and is at least 98% (mol/mol) (Beetz and van Boeckel, 1986; Petitou et al., 1987; van Boeckel et al., 1988). Using the CE conditions established for the heparin disaccharides, the Org 31213 sample gives rise to two peaks at 13.9

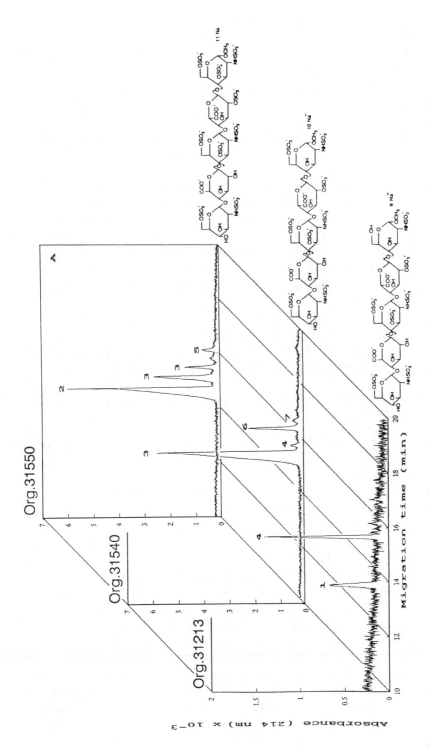

Figure 3. CE of Org 31213, Org 31540 and Org 31550. Conditions as in Fig. 1, except that the concentration of the Org compounds was 0.2–1 mg/ml.

(peak 1 in Fig. 3) and 15.7 (peak 4) min, respectively. Based on its migration time, the main peak at 15.7 min is assigned to Org 31213. Similarly, CE of the Org 31540 preparation yields two peaks at 15.3 (peak 3) and 16.2 (peak 6) min, respectively. According to its migration time, the main peak at 15.3 min can be assigned to Org 31540. In addition, two minor peaks at 15.7 (peak 4) and 16.5 (peak 7) min are visible. The Org 31550 sample yields four peaks, namely, a main peak at 14.7 (peak 2) min, which is assigned to Org 31550, and three smaller peaks at 15.2 (peak 3), 15.5 (peak 3) and 16.1 (peak 5) min, respectively. Apparently, the various pentasaccharide structures can be separated efficiently from each other by CE.

The contaminants in the pentasaccharide preparations (forming < 2% of the preparation) most likely represent synthetic precursors, although at present their identity is not known. Because the synthetic precursors in general contain strong UV absorbing groups (protecting groups), these components are detected with much higher sensitivity than the end products. Evidently, this results in a major overestimation of the amount of contaminants. Consequently, in this case, CE/direct UV detection leads to an incorrect assessment of the purity of the sample.

The previous example demonstrates that for a quantitative determination of the purity of samples that contain constituents with various (unknown) chromophoric groups, the analytical chemist should rely on a uniform detection method, such as indirect UV detection (see section on this topic).

Application of CE for the determination of structural differences between heparin and heparan sulphate

Although heparin and heparan sulphate are structurally related glycosamino-glycans (Guo and Conrad, 1989), these biomolecular entities are synthesized in different cell types and differ in their fine structure and biological function. Heparin, found in mast cell granules, is primarily known for its anti-blood clotting activity, whereas heparan sulphate is an important component of the extracellular matrix and influences cell growth, development and differentiation (Gallagher *et al.*, 1986). Differences also exist between the (ultra)structure of both proteoglycans. Heparin is generally conceived as the GAG fraction from a proteoglycan consisting of a relatively small protein core and multiple GAG chains. As mentioned earlier, the heparin GAGs are mainly composed of uninterrupted stretches of →4)-α-D-GlcNS6S(1-4)-α-L-IdoA2S (1→, that radially extend from the protein core. In contrast, heparan sulphate is envisaged as the GAG moiety from a proteoglycan comprised of a linear protein back-bone, accommodating many GAG side chains that primarily contain →4)-α-D-GlcNAc(1-4)-β-D-GlcA(1→ disaccharide units, forming continuous blocks of N-acetylated disaccharides, interspersed with blocks of N-sulphated disac-charides (Linhardt *et al.*, 1990). Both GAGs are very heterogeneous polymers and consequently, heparin may contain D-GlcNAc and D-GlcA residues and heparan sulphate often contains a substantial amount of D-GlcNS and L-IdoA residues. In addition, both GAGs show variable O-sulphation at the 3 and 6 positions of D-GlcN and the 2 position of L-IdoA (Ampofo *et al.*, 1991).

CE being a powerful method for the analysis of GAG fragments, it can be applied to ascertain the structural differences between heparin and heparan sulphate. A straightforward strategy towards this end is the enzymatic degradation of both GAG preparations by treatment with (a combination of) heparinase I, II and III, and subsequent analysis of the resulting disaccharide constituents by capillary electrophoresis.

Heparinase I, II and III are able to degrade the GAG polysaccharides, leading to oligosaccharide fragments having an unsaturated uronic acid residue as the non-reducing terminus allowing sensitive UV detection at 230 nm. Heparinase I acts only on GlcNS-α-(1-4)-IdoA2S and on GlcNS6S-α-(1-4)-IdoA2S bonds (Silva *et al.*, 1976). Heparinase II also has, apart from heparinase I substrate activity, the ability to degrade all linkages but the GlcNS3S-α-(1-4)-GlcA linkage and the linkages in the GlcA-β-(1-4)-GlcNAc-α-(1-4)-GlcA-β-(1-3)-Gal-β-(1-3)-Gal-β-(1-4)-Xylose sequence linking the GAG chains of heparin and heparan sulphate to the core protein. Therefore, heparinase II is capable of cleaving tetra- and larger oligosaccharides, resulting from a heparinase I digest of heparin to the corresponding disaccharides (Linhardt *et al.*, 1990). Heparinase III (also called heparitinase I) preferably degrades GlcNAc-α-(1-4)-GlcA or GlcNAc6S-α-(1-4)-GlcA bonds. The GlcNS3S-α-(1–4)-GlcA bond is resistant towards Heparinase III action (Silva *et al.*, 1976).

Digestion of heparin and heparan sulphate (isolated from porcine mucosa, Diosynth B.V., Oss, The Netherlands) with all three lyases, and successive CE analysis of the digestion mixture results in the electropherograms shown in Fig. 4A and Fig. 4B, respectively. The electropherogram of heparin obtained after exhaustive digestion (digestion conditions in Fig. 4, CE conditions in Table 1), displays one major peak migrating at 16.1 min, belonging to δUA2S-GlcNS6S and some smaller additional peaks, representing δUA2S-GlcNS (20.4 min), δUA-GlcNS6S (21.2 min), δUA2S-GlcNAc6S (21.8 min), δUA2S-GlcNAc (34.9 min), δUA-GlcNAc6S (35.9 min) and tetra- or larger oligosaccharides (19.0, 19.4 and 23.3 min). In the electropherogram of heparan sulphate, major peaks are observed corresponding to δUA2S-GlcNS6S (16.1 min), δUA2S-GlcNS (20.4 min), δUA-GlcNS6S (21.2 min), δUA2S-GlcNAc (34.9 min) and δUA-GlcNAc6S (35.9 min). Furthermore, the electropherogram reveals minor peaks at 19.0, 19.4, 21.8, 22.3 and at 23.3 min, corresponding to tetra- or larger oligosaccharides, except for the peak observed at 21.8 min, which corresponds to δUA2S-GlcNAc6S. The results are compiled in Table 1. For identification of the constituting disaccharides of the GAGs, well-defined heparin-derived disaccharides were used (see earlier, Fig. 1).

Thus, it is shown that heparin consists mainly of δUA2S-GlcNS6S (60% (mole/mole) of the total disaccharides) and contains only a small amount of GlcNAc(6S) residues (about 10% mole/mole). Within heparan sulphate, the major component is δUA2S-GlcNAc (29% mole/mole), while δUA-GlcNAc6S

J. B. L. Damm

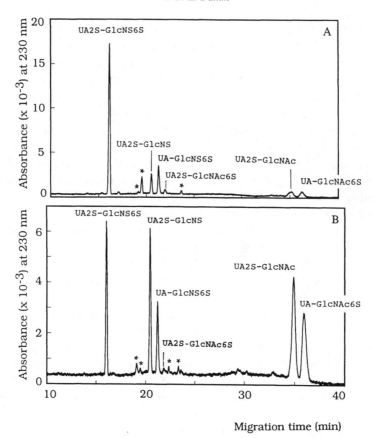

Figure 4. CE electropherograms obtained for heparin (A) and heparan sulphate (B) after a codigestion with heparinase I, II and III. Heparin and heparan sulphate were each dissolved to a concentration of 1 mg/ml in a buffer containing 150 mM sodium acetate, 0.1 mM calcium acetate, adjusted to pH 7 with concentrated acetic acid. Heparinases I, II and III (2U of each enzyme per mg GAG) were added to the solution and incubation was carried out in a waterbath for 24 h at 25°C. A sample of 100 µl was taken for CE analysis. The heparinase digest was analyzed by CE (conditions indicated in Table 1) without further purification. The identified disaccharides are compiled in Table 1. Peaks that could not be identified are labeled with an asterisk.

and δUA2S-GlcNS6S form 22% and 18% (mole/mole), respectively, of the total amount of constituting disaccharides. It should be noted that the peak found at 34.9 min is exclusively assigned to δUA2S-GlcNAc as it is generally understood that heparan sulphate contains no (or a neglible amount of) δUA-GlcNS. This brings the total amount of heparan sulphate – derived disaccharides containing GlcNAc residues to 52% (mole/mole).

The relative amounts of the various disaccharides are deduced from their corresponding peak areas obtained by UV-absorbance detection at 230 nm. This seems a reasonable approach, since at this wavelength the absorption stems mainly from the unique C4–C5 unsaturated bond (one in each disaccharide).

Table 1.

Disaccharides obtained for heparin and heparan sulphate after digestion with heparinases I, II and III[a]

Migration time (min)	Disaccharide	No.	Area percent Heparin	Heparan sulphate
16.1	δUA2S-GlcNS6S	1	60.1	18.0
19.0	[b]		0.4	1.3
19.4	[b]		5.7	0.5
20.4	δUA2S-GlcNS	2	7.5	17.7
21.2	δUA-GlcNS6S	3	10.8	9.4
21.8	δUA2S-GlcNAc6S	4	2.4	0.6
22.3	[b]		0.6	0
23.3	[b]		1.5	0.7
34.9	δUA2S-GlcNAc/	6		
	δUA-GlcNS	7	6.6	29.3
35.9	δUA-GlcNAc6S	8	5.0	22.3
			100	100

[a]CE analysis was performed applying 200 mM NaH_2PO_4 pH 2.5 as electrophoresis buffer, at 40°C using a fused silica capillary of 50 μm × 57 cm (50 cm till detector). The applied voltage was 7.5 kV (131.5 V/cm). Detection was on line at 230 nm. Analyses were performed in the reversed polarity mode, e.g. injections were made at the cathodic side of the capillary. Area percentages are calculated with the software 'System Gold', version 6.0 (Beckman).

[b]Probably representing tetra- or higher oligosaccharides (7.6% for heparin, 3.1% for heparan sulfate). The identity could not be determined.

Therefore, the stated area percentages reflect the molar amounts of the corresponding disaccharides.

Apart from the establishment of structural differences between heparin and heparan sulphate on the basis of CE analysis of their respective disaccharide compositions after (near) complete digestion with heparinases I, II and III, it is possible to get more insight into the arrangement of the disaccharide units in both GAG molecules by partial enzymatic degradation.

Heparinase III treatment of heparin yields no low molecular weight heparin fragments, indicating that stretches of →4)-α-D-GlcNAc(1-4)-β-GlcA(1→ disaccharides are indeed absent in the heparin preparation. It should be borne in mind that under the applied CE conditions, only low molecular weight (< ±4000 Da) GAG fragments will migrate in the specified time interval (Damm *et al.*, 1993). Therefore, absence of peaks indicates that the GAG does not contain heparinase-susceptible bonds and resists degradation to low molecular weight fragments. In contrast, heparinase III digestion of heparan sulphate affords the expected δUA2S-GlcNAc and δUA-GlcNAc6S disaccharides, confirming the presence of continuous regions of these disaccharides in heparan sulphate.

After a combined heparinase I/III digestion, heparin yields the expected δUA2S-GlcNS6S disaccharide, proving that it occurs in uninterrupted stretches in the heparin molecule. Some additional low intensity peaks are observed which can be ascribed to highly charged tetra- or higher oligosaccharides. The same treatment of heparan sulphate leads to the appearence of δUA2S-GlcNS6S, δUA2S-GlcNS6S, δUA-GlcNS6S and δUA2S-GlcNS, next to δUA2S-GlcNAc and δUA-GlcNAc6S. This means that in heparan sulphate, apart from stretches of UA2S(1-4)-D-GlcNAc and/or UA(1-4)-D-GlcNAc6S, also concatamers of UA2S(1-4)-D-GlcNS6S, UA(1-4)-D-GlcNS6S and UA2S (1-4)-D-GlcNS occur.

As discussed, upon treatment with all three enzymes, heparin gives rise to low amounts of δUA-GlcNS6S and δUA2S-GlcNS, δUA2S-GlcNAc and δUA-GlcNAc6S, besides the main disaccharide constituent δUA2S-GlcNS6S. This indicates that the low abundance disaccharide units do not occur as concatamers, since in that case they would have been formed previously upon heparinase I/III treatment. The minor disaccharides are probably derived from the larger heparin fragments that result after heparinase I/III treatment. Heparinase I/II/III treatment of heparan sulphate does not yield additional peaks in comparison to the heparinase I/III incubation. However, after heparinase I/II/III digestion, the relative amount of non-acetylated disaccharides is higher than the amount of acetylated disaccharides, which is the reverse after heparinase I/III treatment. This indicates that part of the non-acetylated disaccharides are combined with other GAG elements in such a way, that is, interspersed, that they are not susceptible to heparinases I and III.

Taken together, it can be concluded that enzymatic hydrolysis of heparin and heparan sulphate, followed by CE analysis of the resulting GAG fragments, enables the determination of structural differences between these GAGs and yields insight into the arrangement of the disaccharide units.

Application of CE for the determination of the species of origin of heparin preparations

Natural heparin and heparin fragments can be prepared from several tissues stemming from various animal species. Well-known examples are bovine and porcine heparin from lung and intestinal mucosa. Recently, the application of bovine heparin preparations for pharmaceutical use in humans has been compromised due to the fact that cows are susceptable to a particular prion infection causing bovine spongiform encephalitis. Since the disease is unknown in pigs, preferably porcine heparin is used for the manufacturing of heparin preparations that are intended for pharmaceutical use. This being so, the availability of an analytical method to determine the origin of species (bovine or porcine) of heparin preparations is desirable. It has been shown (Dabat et al., 1994) that CE is a potential candidate for such an analytical method. In this procedure, the heparin preparation is enzymatically degraded to its constituting disaccharides by heparinases I–III, followed by quantification of the individual disaccharide components by CE.

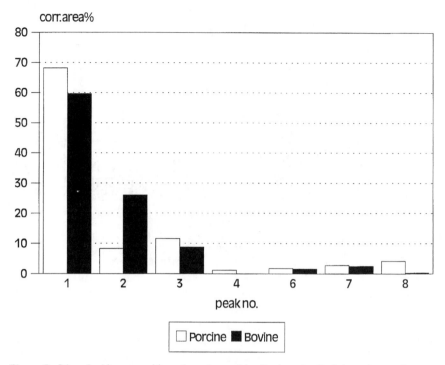

corr.area%

Figure 5. Disaccharide composition of porcine and bovine heparin. Both heparin samples were treated with heparinase I–III followed by analysis of the disaccharide composition by CE. The identity of disaccharides **1–4** and **6–8** is indicated in Table 1. Further details in Fig. 4 and text.

Figure 5 shows the disaccharide composition obtained by CE analysis of heparinase-treated (pure) bovine and porcine heparin. The structures of the disaccharides **1–8** are indicated in Table 1. Evidently, quantitative differences can be detected for the various disaccharides. Particularly, the amount of disaccharides **2** (δUA2S-GlcNS), and to a lesser extent of disaccharides **1** (δUA2S-GlcNS6S) and **3** (δUA-GlcNS6S), as measured by their respective relative peak areas in the electropherograms, differ significantly. In Fig. 6A, the amounts of disaccharides **1–3**, retrieved for several artificial mixtures of bovine and porcine heparin are compiled. Presented in this way, it is apparent that the amount of the constituting disaccharides as detected by CE analysis after enzymatic hydrolysis of the parent heparin preparation, is indicative of the origin/composition of the preparation (pure porcine, pure bovine or a mixture of both). This is even more clear when the ratio of the amount of two of the indicative disaccharides, e.g. amount of disaccharide **1**/amount of disaccharide **2**, is calculated and plotted against the relative amount (w/w) of bovine or porcine heparin in the mixture. To illustrate the principle, the ratios **1/2** and **2/3** are calculated for a series of (artificial) mixtures of bovine and porcine heparin, and plotted against the relative amount (w/w) of bovine heparin in the mixture (Fig. 6B). In this way, a possible contamination of a porcine heparin preparation by bovine heparin can be quantitatively determined. The procedure is illustrated in Fig. 6B for an

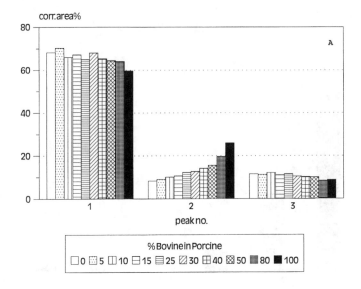

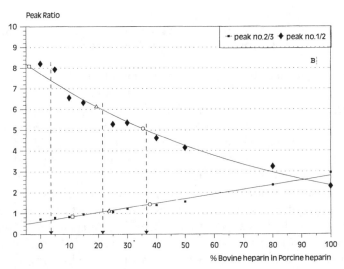

Figure 6. Relative amounts of δUA2S-GlcNS6S (**1**), δUA2S-GlcNS (**2**) and δUA-GlcNS6S (**3**) in pure porcine and bovine heparin and several artificial mixtures of porcine and bovine heparin (varying from 5% (w/w) bovine/95% porcine heparin to 80% bovine/20% porcine heparin). The heparin samples were treated with heparinase I–III followed by analysis of the disaccharide composition by CE. The relative amounts of disaccharides **1**–**3** are plotted (A). Differences between the relative amounts of disaccharides **1**–**3** are most apparent by plotting the ratio of the (relative) amount of these disaccharides retrieved in the various mixtures of porcine and bovine heparin (B). The dashed lines indicate how the composition (in terms of porcine and bovine heparin) of heparin samples can be determined.

unknown heparin sample. After enzymatic breakdown, the relative amounts of disaccharides **1–8** are determined by CE analysis and the ratios **1/2** and **2/3** are calculated and plotted in Fig. 6B (open squares). By extrapolation of the vertical line which best fits the two data points (dashed line) it is clear that the sample contains a small amount of bovine heparin (± 3.5 mass %), next to porcine heparin (± 96.5 mass %). To verify the result, the heparin sample was supplemented with 20% or 40% (w/w) pure bovine heparin upon which the composition of the sample in terms of porcine/bovine heparin was determined again. In this case, the analyses indicate the presence of $\pm 22\%$ (w/w) and $\pm 37\%$ (w/w) bovine heparin in the sample, which is close to the expected value. By the same principle, heparin preparations derived from different tissues from one animal species can be discriminated.

CE OF HEPARIN-LIKE GLYCANS USING INDIRECT UV DETECTION

Capillary electrophoresis in combination with indirect UV detection is carried out using the same materials and experimental conditions as described above for CE/direct UV detection, except that indirect UV detection is realized by quenching of the UV signal at 214 nm using 5 mM 5-sulphosalicylic acid (pK_a carboxylic acid group, 2.27), pH 3.0 or 5 mM 1,2,4-tricarboxy benzoic acid (pK_as 2.28, 3.58 and 4.79), pH 3.5 as electrophoresis buffer. The applied potential is 5 kV (87.7 V/cm) and the thermostatted capillary is kept at 25 °C. Unless stated otherwise, a 50 μm (i.d.) × 57 cm capillary is used. The injection volumes are 1.8 or 9 nl. The heparin disaccharides were purchased from Grampian Enzymes, Aberdeen, UK. The synthetic heparin and heparan sulphate fragments are from N.V. Organon, Oss, Netherlands.

Natural heparin disaccharides

Figure 7A depicts the CE electropherogram obtained for a nearly equimolar mixture of eight heparin disaccharides (denoted **1–8**, structures in Fig. 1) employing 200 mM NaH_2PO_4, pH 2.5 as electrophoresis buffer and using direct UV absorbance at 230 nm as detection method. Injection of ~ 5 pmol of each disaccharide yields a satisfactory signal-to-noise ratio, demonstrating the high mass sensitivity of CE. Figure 7B shows the CE pattern obtained for the same mixture applying 5 mM 1,2,4-tricarboxy benzoic acid, pH 3.5 as electrophoresis buffer and observing the quenching of the UV signal at 214 nm (further referred to as indirect UV detection). In this case only ~ 0.5 pmol of each disaccharide is injected. The disaccharides **2–4** are not completely separated under the conditions of indirect UV detection, but the three disaccharides denoted **6–8**, all having two negative charges, are baseline separated, which is not the case under the conditions used for direct detection.

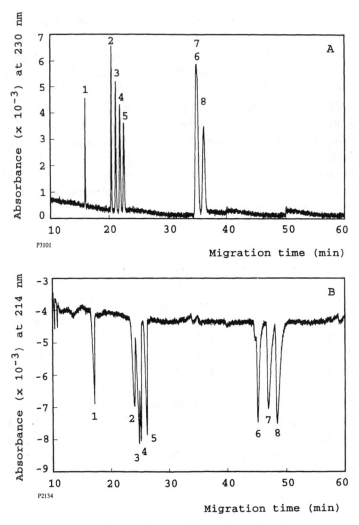

Figure 7. CE of eight heparin disaccharides using direct (A) or indirect (B) UV detection. In case of direct UV detection the electrophoresis buffer was 200 mM Na_2HPO_4, pH 2.5 and the injection volume was 9 nl from a solution containing approximately 0.16 mg/ml of each disaccharide. The potential across the capillary was 7.5 kV (131.5 V/cm) and the thermostatted capillary was kept at 40°C. In case of indirect UV detection the electrophoresis buffer was 5 mM 1,2,4-tricarboxy benzoic acid, pH 3.5 and the injection volume was 1.8 nl from the same solution as in A, except that the concentration was approximately 0.1 mg/ml of each disaccharide. The potential across the capillary was 5 kV (87.7 V/cm) and the thermostatted capillary was kept at 25°C. The structures of disaccharides **1–8** are given in Fig. 1.

Application of CE for the purity determination of synthetic heparin-like oligosaccharides

The practical value of CE in combination with indirect UV detection for the determination of the purity of actual pentasaccharide preparations is demonstrated by the analysis of several synthetic pentasaccharide preparations.

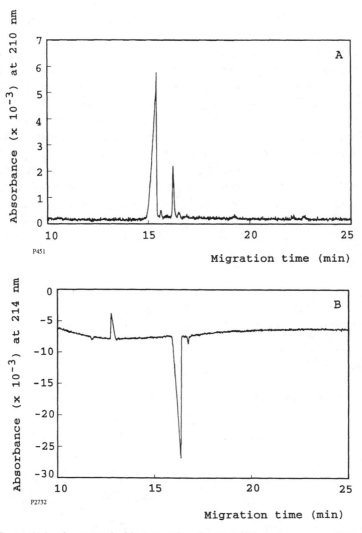

Figure 8. Analysis of pentasaccharide preparation Org 31540E by CE using direct (A) and indirect (B) UV detection. Direct UV detection was carried out using 200 mM Na_2HPO_4, pH 3.0 as electrophoresis buffer. The potential across the capillary (75 μm × 57 cm) was 7.5 kV (131.5 V/cm) and the thermostatted capillary was kept at 40 °C. Indirect UV detection was performed applying 5 mM 5-sulphosalicylic acid, pH 3.0 as electrophoresis buffer. The potential across the capillary (50 μm × 57 cm) was 5 kV (87.7 V/cm) and the thermostatted capillary was kept at 25 °C. Concentration of Org 31540 in the sample solution was 1 mg/ml, injection volume 50 nl (A) and 1.8 nl (B). The structure of Org 31540 is outlined in Table 2.

In Fig. 8 the CE electropherograms obtained for the natural pentasaccharide sequence Org 31540E using direct (A) and indirect (B) UV detection are depicted. Prior to CE, the organic purity of Org 31540E was established by NMR spectroscopy to be 98% (mol/mol). CE in combination with direct UV detection gives rise to a major peak at 15.4 min (79% of total peak area, cf. Fig. 3) representing Org 31540 and three minor peaks at 15.6, 16.2 and 16.5 min stemming from contaminants. In the indirect detection mode, the main peak stemming from Org 31540 migrates at 16.6 min and forms 98% of the total peak area, which exactly agrees with the NMR data. The peak belonging to the main contaminant is observed at 16.9 min and accounts for 1.8% of the total peak area. This peak probably corresponds to the peak at 15.6 min in the direct detection mode. Remarkably, the contaminant that yields a major signal at 16.2 min in the direct detection mode is not observed in the indirect detection mode, implying that it represents less than 0.5% (w/w) relative to Org 31540E (see below). It should be noted that, apart from yielding representative peak areas enabling quantitative analysis, the sensitivity of indirect UV detection is superior to that of direct detection. In the latter experiment, 28 pmol (50 ng) and 1 pmol (1.8 ng) of Org 31540 are injected in the case of direct and indirect UV detection, respectively, yielding a still better signal to noise ratio for the indirect detection mode.

Table 2.
Structures of Org 31540, 31550, 33232, 34275, 34276 and 34277. Org 33271 and Org 33263 are close derivatives of Org 33232 having twelve and eleven negative charges, respectively

GAG	Structure
Org 31540	
Org 31550	
Org 33232	
Org 34277	
Org 34276	
Org 34275	

A second example of the practical value of CE in combination with indirect UV detection for the determination of the purity of pentasaccharide preparations is furnished by the analysis of HH2174. Sample HH2174 represents a batch of raw material of Org 31550 that was deliberately kept from further purification. Org 31550 (structure in Table 2) is a derivative of the unique natural pentasaccharide sequence responsible for the anticoagulant activity of heparin. Relative to the natural sequence, Org 31550 contains one extra 3-O-sulphate group in the first glucosamine-N,6-disulphate residue. NMR spectroscopic analysis proves that HH2174 consists for approximately 85% (mol/mol) of Org 31550. When HH2174 is subjected to CE using direct UV detection, at least nine peaks are discernable in the electropherogram (Fig. 9A). From injection of pure Org 31550 it is known that the peak at 14.7 min can be attributed to Org 31550. The additional peaks, accounting for 75% of the total peak area, belong to minor contaminants. This clearly demonstrates the limitations of direct UV detection for the quantitative analysis of these type of preparations by CE. The contaminants most probably represent synthetic precursors of Org 31550 that contain strong UV absorbing groups, which is the main reason for the overestimation of the amounts present. In contrast, when HH2174 is analyzed by CE applying the indirect detection mode, a pattern is obtained displaying proportionality between peak areas and the amount of the components present (Fig. 9B). The main peak at 24.9 min corresponds to Org 31550 and constitutes 86% of the total peak area, which corresponds with the NMR data. Four minor peaks and three barely discernable peaks, all belonging to contaminants, are visible and account for 14% of the total peak area. Like in the previous example, the contaminant that gives rise to a major peak in the direct detection mode yields only a minor signal in the indirect detection mode. Note that in the latter experiment, the pH of the electrophoresis buffer is 3.5, which explains why the migration times obtained for the analytes in the indirect mode are higher than those obtained using the direct mode.

Also, other types of GAGs are amenable to analysis by CE/indirect UV detection. Figure 10 illustrates the separation of a mixture of GAGs, comprising alkylated and O-sulphated synthetic pentasaccharides, synthetic dermatan sulphate di-, tetra- and hexasaccharides and a natural heparin disaccharide (structures indicated in Table 2). The mixture contains approximately 1 mg/ml of each of the mentioned GAGs. Evidently, all GAGs can be analyzed in a single run, which demonstrates the general applicability of CE/indirect UV detection for the analysis of GAG fragments. In Table 3, the charge and migration time of the various GAG fragments are compiled. The O-sulphated, alkylated pentasaccharide Org 33271, possessing 12 negative charges, migrates at 18.2 min. The N,O-sulphated pentasaccharides, Org 31550 and Org 31540, having 11 and 10 negative charges, respectively, show slightly higher migration times (18.7 and 19.1 min, respectively) in line with their reduced negative charge compared to Org 33271. As discussed earlier, the relationship between the molecular mass and charge of carbohydrate molecules and their electrophoretic

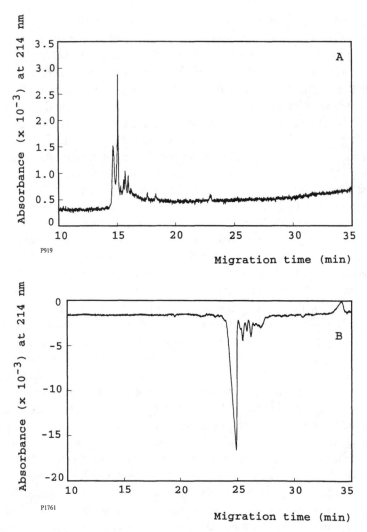

Figure 9. Analysis of pentasaccharide preparation HH2174 by CE using direct (A) and indirect (B) UV detection. The electrophoresis conditions for direct and indirect UV detection are reported in Fig. 8A and 8B, respectively, except that in 9B the pH of the electrophoresis buffer is 2.5. HH2174 represents a sample of Org 31550 (structure indicated in Table 2) that was deliberately kept from purification. The concentration of the pentasaccharide in the sample solution is 1 mg/ml, injection volume 50 nl (A) and 1.8 nl (B).

mobility can be expressed as $\mu_{ep} = Cq(M_r)^{-2/3}$ (Chiesa and Horváth, 1993). Although the results presented in this study are in line with this relationship, for a correct interpretation of the data it should be noted also that the charge distribution of the analyte may influence the migration time (Damm *et al.*, 1992). The alkylated pentasaccharides Org 33263 and Org 33232, both having 11 negative charges, migrate at 19.4 and 19.9 min, respectively. The fact that the latter two compounds have higher migration times than Org 31540,

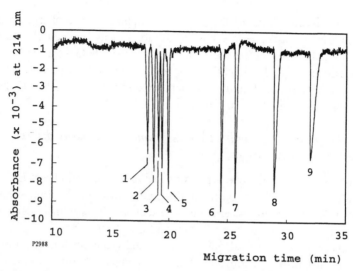

P2988

Figure 10. CE/indirect UV detection of a mixture of various glycosaminoglycans, comprising different synthetic penta-saccharides (1: Org 31550, 2: Org 31540, 3: Org 33271, 4: Org 33263 and 5: Org 33232), synthetic dermatan sulphate hexa-, tetra- and disaccharides (6: Org 34275, 7: Org 34276 and 8: Org 34277, respectively) and a natural heparin disaccharide (9: δUA2S-GlcNAc6S). The concentration of each GAG in the mixture is approximately 1 mg/ml and the injection volume is 1.8 nl. The structures and migration times of the GAGs are compiled in Table 2 and 3, respectively. Electrophoresis conditions as in Fig. 8B.

Table 3.

Migration time of GAGs in CE/indirect UV detection

GAG	Type	Charge	Migration time
Org 33271 (1)	synth. pentasacch.	12	18.2
Org 31550 (2)	synth. pentasacch.	11	18.7
Org 31540 (3)	synth. pentasacch.	10	19.1
Org 33263 (4)	synth. pentasacch.	11	19.4
Org 33232 (5)	synth. pentasacch.	11	19.9
Org 34275 (6)	derm. sulph. hexasacch.	10	24.4
Org 34276 (7)	derm. sulph. tetrasacch.	7	25.6
Org 34277 (8)	derm. sulph. disacch.	3	28.9
δUA2S-GlcNAc6S (9)	heparin disacch.	3	32.0

[a]Experimental details in Fig. 10. Org 33271, Org 33263 and Org 33232 differ from Org 31550 and Org 31540 in that they contain exclusively O-sulphated groups and methylated hydroxyl functions, whereas Org 31550 and Org 31540 possess N- and/or O-sulphated and free hydroxyl functions. The numbers in brackets refer to the peaks in Fig. 10.

in spite of containing one extra negative charge, must be ascribed to their higher molecular mass (Org 31540, 1727 Da; Org 33263, 1917 Da; Org 33232, 1917 Da) and/or differences in charge distribution. Since Org 33263 and Org 33232 have an identical molecular mass and charge, the difference in migration time must be ascribed to the unequal charge distribution. The dermatan sulphate hexasaccharide, bearing ten negative charges, migrates at 24.4 min.

The higher migration time of the dermatan sulphate hexasaccharide compared to the heparin pentasaccharides can be ascribed to its reduced charge and/or higher molecular mass (2108 Da). The dermatan sulphate tetra- and disaccharides, possessing 7 and 3 negative charges, respectively, migrate at 25.6 and 28.9 min, respectively. Finally, the heparin disaccharide δUA2S-GlcNAc6S, having three negative charges, migrates at 32.0 min.

Validation aspects of CE/indirect UV detection for the purity determination of heparin-like oligosaccharides

In the following section, some aspects of the validation of indirect UV absorption as a *quantitative* detection method for CE, affording determination of the purity of GAG preparations, are briefly discussed. This item is dealt with in more detail in a recent study by Damm and Overklift (1994).

The experiments described below are carried out with known amounts of a highly purified batch of the synthetic pentasaccharide D-glucosamine-N,6-disulphate(α1-4)-L-iduronic acid-2-sulphate-(β1-4)-D-glucosamine-N,3,6-trisulphate-(α1-4)-D-glucuronic acid(β1-4)-D-glucosamine-N,3,6-trisulphate (Org 31550: > 99.5% pure (mol/mol) by NMR spectroscopy; residual water content 8.7% (w/w) by Karl Fischer titration). CE is carried out using 5 mM sulphosalicylic acid, pH 3.0, at a thermostatted temperature of 25 °C, applying a potential of 5 kV acrosss the capillary. The injection time is, in each case, 2 s resulting in injection of 1.8 nl of sample solution. These conditions are referred to as standard conditions.

The method shows excellent repeatability as the average migration time of Org 31550 after six successive injections of 1 pmol is 17.9 min ± 0.03 (s.e.m., $n = 6$), whereas the average integrated peak area is 17.4 ± 0.43 (s.e.m., $n = 6$). The day-to-day reproducibility of the migration time is somewhat less favourable. The average migration time found for Org 31550, analyzed on ten different days in a time interval of four weeks, keeping all analysis conditions identical as much as possible, is 17.0 ± 0.81 (s.e.m., $n = 10$), whereas the average peak area is 15.6 ± 2.44 (s.e.m., $n = 10$). Replacement of the capillary reduces the reproducibility. Therefore, a reference mixture and an internal standard should be used when peaks in a sample electropherogram are to be identified on the basis of their migration time.

In Fig. 11, the peak areas obtained after injection of several amounts (see Table 4) of Org 31550 are plotted. Taking into account all data points, except that for the highest injected amount (1960 fmol), a regression line $y = 0.015x + 0.487$ and a correlation coefficient $R = 0.996$ is obtained. When all data points are taken into account, the regression line is $y = 0.018x - 0.581$ with $R = 0.981$. It should be mentioned however, that the data points found for the three lowest concentrations (injected amounts 5, 10 and 25 fmol pentasaccharide) are close together and might be less reproducible. Taken together, it can be stated that the limit of quantitation is better than 25 fmol pentasaccharide and that the method is linear at least from 25 to 1480 fmol injected

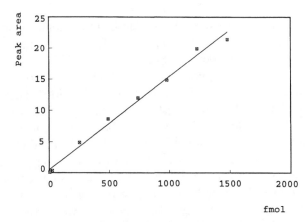

fmol

Figure 11. Peak areas vs. injected amount, obtained for various amounts of Org 31550 by CE/indirect UV detection. Peak areas are given in arbitrary units. Experimental details in Table 3. The obtained peak areas (electropherograms not shown) are compiled Table 4.

Table 4.
LOD, LOQ and linearity of CE/indirect UV detection

Stock solution (%)	pg	fmol	Area	Area %
200	3600	1960	39.43	265
150	2700	1480	21.42	144
125	2250	1230	19.93	134
100	1800	980	14.86	100
75	1350	740	12.00	81
50	900	490	8.63	58
25	450	245	4.90	33
2.5	45	25	0.33	2.2
1.0	18	10	0.26	1.7
0.5	9	5	0.07	0.5

*Four stock solutions containing 1.00 (100%), 1.25 (125%), 1.50 (150%) and 2.00 (200%) mg Org 31550 per ml Milli Q water were prepared. The 100% solution was diluted to the concentrations indicated in the Table. From each solution 1.8 nl was analyzed by CE/indirect UV detection applying the standard conditions (see text). The injected amount of Org 31550 is given in pg and fmol. The resulting peak areas are in arbitrary units (area) and are expressed relative to the peak area obtained for injection from the 100% solution (area %).

pentasaccharide. Consequently, for accurate quantitative analysis, the concentration of the pentasaccharide(s) in the sample solution should be between 0.025 and 1.50 mg/ml. More concentrated solutions may be diluted prior to analysis, whereas sample solutions containing less than 0.25 mg pentasaccharide per ml should be concentrated or, alternatively, the injection volume shoud be increased. From the above data it can be deduced that for quantitative analysis of pentasaccharides in a multicomponent mixture, e.g. main component plus contaminants, the ratio between the compounds should preferably be about

1 : 60 (w/w) or smaller. This means that the LOQ for contaminants is about 1.7% (w/w) relative to the main component. This issue is more accurately addressed in the experiment described below.

For determination of the accuracy of the method, Org 31540 is spiked with decreasing amounts of Org 31550. Prior to CE, the purity and the residual water content of Org 31540 were established to be > 98% (mol/mol) by ^{1}H-NMR spectroscopy and 19.2% (w/w) $\pm$ 0.28 (s.e.m., $n = 3$) by Karl Fischer titration, respectively. Ten mixtures of Org 31550 and Org 31540 are prepared, the total concentration of all mixtures being 1 mg pentasaccharide per ml (defined as 100%, Table 5). Analysis of a mixture containing 50% Org 31540 and 50% Org 31550 (i.e. both compounds 0.5 mg/ml) yields two baseline separated peaks at 15.57 and 16.09 min, followed by a minor peak at 16.47 min (Fig. 12). From injection of Org 31550 and Org 31540 separately (not shown), it is known that the signal at 15.57 min stems from Org 31550, whereas the signals at 16.09 and 16.47 min are derived from Org 31540 and an impurity in Org 31540, respectively. The peak areas found for Org 31540 (impurity not included) and Org 31550 are 8.49 and 7.83, respectively (Table 5). However, since the water content of Org 31550 and Org 31540 are 8.7% and 19.2%, respectively, a correction factor of 1.13 for the area found for Org 31540 must be applied, which gives a corrected peak area of Org 31540 as 8.85. Nine additional mixtures were made, gradually increasing the ratio Org

Table 5.

Accuracy of CE/indirect UV detection

Org 31540			Org 31550		
% Org 31540	Peak area	Area %	% Org 31550	Peak area	Area %
0	0	0	100	16.83	100
50	8.85	51.1	50	8.49	48.9
60	11.45	60.3	40	7.53	39.7
80	15.10	79.9	20	3.81	20.1
90	16.89	88.6	10	2.17	11.4
95	22.14	93.9	5	1.43	6.1
96	18.03	95.9	4	0.77	4.1
97	20.73	96.0	3	0.86	4.0
98	16.80	99.9	2	0.02	0.1
99	16.03	99.8	1	0.03	0.2
99.5	16.03	99.8	0.5	0.03	0.2
100	16.46	100	0	0	0

[a] Ten mixtures of Org 31540 and Org 31550 having an increasing Org 31540/Org 31550 ratio were analyzed by CE/indirect UV detection applying standard conditions. In each mixture the total pentasaccharide concentration (Org 31540 + Org 31550) was 1 mg/ml (defined as 100%). Peak areas were corrected for residual water content (details in text).

[b] % Org 31540: percentage (w/w) of Org 31540 in the Org 31540/Org 31550 mixture. Peak area: observed peak area (in arbitrary units) in CE/indirect UV analysis. Area %: observed peak area for Org 31540 or Org 31550 in CE/indirect UV analysis of a mixture of both Org compounds. The total peak area obtained for Org 31540 + Org 31550 is taken as 100%.

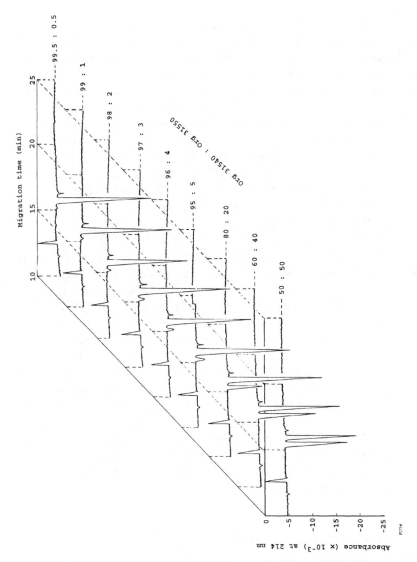

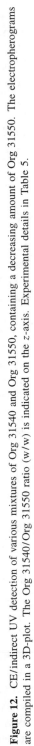

Figure 12. CE/indirect UV detection of various mixtures of Org 31540 and Org 31550, containing a decreasing amount of Org 31550. The electropherograms are compiled in a 3D-plot. The Org 31540/Org 31550 ratio (w/w) is indicated on the z-axis. Experimental details in Table 5.

31540/Org 31550. In fact, in this experiment, Org 31550 can be regarded as a contaminant in the Org 31540 preparation. As is clear from Fig. 12 and Table 5, the theoretical and observed ratios are in accordance with each other, indicating on the one hand that, using indirect UV detection, a nearly identical detector response is obtained for Org 31540 and Org 31550, and that on the other hand the presence of a small amount of Org 31550 (e.g. 9 pg or 4.9 fmol) can be detected in the presence of a large excess of Org 31540 (e.g. 1791 pg or 1036 fmol, Table 5). There is, however, a discrepancy between the limit of detection and the limit of quantitation of Org 31550 in the presence of excess Org 31540. Figure 12 demonstrates that the observed signal for Org 31550 continuously decreases, relative to the signal obtained for Org 31540, as the ratio Org 31540/Org 31550 increases. Therefore, the Org 31550/Org 31540 ratio (w/w) can be assessed in a qualitative way at least up to a ratio of 0.5/99.5, i.e. the limit of detection of Org 31550 in the presence of an excess of Org 31540 is about 0.5% (w/w). However, from Table 5 it will be clear that the limit of quantitation for Org 31550 in a mixture with Org 31540 is about 2% (w/w). This difference with the limit of detection is mainly due to the limited accuracy of the integration software, as it is evident from Fig. 12 that the detection signal does reflect the actual ratio.

The former experiment shows that closely related pentasaccharides yield a (nearly) identical detector response in the indirect detection mode and thus are registered in proportion to their relative (gravimetric) abundance. For a more general application of CE/indirect UV detection to determine the purity of GAG preparations, it is of importance to ascertain that equal amounts of GAGs widely differing in molecular mass and charge are also detected with equal sensitivity. This item is briefly addressed by injection of equal (gravimetric) amounts of Org 31550 (0.9 ng or 0.49 pmol) and δUA2S-GlcNS6S (0.9 ng or 1.35 pmol). CE/indirect UV detection yields two peaks (not shown) at 17.2 min, corresponding to Org 31550 and 20.1 min, corresponding to δUA2S-GlcNS6S having nearly identical peak areas, namely 12.48 (48.2%) and 13.43 (51.8%), respectively. This indicates that injection of equal mass amounts of different types of GAGs yield an equal detector response, enabling their quantitative analysis by CE/indirect UV detection. It is of importance to note that the detector response correlates with the gravimetric amount of the saccharide and not with the molar amount.

CONCLUSIONS AND PERSPECTIVES

Within a short time CE has become an indispensable tool in the analysis of low molecular mass heparin preparations. Due to its high resolution and mass sensitivity, CE complements or sometimes even replaces traditional HPLC methods, relieving previously existing limitations with respect to separation and sensitivity. The establishment of structural differences between heparin and heparan sulphate and the determination of the species of origin of heparin preparations exemplify the application of CE in the analysis of small amounts of natural GAGs. In combination with indirect UV detection, CE enables a reliable qualitative and quantitative analysis of the purity of sub-picomole levels of low molecular weight heparin preparations.

Current efforts to improve the reproducibility, precision of integration and uniformity of (indirect) detection, as well as the scope of the method (e.g. analysis of medium mass GAGs) will undoubtedly further refine and extend the capabilities of CE in analytical chemistry in general, and in the field of GAG analysis in particular. Especially worthwhile mentioning in this respect is the development of commercial equipment for the routine linking of CE with various types of mass spectrometry which may open the way to the accurate and definitive on-line identification of many biologically interesting compounds by CE.

ACKNOWLEDGEMENTS

The author is grateful to Prof. Dr H. Poppe (University of Amsterdam, Netherlands) and P. S. L. Janssen for the stimulating discussions, G. T. Overklift, H. Peters and M. H. J. M. Langenhuizen for the technical assistance in the CE experiments and Prof. Dr C. A. A. van Boeckel (all from N.V. Organon, Oss, Netherlands) for supplying the synthetic pentasaccharides.

REFERENCES

Albada, J., Nieuwenhuis, H. K. and Sixma, J. J. (1989). Pharmacokinetics of standard and low molecular weight heparin. In: *Heparin: Chemical and Biological Properties, Clinical Applications*, Lane, D. A. and Lindahl, U. (Eds). Edward Arnold, London, pp. 417–432.

Al-Hakim, A. and Linhardt, R. J. (1991). Capillary electrophoresis for the analysis of chondroitin sulphate- and dermatan sulphate-derived disaccharides. *Anal. Biochem.* **195**, 68–73.

Ampofo, S. A., Wang, H. M. and Linhardt, R. J. (1991). Disaccharide compositional analysis of heparin and heparan sulfate using capillary zone electrophoresis. *Anal. Biochem.* **199**, 249–255.

Bashkin, P., Doctrow, S., Klagsbrun, M., Svahn, C. M., Folkman, J. and Vlodavsky, I. (1989). Basic fibroblast growth factor binds to subendothelial extracellular matrix and is released by heparanase and heparin-like molecules. *Biochemistry* **28**, 1737–1743.

Beetz, T. and van Boeckel, C. A. A. (1986). Synthesis of an antithrombin binding heparin-like pentasaccharide lacking 6-O sulphate at its reducing end. *Tetrahedron Lett.* **27**, 5889–5892.

Blake, D. A. and McLean, N. V. (1990). High-pressure, anion-exchange chromatography of proteoglycans. *Anal. Biochem.* **190**, 158–164.

van Boeckel, C. A. A., Beetz, T., Vos, J. N., de Jong, A. J. M., van Aelst, S. F., van den Bosch, R. H., Mertens, J. M. R. and van der Vlugt, F. A. (1985). Synthesis of a pentasaccharide corresponding to the antithrombin III binding fragment of heparin. *J. Carbohydr. Chem.* **4**, 293–321.

van Boeckel, C. A. A., Lucas, H., van Aelst, S. F., van den Nieuwenhof, M. W. P., Wagenaars, G. N. and Mellema, J.-R. (1987). Synthesis and conformational analysis of an analogue of the antithrombin-binding region of heparin: the role of the carboxylate function of α-L-idopyranuronate. *Recl. Trav. Chim. Pays-Bas* **106**, 581–591.

van Boeckel, C. A. A., Beetz, T. and van Aelst, S. F. (1988). Synthesis of a potent antithrombin activating pentasaccharide: a new heparin-like fragment containing two 3-O-sulphated glucosamines. *Tetrahedron Lett.* **29**, 803–806.

van Boeckel, C. A. A., van Aelst, S. F., Beetz, T., Meuleman, D. G., van Dinther, Th. G. and Moelker, H. C. T. (1989). Structure–activity relationships of synthetic heparin fragments. Discovery of a very potent ATIII activating pentasaccharide. *Ann. N. Y. Acad. Sci.* **556**, 489–491.

Bruno, A. E., Krattiger, B., Maystre, F. and Widmer, H. M. (1991). On-column laser-based refractive index detector for capillary electrophoresis. *Anal. Chem.* **63**, 2689–2697.

Caprioli, R. M., Moore, W. T., Martin, M., de Gue, B. B., Wilson, K. and Moring, S. (1989). Coupling capillary zone electrophoresis and continuous-flow fast atom bombardment mass spectrometry for the analysis of peptide mixtures. *J. Chromatogr.* **480**, 247–257.

Carney, S. L. and Osborne, D. J. (1991). The separation of chondroitin sulfate disaccharides and hyaluron oligosaccharides by capillary zone electrophoresis. *Anal. Biochem.* **195**, 132–140.

Casu, B. (1989). Methods of structural analysis. In: *Heparin: Chemical and Biological Properties, Clinical Applications*, Lane, D. A. and Lindahl, U. (Eds). Edward Arnold, London, pp. 25–50.

Chen, C. Y. and Morris, D. (1991). On-line multichannel Raman spectroscopic detection system for capillary zone electrophoresis. *J. Chromatogr.* **540**, 355–363.

Chiesa, C. and Horváth, C. (1993). Capillary zone electrophoresis of malto-oligosaccharides derivatized with 8-aminonaphthalene 1,3,6-trisulfonic acid. *J. Chromatogr.* **645**, 337–352.

Choay, J., Lormeau, J. C., Petitou, M., Sinay, P. and Fareed, J. (1981). Structural studies on a biologically active hexasaccharide obtained from heparin. *Ann. N. Y. Acad. Sci.* **370**, 644–649.

Choay, J., Petitou, M., Lormeau, J. C., Sinay, P., Casu, B. and Gatti, G. (1983). Structure–activity relationship in heparin: A synthetic pentasaccharide with high affinity for antithrombin III and eliciting high anti-factor Xa activity. *Biochem. Biophys. Res. Commun. USA* **116**, 492–499.

Colón, L. A., Dadoo, R. and Zare, R. N. (1993). Determination of carbohydrates by capillary zone electrophoresis with amperometric detection at a copper microelectrode. *Anal. Chem.* **65**, 476–481.

Dabat, M. Espejo, J. H., Branellec, J.-F., Damm, J. B. L., Derksen, M. and van Dedem, G. W. K. (1994). Identification of the animal species of origin of purified heparins by enzymatic degradation and quantitative analysis of the degradation products. *Anal. Biochem.* (submitted).

Damm, J. B. L. and Overklift, G. T. (1994). Indirect UV detection as a non-selective detection method in the qualitative and quantitative analysis of heparin fragments by high performance capillary electrophoresis. *J. Chromatogr.* **678**, 151–165.

Damm, J. B. L., Overklift, G. T., Vermeulen, B. W. M., Fluitsma, C. F. and van Dedem, G. W. K. (1992). Separation of natural and synthetic heparin fragments by high-performance capillary electrophoresis. *J. Chromatogr.* **608**, 297–309.

Damm, J. B. L., Overklift, G. T. and van Dedem, G. W. K. (1993). Determination of structural differences in the glycosaminoglycan chains of heparin and heparan sulfate by analysis of the constituting disaccharides with capillary electrophoresis. *Phar. Pharmacol. Lett.* **3**, 156–160.

Folkman, J. and Ingber, D. E. (1989). Angiogenesis: regulatory role of heparin and related molecules. In: *Heparin: Chemical and Biological Properties, Clinical Applications*, Lane, D. A. and Lindahl, U. (Eds). Edward Arnold, London, pp. 317–334.

Foret, F., Fanali, S., Ossicini, L. and Bocek, P. (1989). Indirect photometric detection in capillary zone electrophoresis. *J. Chromatogr.* **470**, 299–308.

Gallagher, J. T., Lyon, M. and Steward, W. P. (1986). Structure and function of heparan sulphate proteoglycans. *Biochem. J.* **236**, 313–325.

Gallus, A. S. and Hirsh, J. (1976). Treatment of venous thromboembolic disease. *Semin. Thromb. Hemostasis* **2**, 291–331.

Garner, T. W. and Yeung, E. S. (1990). Indirect fluorescence detection of sugars separated by capillary zone electrophoresis with visable laser excitation. *J. Chromatogr.* **515**, 639–644.

Gorsky, A., Wasik, M., Nowaczyk, M. and Korczak-Kowalska, G. (1991). Immunomodulating activity of heparin. *FASEB J.* **5**, 2287–2291.

Grootenhuis, P. D. J. and van Boeckel, C. A. A. (1991). Constructing a molecular model of the interaction between antithrombin III and a potent heparin analogue. *J. Am. Chem. Soc.* **113**, 2743–2747.

Guo, Y. C. and Conrad, H. E. (1988). Analysis of oligosaccharides from heparin by reversed-phase ion-pairing high-performance liquid chromatography. *Anal. Biochem.* **168**, 54–62.

Guo, Y. and Conrad, H. E. (1989). The disaccharide composition of heparins and heparan sulfates. *Anal. Biochem.* **176**, 96–104.

Hallen, R. W., Shumate, C. B., Siems, W. F., Tsuda, T. and Hill Jr., H. H. (1989). Preliminary investigation of ion mobility spectrometry after capillary electrophoresis introduction. *J. Chromatogr.* **480**, 233–245.

Hjertén, S., Elenbring, K. Kilar, F., Liao, J. L., Chen, A. J. C., Siebert, C. J. and Zhu, M. D. (1987). Carrier-free zone electrophoresis, displacement electrophoresis and isoelectric focussing in a high-performance electrophoresis apparatus. *J. Chromatogr.* **403**, 47–61.

Holmer, E. (1989). Low-molecular weight heparin. In: *Heparin: Chemical and Biological Properties, Clinical Applications*, Lane, D. A. and Lindahl, U. (Eds). Edward Arnold, London, pp. 575–596.

Huang, X. H., Luckey, J. A., Gordon, M. J. and Zare, R. N. (1989). Quantitative analysis of low molecular weight carboxylic acids by capillary zone electrophoresis/conductivity detection. *Anal. Chem.* **61**, 766–770.

Jandik, P. and Jones, W. R. (1991). Optimization of detection sensitivity in the capillary electrophoresis of inorganic anions. *J. Chromatogr.* **546**, 431–443.

Lider, O., Mekori, Y. A., Miller, T., Bar-Tana, R., Vlodavsky, I., Baharav, E., Cohen, I. R. and Naparstek, Y. (1990). Inhibition of T-lymphocyte heparanase by low dose heparin leads to impaired lymphocyte traffic and reduced cellular immune reactivity in mice. *Eur. J. Immunol.* **20**, 493–499.

Lindahl, U., Bäckström, G., Höök, M., Thunberg, L., Fransson, L. A. and Linker, A. (1979). Structure of the antithrombin-binding site in heparin. *Proc. Natl. Acad. Sci. USA* **76**, 3198–3202.

Linhardt, R. J., Rice, K. G., Zohar, Z. M., Yeong, K. S. and Lohse, D. L. (1986). Structure and activity of a unique heparin-derived hexasaccharide. *J. Biol. Chem.* **261**, 14448–14454.

Linhardt, R. J., Gu, K. N., Loganathan, D. and Carter, S. R. (1989). Analysis of glycosaminoglycan-derived oligosaccharides using reversed-phase ion-pairing and ion-exchange chromatography with suppressed conductivity detection. *Anal. Biochem.* **181**, 288–296.

Linhardt, R. J., Turnbull, J. E., Wang, H. M., Loganathan, D. and Gallagher, J. T. (1990). Examination of the substrate specificity of heparin and heparan sulfate lyases. *Biochemistry* **29**, 2611–2617.

Moseley, M. A., Deterding, L. J., Tomer, K. B. and Jorgenson, J. W. (1989). Capillary zone electrophoresis/fast atom bobardment mass spectrometry: Design of an on-line coaxial continuous-flow interface. *Rapid Commun. Mass Spectrom.* **3**, 87–93.

Mosher, R. A., Dewey, D., Thormann, W., Saville, D. A. and Bier, M. (1989). Computer simulation and experimental validation of the electrophoretic behaviour of proteins. *Anal. Chem.* **61**, 362–366.

Nakajima, M., Irimura, T. and Nicolson, G. L. (1988). Heparanases and tumor metastasis. *J. Cell. Biochem.* **36**, 157–167.

Olefirowicz, T. M. and Ewing, A. G. (1990). Capillary electrophoresis with indirect amperometric detection. *J. Chromatogr.* **499**, 713–719.

O'Shea, T. J., Lunte, S. M. and LaCourse, W. R. (1993). Detection of carbohydrates by capillary electrophoresis with pulsed amperometric detection. *Anal. Chem.* **65**, 948–951.

Pangrazzi, J., Abbadini, M., Zametta, M., Naggi, A., Torri, G., Casu, B. and Donati, M. B. (1985). Antithrombotic and bleeding effects of a low molecular weight heparin fraction. *Biochem. Pharmacol.* **34**, 3305–3308.

Petitou, M., Duchaussoy, P., Lederman, I. and Choay, J. (1987). Synthesis of heparin fragments: A methyl α-pentaoside with high affinity for antithrombin III. *Carbohydr. Res.* **167**, 67–75.

Petitou, M., Lormeau, J. C. and Choay, J. (1988). Interaction of heparin and antithrombin III. The role of O-sulfate groups. *Eur. J. Biochem.* **176**, 637–640.

Ruoslathi, E. (1989). Proteoglycans in cell recognition. *J. Biol. Chem.* **264**, 13369–13372.

Ruoslathi, E. and Yamaguchi, Y. (1991). Proteoglycans as modulators of growth factor activities. *Cell* **64**, 867–869.

Silva, M. E., Dietrich, C. P. and Nader, M. B. (1976). Analyses of the products formed from heparitin sulfates by two heparitinases and a heparanase from *Flavobacterium heparinum*. *Biochim. Biophys. Acta* **437**, 129–141.

Sinay, P., Jacquinet, J.-C., Petitou, M., Duchaussoy, P., Lederman, I., Choay, J. and Torri, G. (1984). Total synthesis of a heparin pentasaccharide fragment having high affinity for antithrombin III. *Carbohydr. Res.* **132**, C5–C9.

Smith, R. D., Olivares, J. A., Nguyen, N. T. and Udseth, H. R. (1988). Capillary zone electrophoresis-mass spectrometry using an electrospray ionization interface. *Anal. Chem.* **60**, 436–441.

Thomas, D. P. and Merton, R. E. (1982). A low molecular weight heparin compared with unfractionated heparin. *Thromb. Res.* **28**, 343–350.

Thunberg, L., Bäckström, G. and Lindahl, U. (1982). Further characterization of the antithrombin-binding sequence in heparin. *Carbohydr. Res.* **100**, 393–410.

Turnbull, J. E. and Gallagher, J. T. (1988). Oligosaccharide mapping of heparan sulphate by polyacrylamide-gradient-gel electrophoresis and electrotransfer to nylon membrane. *Biochem. J.* **251**, 597–608.

Vorndran, A. E., Oefner, P. J., Scherz, H. and Bonn, G. K. (1992). Indirect UV detection of carbohydrates in capillary zone electrophoresis. *Chromatographia* **33**, 163–168.

de Vries, X. J. (1989). Analysis of heparins by size-exclusion and reversed-phase high-performance liquid chromatography with photodiode-array detection. *J. Chromatogr.* **465**, 297–304.

Walenga, J. M., Fareed, J., Petitou, M., Samana, M., Lormeau, J. C. and Choay, J. (1986). Intravenous antithrombotic activity of a synthetic heparin pentasaccharide in a human serum induced stasis thrombosos model. *Thromb. Res.* **43**, 243–248.

Wang, T. and Hartwick, R. A. (1992). Binary buffers for indirect absorption detection in capillary zone electrophoresis. *J. Chromatogr.* **589**, 307–313.

Wright Jr., T. C., Castellot, J. J., Petitou, M., Lormeau, J.-C., Choay, J. and Karnovsky, M. J. (1989). Structural determinants of heparin's growth inhibitory activity. Interdependence of oligosaccharide size and charge. *J. Biol. Chem.* **264**, 1534–1542.

Yayon, A., Klagsbrun, M., Esko, J. D., Leder, P. and Ornitz, D. M. (1991). Cell surface, heparin-like molecules are required for binding of basic fibroblast growth factor to its high affinity receptor. *Cell* **64**, 841–848.

APPENDIX — ABBREVIATIONS

CE	capillary electrophoresis
GAG	glycosaminoglycan
Gal	galactose
GlcA	glucuronic acid
GlcN	glucosamine
GlcNAc	N-acetylglucosamine
GlcNCOEt6S	glucosaminecarboxyethyl-6-sulphate
GlcNS	glucosamine-N-sulphate
GlcNS3S	glucosamine-N,3-disulphate
GlcNS6S	glucosamine-N,6-disulphate
HPAEC	high performance anion exchange chromatography

IdoA	iduronic acid
IdoA2S	iduronic acid-2-sulphate
s.e.m.	standard error of the mean
δUA	4(5)-unsaturated uronic acid
δUA2S	4(5)-unsaturated uronic acid-2-sulphate
δUA2S-GlcNAc	4(5)-unsaturated uronic acid-2-sulphate β1-4 N-acetylglucosamine

Progress in HPLC-HPCE, Vol. 5, pp. 355–382
H. Parvez *et al.* (Eds)
© VSP 1997.

Chiral separation by HPCE with oligosaccharides: Application to acidic compounds

A. D'HULST and N. VERBEKE

*Clinical Pharmacy, Catholic University of Leuven, Gasthuisberg,
B-3000 Leuven, Belgium*

INTRODUCTION

The present paper deals with the possibilities of capillary electrophoresis (CE) for direct enantioselective analyses, as illustrated by its application to acidic compounds. Maltodextrins or corn syrups, i.e. crude maltooligosaccharide mixtures obtained from acidic/enzymatic hydrolysis of corn starch, were added to the background electrolyte as chiral discriminating agents. Some specific mixtures allowed highly efficient and selective chiral separations. The chiral discriminative capacity of the maltooligosaccharides was shown to be dependent on both the qualitative and quantitative composition of the brand used. Enantioselectivity was observed towards a widely varying series of acidic racemic compounds, including 2-arylpropionic acid non-steroidal anti-inflammatory drugs, coumarinic anti-coagulant drugs, a cholecystographic agent, a benzodiazepine and phenoxypropionic acid-type herbicidal substances. Direct chiral CE with maltodextrins as enantioselective electrolyte modifiers also provided a method of choice to measure in a very accurate way enantiomeric excess. It is suggested that direct chiral CE in general may develop into a valuable alternative to currently available chiral HPLC methods for this purpose.

Chirality plays a major role in virtually all disciplines related to biology. The relatively recent and increasing awareness of the importance of stereochemical specificities in applied biological sciences has also created the need for stereoselective analysis methods, especially in the field of pharmacology-pharmacokinetics-toxicology and the agricultural/environmental sciences. Whereas in the past little if any attention was paid to the question of possible enantiomeric differences with respect to activity, disposition or toxicity when racemic drugs were filed for registration, regulatory offices now require stereochemical data for new candidate drugs. Indeed, numerous synthetic drugs have been developed and marketed as racemates and were subsequently shown to display stereoselectivity in their pharmacokinetic, pharmacodynamic or toxicicological behaviour. The group of the 2-arylpropionic acid (2-APA) non-steroidal anti-inflammatory drugs (NSAIDs) is one of the most illustrative examples in this respect: except for naproxen, they are all marketed as racemic mixtures, yet the anti-inflammatory activity entirely resides in the S(+)-enantiomer. However, in vivo inversion from the inactive R(−)-enantiomer to the active S(+)-antipode may occur, depending on the compound and the species studied (Caldwell *et al.*, 1988). The chemically closely related group of the 2-phenoxypropionic acid herbicides (HERBs) shows an analogous enantiospecific bioactivity: the herbicidal activity is due to the R(+)-eutomers, whereas the S(−)-distomers are considered inactive and therefore environmental ballast (Dicks *et al.*, 1985). Another important group of drugs showing enantiomeric differences in both pharmacokinetics and pharmacodynamics are the coumarinic anticoagulant drugs (COUMs): the metabolism is highly stereospecific, resulting in metabolites of different activity (Pohl *et al.*, 1976). Stereospecificity, although demonstrated, has been less extensively studied in the case of some benzodiazepine drugs (Möhler and Okada, 1977) and cholecystographic agents (Cooke and Cooke, 1983).

Enantioselective analysis methods have been developed for all of the above mentioned compounds, using different techniques. Routinely used enantioselective separation methods now include chiral gas chromatography (GC), high-performance liquid chromatography (HPLC) and supercritical fluid chromatography (SFC). Preparative as well as analytical separations can be performed. Chiral HPLC has been used most extensively. Stereoselectivity in HPLC is obtained through different approaches, i.e. either by derivatization of the analyte with a chiral compound and subsequent conventional achiral separation of the diastereoisomers or by incorporating chiral modifiers into the HPLC mobile phase or by using chiral stationary phases. Most chiral HPLC research now focuses on the development and the use of chiral phases. Chiral column filling materials are grouped in ligand-exchange, Pirkle-type, cavity, protein and carbohydrate stationary phases, the majority of which have been successfully applied to the separation of the above mentioned drug and herbicidal substances (De Vries and Völker, 1989; Noctor *et al.*, 1991; Pirkle and Welch, 1991; Tambuté *et al.*, 1991; Fujima *et al.*, 1993). Although high selectivities can be achieved, many columns suffer from one or more major drawbacks, such as lack of general applicability (causing an important cost-increase due to investments in an array of columns necessary to cover a sufficiently wide application range), unpredictable column enantioselectivity and performance (causing a sometimes lengthy method development time) or column instability.

Capillary electrophoresis (CE) is one of the most recent developments in analytical separation techniques and offers great potential. Not only is the theoretically achievable efficiency extremely high as compared to HPLC — 500 000 theoretical plates are very easily obtained in CE whereas only highly efficient HPLC columns reach 15 000 plates — the range of compounds which can be analyzed is also virtually unlimited. Moreover, unlike HPLC, CE does not require major hardware modifications to switch from small ion analysis to high molecular weight compound analysis. In addition, although separations are always based on true or apparent differences in charge density, neutral compounds can still be separated. Using the micellar electrokinetic capillary chromatography (MECC) mode, i.e. performing CE in micellar solutions, neutral solutes can be given an apparent charge and consequently display an apparent electrophoretic mobility. As expected, the high efficiency of CE was rapidly exploited in chiral separations. As early as four years after the first description of CE by Jorgenson and Lukacs (1981), a paper on the direct separation of amino acid enantiomers was presented, using the optically active Cu-L-histidine complex as chiral selector (Gassman *et al.*, 1985). Several reports have appeared since, using a variety of chiral selection approaches. Attempts have been made to mimic the high selectivity of chiral stationary phases in HPLC and GC through coating or derivatization of the capillary wall with a chiral selector, such as cyclodextrins (Armstrong *et al.*, 1993). The most often used approach however is the addition of one or more chiral selectors to the background electrolyte. Micellar electrolytes containing either non-chiral surfactants functionalized with optically active compounds or chiral surfactants were shown to resolve enantiomers without prior diastereomeric derivatization (Otsuka and Terabe, 1990; Dobashi *et al.*, 1989). Cyclodextrins are at present the most frequently used chiral electrolyte modifiers. Initially, cyclodextrins were used in their native form and organic solvents were frequently added to alter the selectivity and expand their application range. Subsequently, numerous derivatives were synthesized, providing an even broader choice of selectivities (Nishi *et al.*, 1991). Different selectivities were also obtained through several combinations of surfactants and cyclodextrins (Sepaniak *et al.*, 1992). Recent comprehensive reviews on chiral separations in CE were presented by Bereuter (1994) and by Terabe *et al.* (1994).

It is well known that cyclodextrin molecules are capable of both hydrophilic and hydrophobic interactions. In the case of linear saccharide polymers however, information on conformation and behaviour in aqueous solutions and potential steric interaction with small organic molecules is sparse. The present study describes the applicability of a selection of linear saccharide polymers in chiral CE.

Maltodextrins and corn syrups are complex malto-oligo- and polysaccharide mixtures, i.e. mixtures of $\alpha(1\text{-}4)$ linked D-glucose oligo- and polymers obtained through a partial acid and/or enzymatic hydrolysis of corn starch. Depending upon the manufacturing procedure and the extent of hydrolysis, the average degree of polymerization (DP) of the maltooligosaccharides obtained may differ and is indirectly measured through the DE value, the dextrose equivalent, defined as the percentage reducing sugars calculated as glucose on a dry substance basis, with glucose given the value 100. Consequently, unhydrolyzed starch will have a DE value of nearly zero. Corn starch subjected to extensive hydrolysis will yield a product composed of primarily maltooligosaccharides in the lower molecular weight range, thus having a low average DP and a high DE value. Maltooligosaccharide mixtures with DE values of 20 or higher are generally referred to as corn syrups, whereas maltodextrins are characterized by a DE value lower than 20. The degree of hydrolysis may thus be fairly accurately monitored through the DE value. However, the DE value gives no indication on the qualitative composition of the product obtained. Indeed, scission of a terminal glucose unit will produce an equal rise of the DE value as a mid-chain splicing of the same polysaccharide, but yield a product with, most likely, entirely different physicochemical properties. So, although the average molecular weight of starch hydrolysates (as evaluated by the DE value) may be identical, the actual molecular weight distribution of the maltooligosaccharides may vary widely, as is observed for instance by gel permeation chromatography or anion exchange HPLC analysis. In addition to the highly variable maltooligosaccharide composition of the maltodextrins and corn syrups caused by different manufacturing procedures, a substantial contribution to variation may also be expected from the origin of the starch used as starting product. Post-hydrolysis modifications (e.g. hydrogenation) may add further variation. Table 1 lists the maltodextrins and corn syrups used in this study (with their respective DE values and the relative contents of maltooligosaccharides higher than maltose).

The general applicability of maltodextrins as chiral selectors was investigated through a chiral selectivity screening of a series of maltodextrin and corn syrup modified electrolytes towards a broad range of acidic drugs. Close to 20 different compounds were tested (Fig. 1), belonging to different chemical and pharmacological groups. Eight racemic 2-APAs, seven of them currently used as NSAIDs and marketed as racemates, were tested and compared with three compounds belonging to the closely related group of the 2-phenoxypropionic acid herbicides, one benzodiazepine and one cholecystographic agent. All these compounds have the acidic carboxyl-group next to the chiral centre in common and differed in the distance and/or nature of the bulky substituent to the chiral carbon atom. Within the group of the COUMs, differing from the former groups in the nature of the acidic function and its distance from the chiral centre, the importance of identical, minor substituent changes in closely related compounds with respect to the extent of the chiral discrimation of the maltodextrins was studied. In addition, the COUMs were run through a series of experiments investigating the effects of chiral selector and background electrolyte modifications. Furthermore, the question of improvement on UV detection limits by wavelength reduction was

Table 1.

DE value and contents of maltooligosaccharides of DP 3 or higher (relative amount in % w/w) of maltodextrins and corn syrups used

Maltodextrin/corn syrup	DE	> DP 2
Maltrin 200	20	89.8
Maltrin 150	15	94.8
Maltrin 100	10	96.8
Maltrin 040	5	99.5
Maldex 30	30	84.0
Maldex 20	20	88.2
Maldex 15	15	90.2
Mylose HDE	60	25
Mylose HM	44	43
Mylose STD33	33	75
Mylose STD40	40	69.5
Mylose STD42	42	68
Mylose STD45	45	67
Glucidex 21	21	90
Glucidex 19	19	91
Glucidex 17	17	93
Glucidex 12	12	97
Glucidex 6	6	98.5
Glucidex 2	2	99
Lycasin 80/85	0	40

addressed. Finally, the possibility of enantiomeric excess determinations was studied.

EXPERIMENTAL

Chemicals

A range of maltodextrin and corn syrup series — kind gifts from different suppliers — were used: Maltrin® maltodextrins (The Grain Processing Corporation, Muscatine, Iowa, USA), Maldex® maltodextrins and Mylose® corn syrups (Amylum, Aalst, Belgium), Glucidex® maltodextrins and Lycasin® corn syrup (Roquette, Lestrem, France). A standard low molecular weight (DP 2 to 6) maltooligosaccharide mixture was obtained from Merck (Darmstadt, Germany). β-, dimethyl-β- and hydroxypropyl-β-cyclodextrin were obtained from Sigma (Deisenhofen, Germany), Avebe (Veendam, The Netherlands) and Janssen Drug Delivery Systems (Beerse, Belgium) respectively.

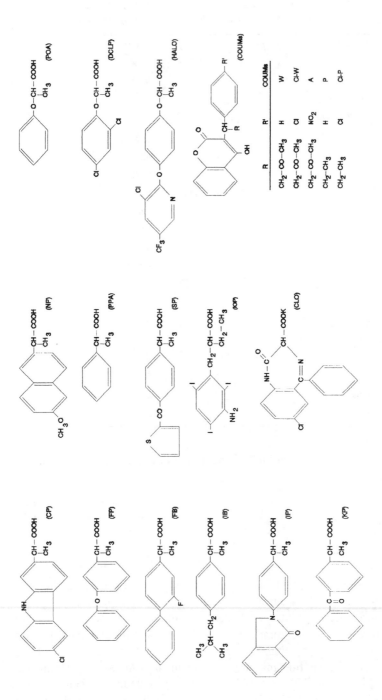

Figure 1. Structure of compounds used. NSAIDs: carprofen (CP), fenoprofen (FP), flurbiprofen (FB), ibuprofen (IB), indoprofen (IP), ketoprofen (KP), naproxen (NP), 2-phenylpropionic acid (PPA), suprofen (SP). COUMs: warfarin (W), p-chlorowarfarin (Cl-W), phenprocoumon (P), p-chlorophenprocoumon (Cl-P), acenocoumarol (A). HERBs: 2-phenoxypropionic acid (POA), dichlorprop (DCLP), haloxyfop (HALO). Iopanoic acid (IOP). Potassium clorazepate (CLO).

Racemic drug substances and analogues included flurbiprofen and ibuprofen (gifts from Boots, Nottingham, UK), ketoprofen (gift from Rhône-Poulenc, Paris, France), carprofen, fenoprofen, indoprofen, 2-phenylpropionic acid and suprofen (obtained from Sigma, Deisenhofen, Germany), warfarin and *p*-chlorowarfarin (Sigma, Deisenhofen, Germany), phenprocoumon and *p*-chlorophenprocoumon (gift from Roche, Basel, Switzerland), acenocoumarol (gift from Ciba-Geigy, Basel, Switzerland), potassium clorazepate (gift from Sanofi-Winthrop, Brussels, Belgium), and iopanoic acid (Sigma, Deisenhofen, Germany). Pure enantiomeric substances included the S(+)-enantiomers of ibuprofen (gift from Profarma, Beerse, Belgium) and naproxen (gift from UCB, Brussels, Belgium) and the R(−)enantiomer of ibuprofen (97.8% enantiomeric purity, obtained from Research Biochemicals Inc., Natick, MA, USA). Analytical standards of racemic herbicides included 2-phenoxypropionic acid and haloxyfop (obtained from Labor Dr. Ehrenstorfer, Augsburg, Germany) and dichlorprop (obtained from the Institute of Organic Industrial Chemistry, Warsaw, Poland).

Other chemicals used were: phosphoric acid and ammonia (obtained from UCB, Leuven, Belgium), sodium dihydrogen phosphate, tris(hydroxymethyl) aminomethane, potassium hydroxide and sodium hydroxide (obtained from Merck, Darmstadt, Germany) and lithium hydroxide (obtained from Sigma, Deisenhofen, Germany), HPLC-grade acetonitrile, ethanol, methanol (obtained from Carlo Erba, Milano, Italy) and isopropanol (obtained from Labscan, Dublin, Ireland). All compounds were used without further purification.

CE operation

Experiments were performed with a Waters Quanta 4000 CE system. Detection occurred with a fixed wavelength UV detector equipped either with a zinc lamp and a 214 nm filter or a mercury lamp and 185 nm filter and window. The system was operated at a constant voltage (30 kV) using normal polarity mode. A fused silica capillary of 50 µm internal diameter and 60 cm length was used. Capillaries were stored overnight filled with water. Each day operation was started by a vacuum purge with NaOH 0.5 M followed by water. When changing electrolytes, the capillary was subjected to an electroosmotic purge following a vacuum purge with the new electrolyte. All runs were preceded by a 3 min purge with the electrolyte used. Samples were introduced by gravity induced siphoning ($\Delta H = 10$ cm). Data were collected through Waters Maxima Software. All sample solutions were prepared in water obtained from a Milli-Q water purification system (Millipore, Bedford, Massachusetts, USA) or HPLC-grade methanol or acetonitrile. Sample concentrations were varied according to detection possibilities (214 versus 185 nm detection) and are mentioned in legends of figures and tables. All CE running buffers were freshly prepared in Milli-Q water, filtered and degassed immediately prior to use. The sodium azide supplemented electrolytes were filtered after preparation and stored overnight refrigerated.

Before use these electrolytes were allowed to warm up to room temperature and degassed. Electrolyte composition is given in legends of figures and tables.

Calculations

In CE, as in HPLC, separation performance is most often measured by the resolution factor R_s. However, due to, for example, non-compliance with approximating conditions, the obtained R_s figures do not always accurately describe the actual separation obtained. We have therefore introduced a relative chiral separation (RCS) factor and a percentage chiral separation (%CS) factor, in order to compare the efficiency and the extent of the enantiomeric resolving capacity of oligosaccharide-modified electrolytes and to discern more easily declining or improving trends of chiral discrimination upon varying electrolyte composition (D'Hulst and Verbeke, 1992, 1994). Briefly, the RCS factor was defined as:

$$RCS = (\Delta T - 2w)/MT\, 100, \qquad (1)$$

with ΔT the time between peak start and peak end of the first and second enantiomeric peaks respectively, w the mean peak width (measured at peak base) and MT the mean migration time. Complete resolution of an enantiomeric pair is reflected in a positive value of the RCS factor. For enantiomers which are not separated to the baseline, a negative value is obtained. As the RCS factor is a measure of overall separation efficiency, more positive RCS values are obtained either through smaller peak widths or shorter migration times. Alternatively, increasing peak overlap as well as decreasing migration times of incompletely resolved peaks may yield a more negative RCS value.

When baseline separation is achieved, mean peak width at peak base can easily be measured. Assuming the peak shapes of enantiomers in chiral CE are identical — an assumption which seems to hold when comparing, for example, peak width and peak height for baseline separated peaks, even when peaks are not symmetrical — the mean peak width at peak base for incompletely resolved enantiomers can be estimated by subtracting the difference in migration time (ΔMT) from ΔT, the time between peak start and peak end of the first and the second enantiomer peaks respectively:

$$w = \Delta T - \Delta MT. \qquad (2)$$

When calculating the RCS values for baseline separated peaks using either the measured or the approximated (as outlined above for incompletely resolved peaks) mean peak width, nearly identical values were obtained.

The percent chiral separation (%CS) relates absolute chiral separation ($\Delta T - 2w$) to peak width and therefore measures the extent of peak separation:

$$\%CS = \left(1 + (\Delta T - 2w)/w\right)100. \qquad (3)$$

%CS values of 100 or higher will thus be indicative of baseline separated peaks, while incompletely resolved peaks will be characterized by %CS values between 0 (100% co-migration) and 100 (baseline separation). The approximating determination of peak width of partially separated peaks according to equation (2) proved to be a workable alternative to extrapolations from peak width at half height, at least in CE, as the calculated %CS is consistent with the visually estimated peak overlap, even when up to 80% co-migration of enantiomers is observed.

RESULTS

Maltodextrin modified electrolytes as chiral selectors towards acidic compounds

The stereoselectivity of maltodextrins as chiral electrolyte modifiers was studied using a range of acidic compounds: eight racemic non-steroidal antiinflammatory drugs (NSAIDs), three coumarinic anticoagulant drugs (COUMs) and two of their chlorinated derivatives, three phenoxypropionic herbicides, one benzodiazepine and one iodinated cholecystographic agent. Since all compounds were acidic, their pK_a values ranging from 3.5 to 5.8, the background electrolyte was buffered at pH 7.5, so all analytes were in their dissociated form. As the electroosmotic flow (EOF) was not suppressed and normal polarity mode of the voltage supply was used, all analytes migrated after the EOF. Except for ketoprofen and the NSAID-like compound 2-phenylpropionic acid, all racemic NSAIDs were resolved using a Glucidex2 modified electrolyte, achieving separation to the baseline for most compounds (Fig. 2a–c). Both warfarin and phenprocoumon as well as their chlorinated derivatives were baseline separated using the same electrolyte, whereas acenocoumarol was not (Fig. 3). Of the three herbicides tested, only the parent compound, 2-phenoxypropionic acid, was not resolved (Fig. 4). Clorazepate (Fig. 5) and iopanoic acid (Fig. 6) enantiomers were equally baseline separated. In addition, an increased non-chiral selectivity of Glucidex2 modified electrolytes was observed as compared to the background electrolyte: co-migration was considerably higher when running samples containing either all the NSAIDs or COUMs listed in Fig. 1, in background electrolyte, as illustrated in Fig. 7a–b. As could be expected from altered selectivity, migration order was also altered upon addition of Glucidex2. When running the combined NSAID sample in background electrolyte, indoprofen migrated shortly after the neutral marker, thus showing the lowest charge density, which was consistent with its relatively high pK_a value. According to the electropherogram of the same sample run in Glucidex2 modified electrolyte (Fig. 2a), carprofen was now the compound with the lowest apparent charge density, indicating a strong interaction with the maltooligosaccharides.

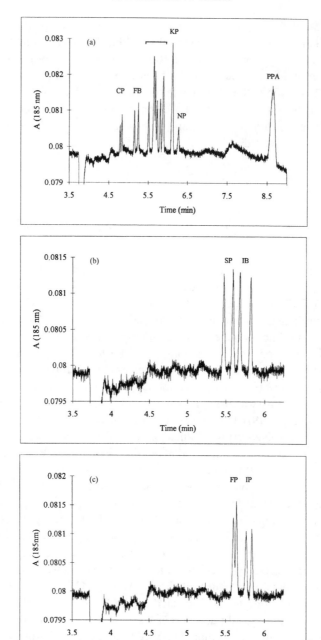

Figure 2. Electropherograms representing chiral separation of NSAIDs. (a) Sample containing all the NSAIDs listed in Fig. 1. (b) Suprofen and ibuprofen. (c) Fenoprofen and indoprofen. Electrolyte: Glucidex2, 3% – Tris phosphate 10 mм – pH 7. Sample concentration 10 µg/ml. Migration order: (a) carprofen (CP), flurbiprofen (FB), peak cluster suprofen-fenoprofen-ibuprofen-indoprofen, ketoprofen (KP, not resolved), S(+)-naproxen (NP), 2-phenylpropionic acid (PPA, not resolved); (b) suprofen (SP) and ibuprofen (IB); (c) fenoprofen (FP) and indoprofen (IP).

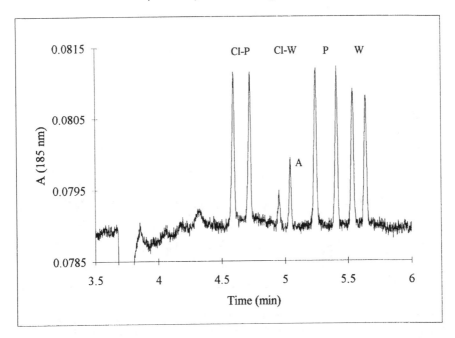

Figure 3. Electropherogram representing chiral separation of COUMs. Electrolyte: Glucidex2, 3% – Tris phosphate 10 mM – pH 7. Sample concentration 10 μg/ml. Order of migration: *p*-chlorophenprocoumon (Cl-P), *p*-chlorowarfarin (Cl-W) with second enantiomer co-migrating with acenocoumarol (A, not resolved), phenprocoumon (P) and warfarin (W).

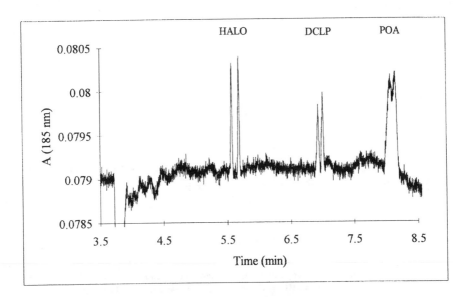

Figure 4. Electropherogram representing chiral separation of HERBs. Electrolyte: Glucidex2, 3% – Tris phosphate 10 mM – pH 7. Sample concentration 10 μg/ml. Order of migration: haloxyfop (HALO), dichlorprop (DCLP), 2-phenoxypropionic acid (POA).

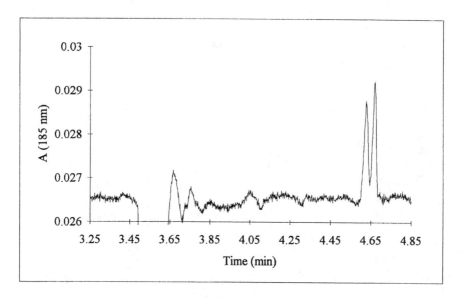

Figure 5. Electropherogram representing chiral separation of iopanoic acid (IOP). Electrolyte: Glucidex2, 2.5% – Tris phosphate 10 mM – pH 7. Sample concentration 25 μg/ml.

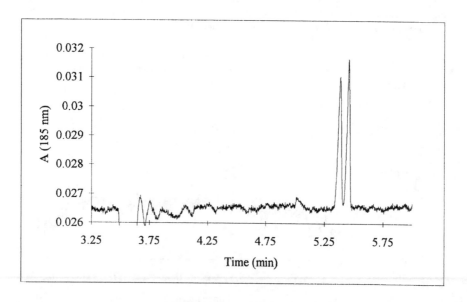

Figure 6. Electropherogram representing chiral separation of clorazepate (CLO). Electrolyte: Glucidex2, 2.5% – Tris phosphate 10 mM – pH 7. Sample concentration 25 μg/ml.

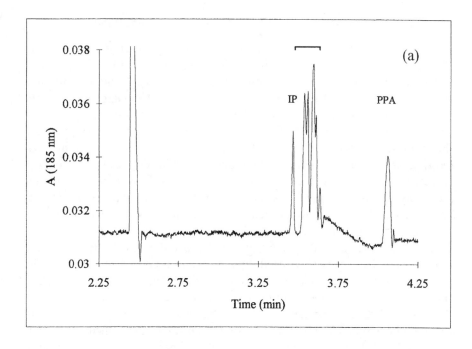

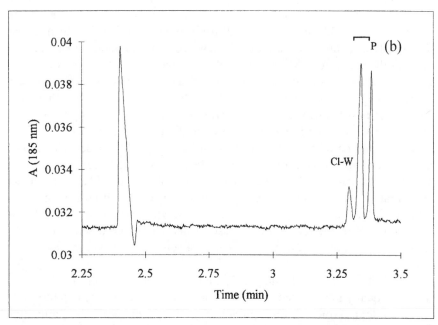

Figure 7. Electropherograms representing non-chiral separation of racemic NSAIDs and COUMs. Electrolyte: Tris phosphate 10 mM – pH 7. Sample concentration 10 μg/ml. Migration order: (a) indoprofen (IP), peak cluster carprofen-ketoprofen-suprofen-fenoprofen-ibuprofen-flurbiprofen-naproxen, 2-phenylpropionic acid (PPA); (b) p-chlorowarfarin (Cl-W), central peak representing co-migrating warfarin + p-chlorophenprocoumon + acenocoumarol, phenprocoumon (P).

Effect of qualitative and quantitative chiral selector modifications

Various maltodextrins and corn syrups covering a range of DE values from 2 to 60 (Table 1), were screened as chiral selectors towards three COUMs. Nearly all electrolyte modifiers allowed complete resolution of phenprocoumon and warfarin enantiomers, while chiral separation of acenocoumarol was not observed. In Figs 8a and 8b, electropherograms (with corresponding plots of the chiral separation factors) of phenprocoumon enantiomeric separations are shown using maltodextrins of the Maltrin and Glucidex series as chiral modifiers. Chiral resolution and efficiency improved with decreasing Maltrin DE value (Fig. 8a) — as reflected by the increasing %CS and RCS values respectively — and this trend was observed within each maltodextrin or corn syrup brand for all compounds tested. The effect of the DE value on migration times was less extensive. Increasing the concentration of the chiral electrolyte modifier also increased the extent of chiral separation, as illustrated by the recordings of Glucidex12 mediated separations, with Glucidex12 added to the background electrolyte at concentrations ranging from 1 to 20% (Fig. 8b). However, at higher maltodextrin modifier concentrations, migration times also showed a substantial increase, as did background absorption and noise. In general, Glucidex maltodextrins in the lower DE range performed best: complete resolution of warfarin and phenprocoumon enantiomers was observed in most instances, while acenocoumarol never showed any separation at all. In addition, the low DE Glucidex maltodextrins allowed for lower concentrations of electrolyte modifier to be used, thereby reducing migration times and increasing the efficiency. Moreover, as the transparency of the electrolyte was increased at lower modifier concentrations, detection sensitivity was increased, as shown in Fig. 9, left panel. Although higher peak areas are also observed when migration velocity is reduced, which is achieved by simply lowering the electric field strength, the increase of peak area is merely a result of the expanded time base and does not reflect a better signal-to-noise ratio, as can be seen from the constant values of peak height and time-corrected peak area (Fig. 9, right panel). At reduced background absorption, i.e. in the lower concentration range of chiral electrolyte modifier, increased peak heights are indicative of a true increase in detection sensitivity and largely override the negative effect of the reduced time base, which can be observed through the negative slope of the time-corrected peak area concentration course (Fig. 9, left panel).

The effect of ageing of Glucidex2 modified electrolytes on chiral separation was equally investigated. Stock solutions were supplemented with sodium azide in order to avoid microbial growth. Chiral separation as well as migration times were unaffected by the addition of sodium azide and were fairly constant over a two-week period of time as shown in Table 2.

Some maltodextrin analogues were also screened. With Lycasin, a hydrogenated corn syrup, as electrolyte modifier no chiral separation was observed. Using a standard DP 2 to 6 maltooligosaccharide mixture at high concentrations,

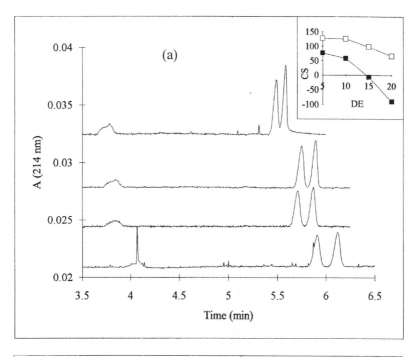

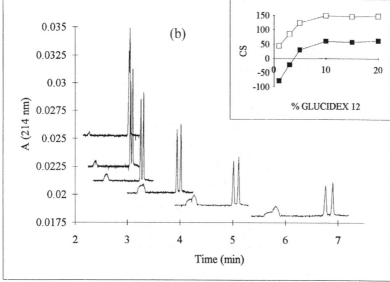

Figure 8. Electropherograms representing chiral separation of phenprocoumon in various maltodextrin modified electrolytes: effect of maltodextrin DE value and concentration on selectivity and efficiency. (a) Maltrin (5%) modified electrolytes with DE values decreasing from top to bottom trace (DE 20, 15, 10 and 5). (b) Glucidex12 modified electrolytes with concentration increasing from top to bottom trace (1, 3, 5, 10, 15 and 20%). Insets: %CS (open markers) and RCS (x100, closed markers) as a function of maltodextrin DE value and concentration respectively. Background electrolyte: sodium phosphate buffer (10 mM – pH 7.5). Sample concentration 50 µg/ml.

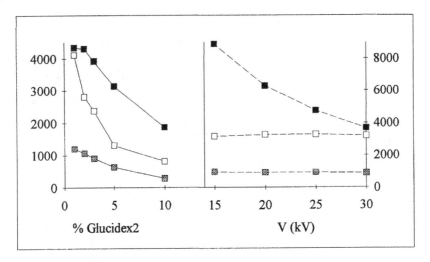

Figure 9. Quantitation dependence on chiral modifier concentration and electric field strength. Peak area, time corrected peak area and peak heigth recorded after an identical sample load of warfarin as a function of Glucidex2 concentration (%) — at 30 kV constant voltage — and as a function of applied voltage (V, in kV) — at fixed 3% Glucidex2 concentration. Electrolyte: Glucidex2-sodium phosphate, 10 mM – pH 7.5. Sample concentration 50 µg/ml. Symbols: peak area, closed markers; time corrected peak area, shaded markers and peak height, open markers.

Table 2.
Chiral separation of ibuprofen by Glucidex2 modified electrolyte. %CS and migration time (MT, in minutes) as a function of electrolyte ageing. Background electrolyte: sodium phosphate 7.43 mM – sodium azide 3 mM – pH 7

Day	%CS (RSD)	MT (RSD)
0	203 (6.8)	7.3 (0.7)
1	176 (3.9)	6.6 (1.7)
2	191 (4.4)	7.0 (2.0)
7	170 (7.3)	6.5 (0.9)
8	198 (3.1)	6.6 (0.3)
9	167 (6.8)	6.5 (0.9)
12	198 (4.2)	6.5 (0.2)

partial separation of acenocoumarol could be observed. However, this modifier was unable to resolve phenprocoumon and warfarin enantiomers. When both a Glucidex — allowing chiral separation of warfarin and phenprocoumon — and the maltooligosaccharide mixture were added to the background electrolyte, chiral separation of acenocoumarol was completely lost, while warfarin and phenprocoumon showed reduced resolution as compared to the pure Glucidex modified electrolyte (Figs 10a to 10c).

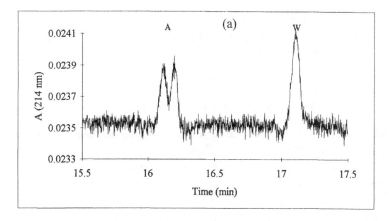

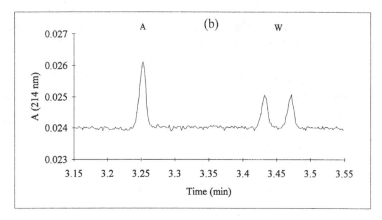

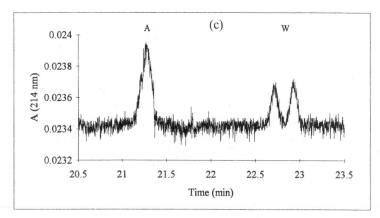

Figure 10. Electropherograms representing chiral separation of warfarin (W) and acenocoumarol (A) in maltodextrin modified electrolytes: effect of maltooligosaccharide composition. Background electrolyte (sodium phosphate buffer, 10 mM – pH 7.5) was supplemented with either a high concentration (40%) of a standard DP 2 to 6 maltooligosaccharide mixture (a), 6.7% of Glucidex12 (b) or both (c). %CS values were 81 (A), 138 (W) and 100 (W) respectively. Sample concentration 50 µg/ml.

Cyclodextrins were equally tested as chiral selectors towards COUMs. β-Cyclodextrin, dimethyl-β-cyclodextrin or hydroxypropyl-β-cyclodextrin were added to the background electrolyte, but in no instance was chiral separation observed.

Effect of background electrolyte cation on chiral separation

In order to assess the importance of the background electrolyte, several phosphate-based buffers were screened. Phosphoric acid was neutralized with either Tris, ammonia, or the hydroxides of sodium, potassium or lithium. A negligable to substantial influence of the buffer cation on chiral separation could be observed, depending on the compound studied. A definitely superior enantioselectivity towards most compounds studied was observed in Tris phosphate background electrolyte. Substituting Tris for lithium reduced chiral resolution and, in the case of flurbiprofen, even caused loss of baseline separation, as shown in Fig. 11. Comparable effects were observed for warfarin, suprofen, haloxyfop and clorazepate. For fenoprofen enantiomers, which were partially separated using Tris phosphate as a background electrolyte, chiral separation was completely lost. Ibuprofen chiral separation was only slightly affected

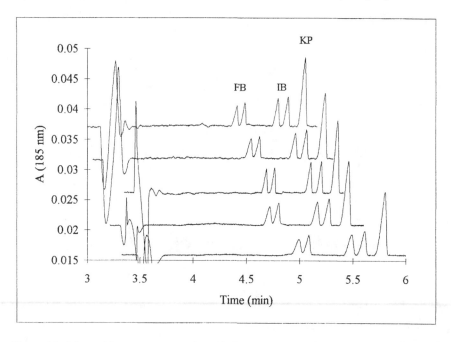

Figure 11. Electropherograms representing chiral separation of NSAIDs in various Glucidex2 modified phosphate buffers: effect of buffer cation. Samples containing flurbiprofen (FB), ibuprofen (IB) and ketoprofen (KP) were run in phosphate-based electrolyte (10 mM – pH 7) containing 2.5% chiral modifier with buffer cation varying from top to bottom trace: NH_4^+, K^+, $Tris^+$, Na^+ and Li^+. Sample concentration 25 μg/ml.

upon varying electrolyte cation (Fig. 11). In addition to their influence on the extent of chiral separation, background electrolyte cations also showed a marked and consistent effect on migration times: ammonium phosphate based buffers allowed the fastest separations, followed by potassium, Tris, sodium and lithium phosphates.

Influence of organic modifiers on chiral separation

Maltodextrins were used in combination with organic modifiers such as methanol, ethanol, isopropanol or acetonitrile. Chiral separation as well as overall analyte mobility decreased upon addition of acetonitrile as organic modifier, as illustrated in Fig. 12. The addition of acetonitrile caused a substantial reduction of conductivity — the measured current dropped by 40% over a concentration range of 0 to 25% — which was parallelled by decreased mobilities of both EOF and warfarin. In addition, when comparing the migration time of warfarin with the neutral marker (EOF signal), a

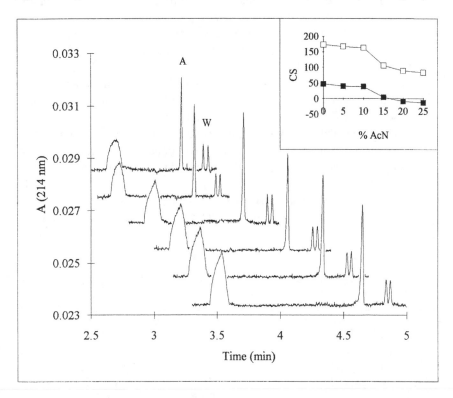

Figure 12. Electropherograms representing chiral separation of warfarin (W) and acenocoumarol (A) in maltodextrin modified electrolytes: effect of organic modifier addition. Acetonitrile was added to chiral electrolyte consisting of 7.5% Glucidex12 in sodium phosphate buffer (10 mM – pH 7.5) — with concentration increasing from top to bottom trace (0, 5, 10, 15, 20 and 25%). Inset: %CS (open markers) and RCS ($\times$100, closed markers) as a function of acetonitrile concentration. Sample concentration 20 µg/ml.

gradual rise in the migration time ratio could be observed, indicating an increased electrophoretic mobility of the analytes and hence a higher apparent charge density. Analogous observations were made when using the lower alcohols as organic modifiers. Conductivity dropped by 45 to 55% over a concentration range of 0 to 20% when methanol, ethanol or isopropanol was used, causing an even more pronounced effect on mobilities. When compared to acetonitrile, chiral selectivity was slightly less affected by methanol, while ethanol and isopropanol were considerably more deleterious to separation.

Improving sensitivity limits by low UV detection in CE

Detection of UV absorption at 214 and 185 nm was compared. As shown in Table 3, a moderate to substantial improvement on detection limits can be expected by reducing the wavelength of detection. The importance of reducing detection limits is of special interest for enantiomeric excess determinations, as shown for ibuprofen. At 214 nm, sample overloading was needed in order to reach detectable levels of the minor enantiomer, however, at the expense of baseline separation, thereby disabling accurate quantitation. At 185 nm, the sample load could be substantially reduced, so the minor and major enantiomers could be baseline separated. As illustrated in Fig. 13, enantiomeric excess determinations with either S(+)- or R(−)-ibuprofen as the minor component are possible. Addition of increasing amounts of the optical antipode to a supposedly pure enantiomer yielded an excellent correlation between peak area of the minor component and the enantiomeric concentration ratio (Fig. 14). The correlation obtained for R(−)-ibuprofen added as a contaminant to the S(+)-ibuprofen enantiomer was 0.9998. In the reversed case, the correlation factor was 0.9861. The enantiomeric excess in the pure R(−)-ibuprofen sample was found to be 98%, which was in excellent agreement with the suppliers' specifications.

Table 3.

UV detector response in chiral CE with Glucidex2 modified electrolyte as a function of detection wavelength, given as the peak area ratio at 185 and 214 nm

Compound	185/214 ratio	Compound	185/214 ratio
NSAIDs		HERBs	
Carprofen	0.55	Haloxyfop	2.3
Fenoprofen	1.06	Dichlorprop	1.2
Flurbiprofen	1.59	COUMs	
Ibuprofen	2.72	Warfarin	1.56
Indoprofen	1.96	p-Chloro-warfarin	1.23
Ketoprofen	1.80	Phenprocoumon	1.56
Naproxen	0.23	p-Chlorophenprocoumon	1.62
Suprofen	3.51	Acenocoumarol	0.92
Iopanoic acid	1.38	Clorazepate	1.27

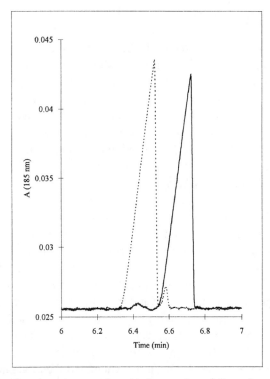

Figure 13. Electropherograms representing chiral separation of ibuprofen. Electrolyte: Glucidex2, 2.5% – Tris phosphate, 10 mM – pH 7. Sample concentration 100 μg/ml. Main graph (solid line) is S(+)-ibuprofen (unknown optical purity) and overlay chart (dotted line) is R(−)-ibuprofen (98% optically pure, as specified by the supplier).

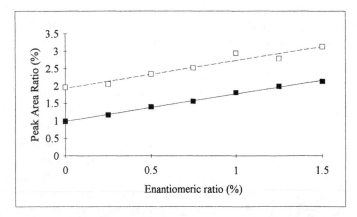

Figure 14. Determination of optical purity. Peak area ratio (enantiomeric contaminant versus major enantiomer) as a function of enantiomeric concentration ratio. Increasing enantiomeric concentration ratio obtained through addition of 0 to 1.5 μg/ml contaminant to 100 μg/ml pure enantiomer. Symbols: solid line, R(−)-IB contaminant added to S(+)-IB pure enantiomer, striped line, S(+)-IB contaminant added to R(−)-IB pure enantiomer.

DISCUSSION

The applicability of a direct stereoselective CE separation method using linear saccharide polymers as chiral electrolyte modifiers was studied. Maltodextrins and corn syrups were subjected to a stereoselectivity screening towards a range of racemic drug substances with different chemical structures and belonging to widely varying, pharmacologically important classes. Seven samples of currently used racemic 2-APA NSAIDs were run in a maltodextrin (Glucidex2) modified electrolyte and, with the exception of ketoprofen, shown to be baseline or near-baseline resolved within very short analysis times. The HERBs dichlorprop and haloxyfop were also included in the screening because of their structural resemblance to the 2-APA NSAIDs. Chiral separation was also easily obtained. The two parent compounds of both groups however could not be resolved. This could be attributed to a lack of steric 'bulkiness'. An identical explanation was given to an observed decline of chiral separation of amino acids with γ-cyclodextrin, while β-cyclodextrin did allow separation (Guttman et al., 1988). Using identical experimental conditions, the enantioselectivity of the same maltodextrin modifier was shown to be very high as well towards warfarin, phenprocoumon and their respective chlorinated derivatives. Acenocoumarol enantiomers could be resolved, however incompletely, upon changing the electrolyte modifier. Baseline separation of enantiomers was also shown for clorazepate and iopanoic acid. Given the extent and the efficiency of the separations and the wide range of racemic drug substances which were shown to be resolved, maltodextrins perform well as chiral selectors in a broad application range.

   Using COUMs as test solutes, the composition of the saccharide modifier as well as the concentration in the electrolyte was shown to affect the migration times and enantioselectivity. Except for acenocoumarol, all COUMs could be baseline separated. Migration time and chiral separation increased as the concentration of the maltodextrin or corn syrup was increased. These effects can easily be explained in terms of increased viscosity and complexing and/or interacting possibilities respectively. Within a selected maltodextrin brand, lowering of the DE value of the chiral selector caused limited changes in apparent mobility, especially in the higher DE range, an effect which was to be expected based on the relatively small viscosity differences between maltodextrins (technical documentation of maltodextrins and corn syrups by Grain Processing Corporation, Muscatine, Iowa, USA; Amylum N.V., Aalst (B); and Roquette Frères, Lestrem (F)). However, substantial improvements in chiral separation were observed. As lower DE maltodextrins contain more higher molecular weight complexing agents, the apparent electrophoretic mobility ratio of complexed and free analytes is expected to decrease, improving the separation. In addition, the increased resolution may be explained by enhanced enantioselectivity, i.e. an increased complexing constant difference between the optical isomers, of higher saccharide polymers, or a higher fraction thereof. Overall, the low DE

Glucidex maltodextrins proved to be the chiral electrolyte modifiers of choice both in terms of selectivity/efficiency and detection sensitivity.

A standard maltooligosaccharide mixture was able to separate enantiomers of acenocoumarol, a compound which could not be resolved using maltodextrins. At the same time, separation of the analogous COUMs was lost. Addition of maltodextrins to an electrolyte already containing the standard maltooligosaccharide mixture caused the loss of acenocoumarol separation, whereas separation of the analogous compounds was partially restored. A similar observation with NSAIDs was reported earlier (D'Hulst and Verbeke, 1992). Indeed, when ketoprofen was subjected to the same tests, it was shown that a separation occurred in pure maltooligosaccharide modified electrolytes, while electrolytes containing maltodextrins or a combination of both modifiers were unable to resolve ketoprofen enantiomers. This is similar to our present observation with acenocoumarol, when comparing the enantioselectivity of electrolytes containing either a highly concentrated standard maltooligosaccharide mixture or maltodextrins or the combination of both chiral modifiers. We have also demonstrated in the same report that ibuprofen and flurbiprofen enantiomers were separated using either purified, separate maltooligosaccharides (DP 4 up to 6) or maltodextrins as electrolyte modifiers. However, electrolytes containing the purified maltooligosaccharide as well as the maltodextrin failed to resolve the same racemates. Sepaniak *et al.* (1992) observed a comparable phenomenon when α- and β-cyclodextrins were mixed for the enantioseparation of two racemic binaphthyl compounds: while resolution for one binaphthyl compound was observed when one single cyclodextrin was used, selectivity towards both compounds was slightly decreased upon addition of the other cyclodextrin enabling the separation of the second compound. It was suggested that mixed cyclodextrin modified electrolytes might in fact be of use.

The results obtained with the COUMs and the NSAIDs may be indicative of multiple interactions resulting in different enantioselectivities, which may be qualitative as well as quantitative. Moreover, the different interactions may not behave independently of each other. The former hypothesis is explained when assuming, for instance, identical separation mechanisms with opposing enantioselectivity. Assuming reversal of migration order when changing from the single maltooligosaccharide modified to the maltodextrin modified electrolyte, an annihilation of opposing enantioselectivities may be the cause of the loss of chiral separation in mixed electrolytes, as observed in the case of the NSAIDs. However, this model, although attractive by its simplicity, may not explain all observations made. Alternatively, a simple quantitative difference in complexing constants between maltooligosaccharides might also account for the observed loss of separation of ketoprofen and acenocoumarol when adding maltodextrins to the separation medium. Aoyama *et al.* (1992) studied complex formation of hydrophobic fluorescent probes with maltooligosaccharides and found binding constants to increase along the DP range (from maltotetraose to maltoheptaose). This is in agreement with our results demonstrating an increase in enantioselectivity with decreasing maltodextrin DE value. Extrapolating the same findings to ketoprofen or acenocoumarol, very high stability constants might be found for complexes with low DE maltodextrins. Although a definite complexation is required for enantiorecognition, a tight fit may actually be deleterious to chiral separation. Indeed, when stable complexes are formed, migration time ratios of free *versus* complexed analyte species are unlikely to differ significantly between enantiomers. This reasoning might apply to the case of acenocoumarol or ketoprofen and account for the lack of enantioseparation in the presence of high DP maltooligosaccharides. Another explanation for the apparent paradox discovered when combining low molecular weight oligosaccharides and maltodextrins can be provided when considering the viscosity of the electrolytes. For enantioselectivity to be discerned, a sufficient difference in migration times between free and complexed form of the compound studied is required. This difference might be compromised in highly viscous electrolytes. Actually, the viscosity increase accompanying the increasing maltodextrin concentrations did not cause a decline in chiral selectivity. Instead, the chiral separations were shown to improve with concentration, levelling off towards the end of the concentration range. In the case of Glucidex 12, the initial sharp rise in resolution levelled off at electrolyte concentrations around 10%, turning into a small decline at approximately 20%. The addition of high concentrations of low molecular weight oligosaccharides to maltodextrin mixtures increased viscosity and caused an abrupt loss of enantioselectivity. So there may be an important difference between the effect on enantiomer separations of a rise in viscosity due to either small molecules (e.g. oligosaccharides) or high molecular weight compounds (e.g. low DE maltodextrins). This could be due to the fact that the enantiomers are offered different diffusion paths in a maltodextrin modified medium and in an oligosaccharide solution, the latter suppressing the enantioselectivity of the former when used together.

The multiplicity of possible interactions, which can be ascribed to the complexity of the maltodextrin preparations, may thus offer a wide applicability. However, the use of such a poorly defined or crude chemical may represent a serious drawback for method validation. Indeed, due to the as-yet unsolved problem of obtaining complete information on the qualitative and quantitative composition of maltodextrins — caused by the lack of available analytical methods — one has to rely on the reproducibility of the maltodextrin production process. This is not unique. Commercial derivatized cyclodextrin preparations were also shown to be heterogeneous: substitution patterns varied considerably depending upon the source (Liu *et al.*, 1993) and the degree of substitution of cyclodextrin derivatives has been demonstrated to cause different complexing behaviour (Müller and Brauns, 1986), leading to different enantioselectivities in chiral cyclodextrin mediated CE separations (Cruzado *et al.*, 1992; Nielen, 1993; Belder and Schomburg, 1994). So, there is clearly a need for as complete as possible information on chiral modifiers, whatever their application in analytical separation techniques.

The possibility of an 'ageing' effect on the chiral separation properties of saccharide solutions was investigated. These experiments necessitated either the use of sterile solutions or a suitable antimicrobial agent. Indeed, dilute solutions are highly susceptible to microbial growth, possibly inducing a breakdown of saccharide polymers, so that even during short term storage without any precautions the DP spectrum could be altered. To avoid microbial growth, the addition of sodium azide was found to be the most convenient approach. Sodium azide is active at very low concentrations and does not interfere with UV detection. The presence of sodium azide was also found not to affect chiral separation. Over the storage period studied, chiral separation was constant, proving both repeatability and the possibility of preparing large volume stocks.

In addition to stereoselectivity, the presence of saccharide modifiers also influenced non-chiral separation. Selectivities were highly improved upon addition of maltodextrins to the background electrolyte, an effect which was analogous to the enhanced selectivity obtained with cyclodextrins, as observed in numerous studies in CE as well as in HPLC.

A number of reports on chiral selectivity tuning using organic modifiers in combination with cyclodextrins prompted us to assess the effect of various lower alcohols (methanol, ethanol and isopropanol) as well as acetonitrile on the chiral selectivity of maltodextrins (Guttman *et al.*, 1988; Fanali, 1991; Busch *et al.*, 1993). Whereas the addition of organic modifiers often proved to be beneficial to resolution for cyclodextrin-based chiral separations as well as for protein-based separations, this was obviously not the case for maltodextrin electrolyte additives. Both resolution and mobility declined with solvent concentration, to a different extent depending on the solvent used. The observed rise in migration time was both absolute and relative compared to the neutral marker, indicating an increased apparent charge density and therefore a decreased interaction with the saccharide polymers. The effect of the addition of organic modifiers on mobilities is explained by the decreased dielectric constant and the increased viscosity of the electrolyte. From the expression for resolution of two analytes in CE derived by Jorgenson and Lukacs (1981) it can be predicted that a general reduction of mobilities and — in the case of anionic analytes, run under normal operating conditions — a relative increase of the analyte/EOF mobility ratio will produce a smaller resolution. The adverse effect of organic modifiers on enantioselectivity when used with maltodextrins as chiral selectors as opposed to cyclodextrins may also be explained in terms of complexation constants. Balasubramanian and co-workers studied the interaction of two surfactants, cetyltrimethylammonium bromide and Triton X-100, with several oligosaccharides (Sivakama Sundari *et al.*, 1991). Both linear and circular α(1-4) D-glucose oligomers increased the critical micellar concentration of the surfactants. The highest increase was noted for α-cyclodextrin while maltodextrins were considerably less effective. Aoyama *et al.* (1992) also reported significantly more stable complexes of hydrophobic fluorescent probes with β-cyclodextrin than with maltooligosaccharides. As already outlined above, metastable associations are preferred to highly stable complexes for chiral resolution. When the latter is the case, the addition of organic modifiers may be used to adjust complexation equilibria and eventually generate the appropriate association strength, as frequently observed in cyclodextrin-mediated chiral CE separations. Maltodextrin complexes are comparatively weak, so any addition of organic solvents may considerably disturb complex formation and eventually interfere with separation.

Further modifications to the background electrolyte as, for example, the nature of the cation of the buffer, also affected enantioselectivity as well as migration times. Whereas the order of mobilities as a function of buffer cation could be predicted based on conductivity data of the electrolyte ions present, results with respect to enantioselectivity were considerably less conclusive. In general, electrolytes containing Tris phosphate tended to show better separations than the other phosphate based buffers, while migration times were average. Maltodextrins dissolved in potassium phosphate showed somewhat less chiral selectivity. Although potassium phosphate caused a higher current, the negative effect on chiral separation could hardly entirely be ascribed to Joule effects — known to decrease separation efficiency — as demonstrated by comparison with lithium buffers. The latter allowed a reduction of the current but did not show substantial improvements on chiral separation factors. If complexation of the analytes by maltodextrins, whether enantioselective or not, involves ion pair formation, the effect of buffer cation on resolution may be explained. It could be speculated, for instance, that complexation would be enhanced with increasing ionic radius and/or 'organic' character of the counter ion. The nearly consistent observation of superior separations in TRIS phosphate and a considerably poorer resolution when lithium phosphate was used may strengthen this hypothesis, but a more extensive study is obviously needed.

As the present method proved to be highly stereoselective, the possibility of enantiomeric excess determinations was also investigated. For reliable enantiomeric excess determinations by chiral HPLC, the minor optical isomer should preferentially elute in front of the major compound. Indeed, peak tailing is otherwise a frequently encountered problem because of the obvious overload of one compound relative to the other. A second difficulty arises from the peak broadening due to diffusion, which affects the second eluting compound more extensively. Both effects add up to the difficulties often experienced in accurate peak quantitation. To avoid this problem, the possibilities of reversing peak order are studied, which comes down to a new method development. In CE, diffusion effects are very limited, if present at all. When baseline separation can be obtained, the order of migration is of no concern. Using maltodextrin modified electrolytes, baseline separation of ibuprofen enantiomers is easily obtained. However, to determine enantiomeric excess a sufficient detection sensitivity is required, which is a drawback of CE with UV detection. The UV detection cell is the capillary itself, giving an optical path length of usually 50 μm, which is an order of magnitude lower than in HPLC UV detection cells. This necessitated higher sample loads at the expense of baseline resolution. Because of the short optical path length and, in most instances, the low absorbance of the running electrolyte, CE is compatible with very low wavelength detection — an insufficiently exploited option which can sometimes lower the detection limits up to tenfold. By detection at 185 nm, baseline resolution of the two nearly pure enantiomeric preparations of ibuprofen and their antipode contaminants were obtained, proving enantiomeric excess determinations to be possible regardless of the migration order. Enantiomeric excess could be determined at least down to the 98% excess level.

In conclusion, linear maltooligosaccharide mixtures were shown to be broadly applicable as electrolyte modifiers for the chiral CE analysis of acidic racemic drug substances. A baseline separation was achieved for most compounds within very short times. Furthermore, the method was both sufficiently selective and sensitive to allow enantiomeric excess determinations at least up to the 98% level. The predictability of enantioselectivity of the maltodextrins or corn syrups towards the test solutes however was limited, as the chiral separation mechanism seemed to involve multiple and mutually influencing interactions.

REFERENCES

Aoyama, Y., Otsuki, J.-I., Nagai, Y., Kobayashi, K. and Toi, H. (1992). Host–guest complexation of oligosaccharides: interaction of maltodextrins with hydrophobic fluorescence probes in water. *Tetrah. Lett.* **33**, 3775–3778.

Armstrong, D., Tang, Y., Ward, T. and Nichols, M. (1993). Derivatized cyclodextrins immobilized on fused-silica capillaries for enantiomeric separations via capillary electrophoresis, gas chromatography, or supercritical fluid chromatography. *Anal. Chem.* **65**, 1114–1117.

Belder, D. and Schomburg, G. (1994). Chiral separations of basic and acidic compounds in modified capillaries using cyclodextrin-modified capillary zone electrophoresis. *J. Chromatogr.* **666**, 351–365.

Bereuter, T. (1994). Enantioseparation by capillary electrophoresis. *LC/GC Int.* **7**, 78–93.

Busch, S., Kraak, J. and Poppe, H. (1993). Chiral separations by complexation with proteins in capillary zone electrophoresis. *J. Chromatogr.* **635**, 119–126.

Caldwell, J., Hutt, J. and Fournal-Gigleux, S. (1988). The metabolic chiral inversion and dispositional enantioselectivity of the 2-arylpropionic acids and their biological consequences. *Biochem. Pharmacol.* **37**, 105–114.

Cooke, W. and Cooke, L. (1993). Effects of anaesthetics on the hepatic metabolism and biliary secretion of iopanoic acid enantiomers in rat. *J. Pharmacol. Exp. Ther.* **225**, 85–93.

Cruzado, I., Rawjee, Y. and Vigh, G. (1992). Chiral separations by capillary electrophoresis using cyclodextrin derivatives. Presented at the *Fourth International Symposium on High Performance Capillary Electrophoresis*.

De Vries, J. and Völker, U. (1989). Separation of the enantiomers of phenprocoumon and warfarin by high-performance liquid chromatography using a chiral stationary phase. Determination of the enantiomeric ratio of phenprocoumon in human plasma and urine. *J. Chromatogr.* **493**, 149–156.

D'Hulst, A. and Verbeke, N. (1992). Chiral separation by capillary electrophoresis with oligosaccharides. *J. Chromatogr.* **608**, 275–287.

D'Hulst, A. and Verbeke, N. (1994). Quantitation in chiral capillary electrophoresis. Theoretical and practical considerations. *Electrophoresis* **15**, 854–863.

Dicks, J., Slater, J. and Bewick, D. (1985). The R-enantiomer of fluazifop-butyl. In: *Proc. 1985 Br. Crop Protect. Conf.–Weeds*, pp. 271–280.

Dobashi, A., Ono, T., Hara, S. and Yamaguchi, J. (1989). Enantioselective hydrophobic entanglement of enantiomeric solutes with chiral functionalized micelles by electrokinetic chromatography. *J. Chromatogr.* **480**, 413–420.

Fanali, S. (1991). Use of cyclodextrins in capillary zone electrophoresis resolution of terbutaline and propranolol enantiomers. *J. Chromatogr.* **545**, 437–444.

Fujima, H., Wada, H., Miwa, T. and Haginaka, J. (1993). Chiral separation of lorazepam on ovomucoid-bonded columns: peak coalescence due to racemization. *J. Liq. Chromatogr.* **16**, 879–891.

Gassman, E., Kuo, J. and Zare, R. (1985). Electrokinetic separation of chiral compounds. *Science* **230**, 813–814.

Guttman, A., Paulus, A., Cohen, A., Grinberg, N. and Karger, B. (1988). Use of complexing agents for selective separation in high-performance capillary electrophoresis. Chiral resolution via cyclodextrins incorporated within polyacrylamide gel columns. *J. Chromatogr.* **488**, 41–53.

Jorgenson, J. and Lukacs, K. D. (1981). Zone electrophoresis in open-tubular glass capillaries. *Anal. Chem.* **53**, 1298–1302.

Liu, G., Goodall, D. and Loran, J. (1993). Quantification of trace levels of beta-cyclodextrin and substitution patterns in hydroxypropyl-beta-cyclodextrin using high-performance liquid chromatography with polarimetric detection. *Chirality* **5**, 220–223.

Möhler, H. and Okada, T. (1977). Benzodiazepine receptor: demonstration in the central nervous system. *Science* **198**, 849–851.

Muller, B. and Brauns, U. (1986). Hydroxypropyl-beta-cyclodextrin derivatives: influence of average degree of substitution on complexing ability and surface activity. *J. Pharm. Sci.* **75**, 571–572.

Nielen, M. (1993). Chiral separation of basic drugs using cyclodextrin-modified capillary zone electrophoresis. *Anal. Chem.* **65**, 885–893.

Nishi, H., Fukuyama, T. and Terabe, S. (1991). Chiral separation by cyclodextrin-modified micellar electrokinetic chromatography. *J. Chromatogr.* **553**, 503–516.

Noctor, T., Felix, G. and Wainer, I. (1991). Stereochemical resolution of 2-aryl propionic acid non-steroidal anti-inflammatory drugs on a human serum albumin based high-performance liquid chromatographic chiral stationary phase. *Chromatographia* **31**, 55–59.

Otsuka, K. and Terabe, S. (1990). Enantiomeric resolution by micellar electrokinetic chromatography with chiral surfactants. *J. Chromatogr.* **515**, 221–226.

Pirkle, W. and Welch, C. (1991). Chromatographic separation of underivatized naproxen enantiomers. *J. Liq. Chromatogr.* **14**, 3387–3396.

Pohl, L., Nelson, S., Porter, W., Trager, W., Fasco, M., Baker, F. and Fenton, J. (1976). Warfarin — Stereochemical aspects of its metabolism by rat liver microsomes. *Biochem. Pharmacol.* **25**, 2153–2162.

Sepaniak, M., Cole, R. and Clark, B. (1992). Use of native and chemically modified cyclodextrins for the capillary electrophoretic separation of enantiomers. *J. Liq. Chromatogr.* **15**, 1023–1040.

Sivakama Sundari, C., Raman, B. and Balasubramanian, D. (1991). Hydrophobic surfaces in oligosaccharides: linear dextrins are amphiphilic chains. *Biochim. Biophys. Acta* **1065**, 35–41.

Tambuté, A., Siret, L., Caude, M. and Rosset, R. (1991). Enantiomeric separation of substituted 2-aryloxy propionic esters. Application to the determination of the enantiomeric excess in herbicide formulations. *J. Chromatogr.* **541**, 349–357.

Terabe, S., Otsuka, K. and Nishi, H. (1994). Separation of enantiomers by capillary electrophoresis. *J. Chromatogr.* **666**, 295–319.

Progress in HPLC-HPCE, Vol. 5, pp. 383–441
H. Parvez *et al.* (Eds)
© VSP 1997.

Separation of enantiomers in capillary electrophoresis

THOMAS SCHMITT

*University of the Saarland, Institute of Applied Physical Chemistry,
Saarbrücken, Germany*

INTRODUCTION

Increasingly, many areas in physical and life sciences require reliable analytical methodologies for identifying optically active enantiomeric compounds. The most important applications of stereoselective techniques are in organic chemistry (organic synthetic processes), pharmaceutical manufacturing (product and purity control), and medical research (metabolism studies in patients).

The principal importance of stereochemistry is in its significant influence upon biological activity. Racemic mixtures may introduce some biological side effects caused by the optical antipode in addition to the desired activity (Drayer, 1988). This is explainable by the ability of living matter to distinguish between enantiomers. This ability is based on the fact that all living organisms are comprised of optically active components such as hormones, carbohydrates, and proteins. The least desirable case arises when one of the optical antipodes produces an intended medical effect, while the complimentary enantiomer produces a toxic or even teratogenic effect, as for example in the case of Thalidomide. Providing analytical methods for stereochemical characterizations is thus very important.

Enantiomeric molecules arise when molecular substitutions in a molecule exhibit a different spatial orientation. The existence of enantiomers in nature is based upon the presence of an asymmetric carbon atom. Chirality can be classified with the help of different categories of isomers (atropisomers and allenisomers). In most cases, there is an asymmetric carbon atom, characterized by four different substituents arranged in a tetrahedron. The chiral center thus resides at the central carbon atom (located in the middle of this tetrahedron). This is shown schematically in Fig. 1. Today, many analytical and preparative methods have been developed for enantiomeric separations employing various chromatographic techniques (Allenmark, 1991; Schurig and Jung, 1990).

T. Schmitt

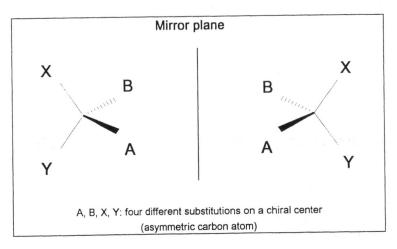

Figure 1. Chirality caused by an asymmetric carbon atom as the chiral center.

In 1853, Pasteur suggested the separation of racemic mixtures of enantiomeric molecules by transforming them into diastereomers (Pasteur, 1853). The addition of pure chiral reagents was suggested to produce a difference in the physical properties of enantiomers. Today, in principle, little has changed in this procedure and it is the basis for all chromatographic and synthetic separation techniques.

There are two methods for enantiomeric separations.

1. The direct method consists of enantiomer separations without previous derivatization. The enantiomeric mixture has to be exposed to an optically active environment, where one of the enantiomers shows a stronger interaction within the optically active environment (i.e. chiral recognition). The common procedures for enantiomeric separations are the application of chiral phases in High Performance Liquid Chromatography (HPLC), Gas Chromatography (GC), Supercritical Fluid Chromatopgraphy (SFC) and Capillary Electrophoresis (CE). The most suitable separation method of these four, for preparative purposes, is HPLC because it lends itself to an appropriate mass throughput.

2. The indirect method transforms pairs of enantiomers into diastereomers by derivatization with a pure chiral molecule. Due to different physical properties (e.g. melting point, crystallization behavior, hydrophobicity) the diastereomers can then be easily separated. A chiral environment may be advantageous, but it is not always a requirement for diasteromer separations.

Since diastereomers often differ in their physical properties, it is possible to separate them by reversed phase HPLC or by MEKC in capillary electrophoresis. This is illustrated in Fig. 2, where the separation can be achieved by using SDS, a micelle-forming detergent in the carrier electrolyte. A nonchiral polymer can also be used instead of SDS, provided it is capable of interacting with

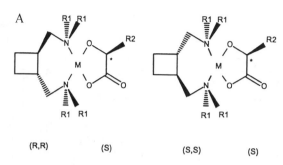

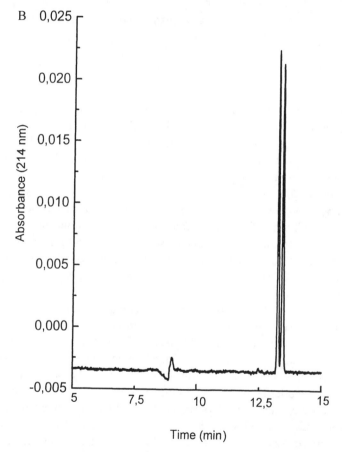

Figure 2. Separation of diastereomeric metallic complexes. A: R^1 = H; R^2 = Methyl; M = Pt. B: Run conditions: L = 60/67 cm, 75 μm i.d., E = 370 V/cm, T = 25°C; buffer: 16 mM phosphate, pH 7.0, 100 mM SDS, 15% (v/v) methanol; detection at 214 nm.

the analyte in a hydrophilic or a hydrophobic way. Schutzner *et al.* (1993) used polyvinylpyrrolidone as a buffer additive for the separation of D,L-tryptophan treated with anhydrous (+)-diacetyl-L-tartaric acid. It is interesting to note that, in this case, the polymer does not effect any molecular size separation.

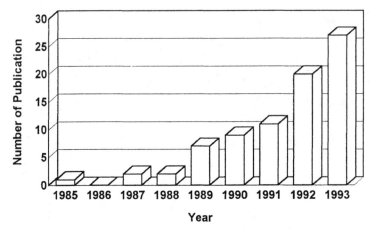

Figure 3. Number of publications per year in the field of enantiomeric separations in capillary electrophoresis.

Recently, CE has gained importance as an analytical tool. This is reflected in the increasing number of commercially available instruments, as well as in the growing number of scientific publications. One great advantage of CE is the fact that charged substances can be easily and efficiently separated with high resolution within a short period of time. This is in contrast to established analytical methodologies where strong acids and bases often create problems (peak tailing, poor efficiencies and resolution). In other methods, GC for example, the analysis of water-containing samples is complicated and requires laborious sample pretreatment. CE is an excellent method for those substances.

The first enantiomeric separation in an electrically driven separation system was performed in 1970 by Yoneda and co-workers using paper electrophoresis (Yoneda and Miura, 1970). Since the separation efficiencies were too small and the analysis time too long, this separation technique did not gain acceptance for routine analysis. With the advent of narrow inner diameter capillaries (< 100 μm), the high electrical fields required could be consistently applied, thereby dramatically shortening the analysis time. The number of scientific contributions addressing enantiomeric separations steadily has grown since the introduction of modern CE. As Fig. 3 illustrates, the number of CE publications has increased exponentially. Consequently, rapid development of CE-based enantiomeric methodologies can be expected. For this reason, it is useful to discuss the present state of the art, as well as separation techniques in the CE field as they exist today.

Chiral selectors in capillary electrophoresis

In optimizing enantiomeric separations, most attention is directed to the choice of an appropriate chiral selector. Generally, this is the case in all enantiomeric separations, since there are no universally applicable chiral selectors available. Chiral selectors in HPLC and GC must be liganded to different packing mate-

Table 1.

Survey of chiral selectors used in capillary electrophoresis

Types of chiral selectors	Chiral selectors	Problems
Cyclodextrins	α-, β-, γ-CDs, methylated CDs, hydroxypropylated CDs, carboxymethylated CDs, carboxyethylated CDs, carboxymethylethylated CDs, succinylated CDs, phosphatated CDs, sulfobutylether CDs, aminated and diaminated CDs	Samples which can form inclusion complexes: aromatic or partially hydrophobic analytes
Chiral micelle-forming detergents (often used as mixed micelles in SDS systems)	SDS-Digitonin, SDVal, SDAla, SDGlu, Cholates	In many cases no baseline separation of the enantiomers; poor efficiencies when SDVal and SDAla are used without additives
Chiral metal complexes	Copper-L-histidine complexes Copper-aspartame complexes	Only used for dansylated-amino acids; fluorescence detection
Chiral crown ether	18-Crown-6 tetracarboxylic acid	For basic, primary amines only
Pure enantiomers as chiral ligand exchange reagents	L-Tartrate	Used for chiral metal complexes only; only a few applications as yet
Proteins and glycoproteins	BSA, human albumin, orosomucoid, ovomucoid and fungal cellulose, cellulases	Feasibility is demonstrated, detection problems when used in the background electrolyte
Capillary coated with a chiral stationary phase	Capillary coated with a 'Chirasil-Dex-Phase'	Feasibility shown, low capacity of the phase, problems with overloading

rials by time-consuming synthetic processes. In contrast, most chiral selectors used in CE can easily be added to a carrier electrolyte. The ease of method development facilitates the search for the selector with the highest separating power. The most common CE chiral selectors, tabulated in Table 1, can be divided into different classes and will be discussed separately later in this chapter. Note that most of the selector classes have long, well-documented histories in HPLC and GC applications. To give the most important example, the molecular structure of many pharmaceuticals enables cyclodextrins to form inclusion complexes with these substances. Thus, of the selectors in Table 1, cyclodextrins and their derivatives offer the broadest, most promising application range in capillary electrophoresis, and will be discussed in detail later.

Apart from choosing an appropriate CE chiral selector, further optimization requires consideration of many parameters that play an important role for enantiomeric separations. Since optimization in CE is based upon mobility differences, analytes must be charged. Ion formation is often induced by working at an appropriate pH. Alternatively, neutral and hydrophobic molecules may be separated by MEKC if they partition into a phase such as the hydrophobic interior of a charged SDS micelle. The four most important optimization parameters are: (1) pH; (2) chiral selector concentration; (3) buffer electrolyte; (4) other buffer additives (SDS, organic solvents, urea) (Altria and Simpson, 1987).

The first successful separations of enantiomers in micellar electrokinetic chromatography triggered a search for new chiral selectors. The use of chiral micelle-forming agents increases the applicability of the technique by facilitating the separation of neutral chiral analytes. Cyclodextrins, chiral detergents, and crown ethers are used in more than ninety percent of all CE chiral applications. CE, with around forty selectors, lacks the large number of useable chiral selectors compared to HPLC with more than four hundred different chiral stationary phases. Many selectors used in HPLC are UV-absorbing and are therefore not applicable in CE as regular buffer additives. Moreover, it was shown that the chiral selectors can support not only the separation of diastereomers, but also the separation of positional isomers (Terabe et al., 1990). The application of chiral selectors is thus not limited to separations of enantiomers, but also includes separations of stereoisomers. In fact, general separation selectivity and optimization of many other non-chiral and non-stereoisomeric separations can be improved by including chiral selectors.

ENANTIOMERIC SEPARATIONS BASED UPON INCLUSION COMPLEXATION: SYSTEMS WITH CYCLODEXTRINS AND CROWN ETHERS

Aside from the purely micellar systems for enantiomeric separations, cyclodextrins and crown ethers are the primary chiral selectors in CE. The chemical structures of the optically active target molecules readily lend themselves to complex formations with cyclodextrins and crown ethers. Pharmacologically active drugs, for example, often contain amines or carboxylic acids on aromatic groups.

CYCLODEXTRINS

Cyclodextrins consist of cyclic oligosaccharides composed of D-glucose units connected to a ring with $\alpha(1\text{-}4)$-glycosidic bonds. Some physical properties of cyclodextrins are shown in Table 2. This arrangement of the D-glucopyranose units forms a bucket or cone-like structure surrounding an inner space, reminiscent of a cage. This is schematically shown in Fig. 4. The carbon atoms of the glucose units are oriented toward the inner space of the cyclodextrin ring, while the sugar's hydroxyl groups are directed toward the exterior. Because glucose is optically active, constraint of the hexoses into a ring provides an optically active surface upon which chiral recognition can occur. The interior of this 'cage' is more hydrophobic than the surrounding solvent. Hydrophobic groups such as aromatic compounds or cycloalkanes can penetrate the cone

Table 2.

Some physical properties of cyclodextrins

Cyclodextrin	UV-transparency	Optical rotation	Solubility (in % (w/v) at 25°C)	Diameter of inner cavity
α-CD	+	+150°	14.5	5.2 Å
β-CD	+	+162°	1.8	6.4 Å
γ-CD	+	+177°	23.2	8.3 Å

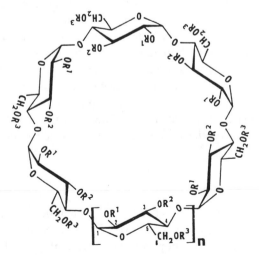

Figure 4. Structure of cyclodextrins. $n = 1$ (α-cyclodextrin); $n = 2$ (β-cyclodextrin); $n = 3$ (γ-cyclodextrin), R^1, R^2, and R^3 are possible positions for the derivatization of the cyclodextrin.

and form an inclusion or host-guest complex. Alternatively, the surface selectivity of the optically active cyclodextrin facilatates a preferential interaction with one of the enantiomers. Such chiral interaction allows separation of the enantiomer-cyclodextrin complex because the mobility of that enantiomer has now changed.

A recent publication of Li and Purdy (1992) discussed an investigation of chiral recognition in enantiomeric separations by means of spectroscopic methods. DNP-derivatives of D,L-valine, -leucine and -methionine were used with β-CD as chiral selector. The changes in circular dichroism, UV spectra, and proton-shifts in the ^{1}H-NMR allow identification of the participating active groups in the chiral recognition of the inclusion complex. Moreover, these spectroscopic data permit estimation of dissociation constants of the cyclodextrin-guest complexes. Generally, all DNP-L-amino acids have smaller dissociation constants than the D-enantiomers. The DNP-group of the amino acid forms a stable inclusion complex with the inner surface of the cyclodextrin ring. This allows the amino acyl moiety to interact with the hydroxyl groups at the chiral centers of the cyclodextrin rim. This arrangement is responsible for the quality of the resultant steric fit, and thus is decisive for chiral recognition. The enantiomer with the poorer steric fit in the cyclodextrin ring will have a smaller binding constant than that of its optical antipode. The difference between the binding constants leads to different residence times of the analytes in the complex. The observable difference in mobilities in the electrophoretic system yields a separation. The formation of a host-guest complex is a necessary criterion for chiral recognition, but it is not sufficient alone. Other functional groups at the chiral center of an analyte may cause a formation of yet another inclusion complex with another cyclodextrin molecule, and thus affect a separation. The preferential stabilization of the binding of one enantiomer by hydrogen bonding or the steric hindrance by cyclodextrin's hydroxy groups are the most important factors determining stability of an inclusion complex.

When an inclusion complex is formed, the change in environment from the bulk solvent to the more hydrophobic cyclodextrin internal surface may result in a change in the UV spectra of the analyte (Szejtli, 1982). This must be considered when quantifying the separated enantiomers, since equimolar amounts of an enantiomeric mixture (racemate) may give rise to different peak areas for the two components. In these cases, a correction factor must be established which will account for the differences in the UV absorption of one of the enantiomers.

Chiral separation models

Several theories have been proposed for the separation of enantiomers in CE. Wren *et al.* analyzed the influence of cyclodextrin concentration. They offered a simple model emphasizing the influence of the electroosmotic mobility as an important criterion for resolution, in addition to maximizing differences in the electrophoretic mobilities of the enantiomers (Wren, 1993). The resolution in a separation strongly depends upon the cyclodextrin concentration. This conclusion was based upon the behaviour of propranolol in a buffer system modified with β- or methylated β-CD (Wren and Rowe, 1992a).

CE enantiomeric separations were modified (increased or decreased resolution) when cyclodextrins were used as chiral selectors together with other buffer additives. Wren and co-workers tried to account for this in their proposed mathematical model (Wren and Rowe, 1992b). For the separation of propranolol with β- and methylated β-CD, the addition of methanol increased migration times. However, the addition of acetonitrile decreased migration times. In both cases, the mobility differences between the enantiomers decreased with increased concentration of the organic component in the buffer and the resolution declined.

Other models were used to determine the optimal cyclodextrin concentration by mathematical approximation (Wren and Rowe, 1993). The latter model was successfully applied to methylated β-CD applications. Some correlation could be made between the optimum chiral selector concentration, hydrophobicity data and log P value [log (octanol/water distribution coefficient)].

Vigh and co-workers introduced a more complex model to describe a separation of enantiomers using cyclodextrins. Ibuprofen and fenoprofen were used as enantiomeric test substances with a β-cyclodextrin chiral selector (Rawjee *et al.*, 1993a). Their model was limited to chiral acids and is based upon three relationships: the pH dependence of enantiomeric separations, the dependence upon the chiral selector concentration, and the dependence of the mobility of the analyte upon pH.

A comparable model based on inclusion complexation equilibria described the dependence of electrophoretic mobility of basic amines and resulting selectivities on experimental parameters, such as pH and cyclodextrin concentration (Rawjee *et al.*, 1993b). Parameters necessary for both the chiral acid and amine models can be determined by three sets of measurements: (1) the change of analyte mobilities with pH (without adding cyclodextrins to the buffer); (2) the dependence of mobilities on cyclodextrin concentration at low pH values; and, (3) the dependence of mobilities on cyclodextrin concentration at high pH values.

A good correlation could be drawn between the observed and the calculated data for homatropine. Additionally, this model described possible changes in selectivity between analytes and cyclodextrin at pHs similar to the analyte's pK. At these pH values, only a portion of the analyte is ionized. This may result in different binding constants of the D- *vs* L-analyte to a chiral selector. Generally, the mobility of an analyte will decrease as the pH approaches its pK.

Karger and co-workers introduced yet another model using cyclodextrin chiral selectors (Guttman *et al.*, 1988). This model correlated the influence of binding constants and cyclodextrin concentrations to resolution in the system.

The three common parameters considered in the above theoretical treatments of chiral separations are: (1) carrier electrolyte pH, which defines the analyte

Table 3.

Description of cyclodextrins used in capillary electrophoresis

Type of cyclodextrin (CD)	Degree of derivatization (molecules per CD-ring)	Average MW (g/mol)	Average formula
Regular cyclodextrins			
α-CD	—	973	$C_{36}H_{60}O_{30}$
β-CD	—	1135	$C_{42}H_{70}O_{35}$
γ-CD	—	1297	$C_{48}H_{80}O_{40}$
Derivatized cyclodextrins			
Hydroxypropylated α-CD	3.6	~1180	
Hydroxypropylated β-CD	6.3	~1500	
Hydroxypropylated γ-CD	4.8	~1580	
β-Methylated CD	12.6	~1311	
Heptakis (di-O-methyl) β-CD	14	1331	$C_{56}H_{98}O_{35}$
Heptakis (tri-O-methyl) β-CD	21	1429	$C_{63}H_{112}O_{35}$
Ionizable cyclodextrins			
Carboxymethylated β-CD	3.6	~1180	—
Carboxyethylated β-CD	~6	~1570	—
Succinylated β-CD	3.5	~1500	—
Sulfobutylether β-CD	4	1730	$C_{58}H_{98}O_{47}S_4Na_4$
Carboxymethylethyl β-CD	—	—	—
6A-Methylamino β-CD	1	1165	$C_{43}H_{73}O_{35}N$
6A,6D-Dimethylamino β-CD	2	1195	$C_{44}H_{76}O_{35}N_2$
2-O-Carboxymethyl β-cyclodextrin	1	1193	$C_{43}H_{72}O_{37}$
6[(3-aminoethyl)amino]-6-deoxy β-cyclodextrin	—	—	

ionic form; (2) chiral selector concentration; and, (3) electrophoretic mobilities of the free, non-complexed D- and L-enantiomers. The observed electrophoretic mobilities in a capillary are affected by the stabilities of the analyte-cyclodextrin complexes formed. This complex stability is, in turn, affected by the fit of an analyte into the cyclodextrin cavity. Analyte considerations are both structural and conformational. However, none of these models can predict conditions under which chiral recognition will be possible and selectivity will be demonstrated. Based on our current knowledge, these conditions can only be determined empirically.

Underivatized cyclodextrins have been used in many applications (Fanali, 1991; Altria *et al.*, 1992; Nardi *et al.*, 1992). Cyclodextrins are also very useful for the separation of diastereomers. The formation of inclusion complexes allows, for example, a baseline separation of pilocarpine and isopilocarpine epimers (Baeyens *et al.*, 1993). Without the presence of a cyclodextrin, these compounds could only be partially resolved. Such increase in selectivity has been observed for many other separations.

A great number of separation systems using underivatized cyclodextrins (α, β, γ) has been described in literature. However, all such systems have only been optimized with respect to a specific separation problem. For this reason, examples detailed in the literature are only valid for the buffer systems and the analytes profiled. Table 3 contains those cyclodextrins that have been applied in capillary electrophoresis. The cyclodextrins α, β, γ are referred to as normal cyclodextrins, because they are used without any derivatization. Some concrete examples from the literature will be discussed below. In addition to the regular optimization procedure, i.e. which cyclodextrin is the best for a specific separation problem and what cyclodextrin concentration and pH value should be used, the main focus will be on specific strategies for the optimization of a chiral separation.

Examples of separations using normal cyclodextrins

A recent work of Nielen dealt with the application of commercially available cyclodextrins to the separation of cationic, pharmacologically relevant enantiomers (Nielen, 1993a). Apart from some purely empirical aspects (matching chiral compounds with appropriate cyclodextrins), Nielen also discussed the influence of field strength on resolution. He attributed this influence to changing separation efficiencies with the increasing field strength. He also concluded that the observed efficiencies depended upon both the analyte mobility and electroosmotic flow. Both these factors contributed, but with different weights, to the efficiencies when increasing the electric field strength. Quantitative analysis and determination of the enantiomeric excess of approximately 1% (relative peak area) demonstrated the applicability of CE to the separation of enantiomers. To increase the sample throughput, multiple consecutive injections were evaluated; the influence of interrupting the separation process during consecutive injections was also discussed. Only a small decrease in resolution was observed, demonstrating the suitability of this method for increasing the sample throughput by multiple consecutive injections.

In addition to pH and cyclodextrin concentration effects upon resolution, Peterson explored the influence of buffer concentration and capillary diameter on the separation performance of the CE system (Peterson, 1993). He used hydroxypropyl-β-CD, DM-β-CD and hydroxyethyl-β-CD in addition to the more common α-, β-, and γ-cyclodextrins. Peterson also found that smaller inner-diameter capillaries (50 and 25 μm) gave higher resolution. Utilizing constant, small-volume injections led to an improved efficiency. These efficiency improvements resulted from the decreased Joule heat generated in the system, a consequence of the decreased electric current. The following racemic mixtures were utilized as samples: epinephrine, atenolol, propranolol, pilocarpine, isopilocarpine, betaxolol, dipivefrine, 2,7-difluoro-4-methoxyspiro(9Hfluorene-9,4'-imidazolidine)-2',5'dione, and 2,7-difluoro-4-hydroxyspiro(9H-fluorene-9,4'-imidazolidine)-2',5'dione.

Snopek *et al.* (1991) also reported enantiomeric separations of basic substances. To avoid adsorption of an analyte to the capillary wall, cellulose derivatives were used as buffer additives. As a side effect, these cellulose derivatives also strongly suppressed the electroosmotic flow.

Schutzner and Fanali (1992) discussed the influence of the structure and concentration of cyclodextrins, as well as the role of the structure of the analyte and the effect of temperature on resolution.

Shibukawa *et al.* (1993) investigated the influence of α-, β-, γ-CD on the diastereomeric and enantiomeric separations of leucovorin and of its metabolite, a tetrahydrofolic acid derivative. The fact that the separation of enantiomers was dependent upon the concentration of a chiral selector allowed a quantitative determination of binding constants. The binding constants of these complexes, as measured by CE, were much smaller than those determined by HPLC. Only leucovorin should form an inclusion complex with the cyclodextrin, based upon complex stability values obtained by CE. The tetrahydrofolic acid derivative showed an affinity to the CD, but did not form an inclusion complex. The influence of changing the ionic strength of the buffer, pH, urea concentration and temperature was also discussed.

In another, more practical study, an example of detecting enantiomeric impurities was briefly discussed with respect to the separation of ergot alkaloids by CD-modified buffer systems in CE (Fanali *et al.*, 1992).

CYCLODEXTRIN DERIVATIVES

Cyclodextrins and their derivatives are an important class of compounds. For instance, a derivatization of DM-β-CD with imidazole at position 3 of the CD can improve the catalytic activity of the hydrolysis of *p*-nitrophenyl acetate a thousandfold under neutral conditions when compared to the unmodified CD (Chen and Pardue, 1993).

The first application of derivatized cyclodextrins was described in isotachophoresis (ITP). The influence of complexation of counter ions with cyclodextrin derivatives (DM- and TM-β-cyclodextrin) on separation efficiencies in an ITP electrolyte system was investigated (Jelinek *et al.*, 1991).

The use of derivatized cyclodextrins in capillary electrophoresis has opened a broad area of application. Derivatized cyclodextrins utilized in CE are listed in Table 3. Those cyclodextrins that undergo ionization as a function of pH will be discussed separately.

The derivatization of a cyclodextrin ring at the hydroxy groups produces changes in their physical properties. Any modification at the cyclodextrin ring has a decisive influence on the steric fit, and thus on the affinity of an analyte to the cyclodextrin. Derivatization strongly influences chiral recognition, which may have both positive and negative consequences for a specific separation problem. Differences in the steric fit result in different binding constants for the (+) and the (−) enantiomer to the chiral selector. The resulting different residence times of an analyte in cyclodextrin lead to different mobilities of the D- and L-forms of the same analyte, and thus to separation.

Penn *et al.* (1993) examined the binding constant of tioconazole enantiomers to hydroxypropylated β-CD. They determined that the maximum separation selectivity occurs at the chiral selector concentration which is equal to the reciprocol value of the average binding constant. Diverging from this optimal concentration diminishes the selectivity and thus the resolution. Additionally, these experiments demonstrated that the analyte-cyclodextrin binding constants can be determined by capillary electrophoresis. Derivatized cyclodextrins may also produce an improvement in the resolution of diastereomers. Thus, d-L- and l-L-folinic acid can also be separated with the help of heptakis (2,6-di-*O*-methyl)-β-CD for quantitative determinations. Short analytical run times and good resolution were obtained with a coated capillary (Cellai *et al.*, 1993).

Nielen used different cyclodextrins to improve the selectivity for various herbicides (phenoxy acid derivatives) and some of their enantiomers (Nielen, 1993b). In addition to the normal cyclodextrins (α-, β-, and γ-CD), he also utilized heptakis (2,6-di-*O*-methyl)-β-CD. The latter cyclodextrin allowed the separation of all components, including minor ones, with a high sensitivity of detection. Additionally, the separation of enantiomers was achieved in two cases. The results of the purity determination of the enantiomers closely agreed with those of HPLC data obtained from a Chiral AGP column.

Soini *et al.* (1992) used trimethylated cyclodextrin (TM-β-CD) for the separation of some test compounds. They investigated the separation power and reproducibility of coated and uncoated capillaries. Their method was characterized by good linearity, sensitivity and precision. It provided reliable quantification and was thus found useful for pharmacokinetic and clinical studies.

Schmitt and Engelhardt (1993a) applied hydroxypropylated α-, β-, and γ-CD as well as methylated and carboxymethylated β-CD for the separation of pharmacologically relevant substances. For dansylated phenylalanine, the migration order inverted as the concentration of hydroxypropylated β-CD increased. This was ascribed to two different separation mechanisms in capillary electrophoresis: first, to different binding constants, and secondly, to the formation of stable diastereomeric complexes at high concentrations of the chiral selector. Fast separations (< 2 min) could be achieved under optimized conditions and with a suppressed EOF in a capillary coated with linear polyacrylamide. The choice of background electrolyte had an important influence on peak symmetries and efficiencies, and thus on the resolution of analytes. The application of ionizable cyclodextrins was especially useful for the separation of basic enantiomers. In the uncharged mode, i.e. at pH < 3, these cyclodextrins are suitable for the separation of sympathomimetic drugs using a capillary with suppressed EOF.

A large number of published applications confirms the potential of capillary electrophoresis as a fast and simple method for the separation of enantiomers. In an outstanding paper, Heumann and Blaschke reported the use

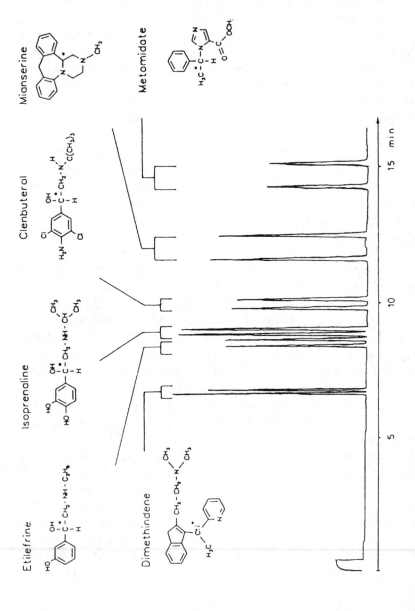

Figure 5. Enantiomeric separation of different basic drugs using hydroxypropylated β-CD. Run conditions: $L = 40/47$ cm, 50 μm i.d., $E = 400$ V/cm, $T = 20°C$; detection at 200 nm; buffer: 50 mM phosphate buffer, pH 3.3, with 30 mg/ml hydroxypropylated β-cyclodextrin. Reproduced with permission of Heumann and Blaschke, 1993.

of different cyclodextrins (hydroxypropylated, methylated and underivatized β-cyclodextrin) for separation of pharmacologically relevant substances like dimethindene derivatives and other aromatic amines (Fig. 5). They discussed the sample selectivity dependence upon CD concentration, temperature and pH. As we know from the preceding discussion, these are indeed the most important parameters when using cyclodextrins as chiral selectors in capillary electrophoresis.

Fanali (1989) used different cyclodextrins for the separation of sympathomimetic drugs (ephedrine, norephedrine, etc.). Although a host-guest complexation of the analyte was observed with β-CD and heptakis (2,3,6-tri-*O*-methyl)-β-CD detectable by a decreased mobility of the analyte, no separation into the enantiomers could be observed for these selectors. However, when using heptakis (2,6-di-*O*-methyl)-β-CD, all the components could be separated (Fig. 6). A later publication shows the separation of (±)-ephedrine using heptakis (2,6-di-*O*-methyl)-β-CD as chiral selector (Fanali and Bocek, 1990). The authors recognized the fact that the absorption spectra of the (+)-form and the (−)-form were not the same for the enantiomers in the inclusion complex. The peak areas obtained for the resolved enantiomers were different from those of the resolved racemic mixture. A correction factor was introduced, allowing an exact quantitative analysis. The report also described an optimized separation of D,L-tryptophan using α-cyclodextrin.

There is a variety of methods in CE for the chiral separation of ephedrine. Swartz (1991) described the influence of different cyclodextrins at a constant concentration on the separation of ephedrine. He also investigated the influence of pH for the use of heptakis (2,6-di-*O*-methyl)-β-CD as a chiral modifier.

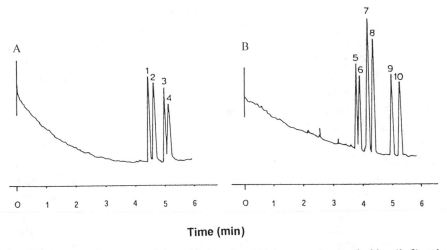

Figure 6. Electropherograms of the resolution of optical isomers. A: norephedrine (1, 2) and ephedrine (3, 4) B: norepinephrine (5, 6), epinephrine (7, 8) and isoproterenol (9, 10). Run conditions: $L = 20$ cm, 25 μm i.d., coated capillary, $E = 400$ V/cm; buffer: 10 mM Tris-phosphate, pH 2.4, containing 18 mM DM-β-CD. Reproduced with permission of Fanali, 1989.

Peterson and Trowbridge (1992) also used heptakis-(2,6-di-O-methyl)-β-CD for the quantitative analysis of L-epinephrine in pharmaceutical drugs. An internal standard was added to the sample, compensating for a poor injection system. This improved reproducibility for all cases.

IONIC CYCLODEXTRINS

The introduction of ionic groups on the cyclodextrin rim led to a further increase in the utility of cyclodextrins. Functional groups such as amino-, sulphonate-, or carboxyl-derivatives are ionized (protonated or deprotonated) according to the buffer pH. These chiral selectors then migrate in an electric field as charged species. The formation of inclusion complexes with the charged cyclodextrins leads to many new possibilities, similar to those described in MEKC.

Terabe et al. (1985b) reported the use of a carboxymethylethylated β-cyclo-dextrin for the separation of aromatic isomers of xylidine, xylenol, nitrophe-nol, nitroaniline, and chloroaniline. They confirmed the compatibility of these charged cyclodextrins with an MEKC system. This compatibility is not based just on the comparable chemical structures of the buffer additives. Rather it stems from their hydrophobic properties and the transport mechanism of un-charged analytes to the detector. Terabe and co-workers used the steric prop-erties of cyclodextrins for the separation of positional isomers, which could be transported toward the detector only when complexed to a charged cyclodextrin. This is comparable to the separation mechanisms occurring in MEKC. Terabe and co-workers also described the stereospecific synthesis of the carboxymethy-lated cyclodextrin in addition to different aspects of a host-guest-complexation of an analyte with that compound. The separation of enantiomers was not attempted with the carboxymethylated β-cyclodextrin.

In a later publication, Terabe (1989) used a positively charged cyclodextrin for the enantiomeric separation of dansylated D,L-amino acids. Here, working at pH 3 and suppressing the EOF by using hydroxypropyl cellulose, a neutral additive with no inherent mobility was obtained. The dansylated amino acids could only be separated while interacting with the charged cyclodextrin carrying the analytes to the detector. 6[(3-aminoethyl) amino]-6-deoxy-β-CD was used as a chiral selector here.

Fanali and co-workers described an enantiomeric separation of aromatic 2-hydroxy acids with derivatized cyclodextrins (Nardi et al., 1993). Neu-tral cyclodextrins and charged derivatized cyclodextrins were used. The ionic cyclodextrins allowed a more efficient chiral recognition by interacting elec-trostatically with the analyte (Fig. 7). Apart from 2-hydroxypropyl-β-CD, which is an uncharged cyclodextrin, positively charged cyclodextrins (6A-methylamino- and 6A,6D-dimethylamino-β-CD) were utilized. The negatively charged chiral analytes exhibited a much higher affinity for the positively charged cyclodextrin-amine than for neutral cyclodextrins. This observation of stronger interactions was offered as explanation of a greater concentra-

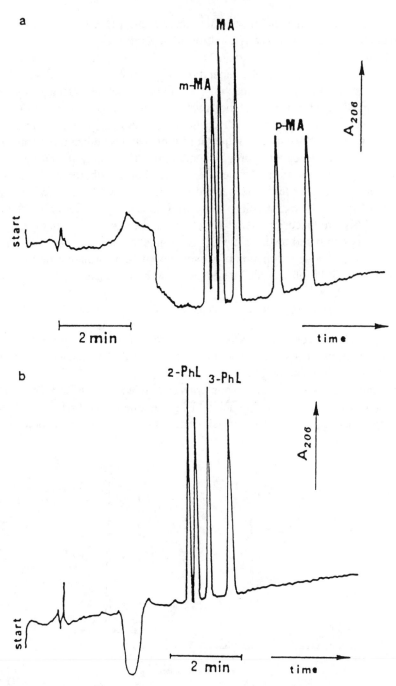

Figure 7. Chiral resolution of 2-hydroxy acids by CZE. Run conditions: $L = 20$ cm, 25 μm i.d., coated capillary; buffer: 10 mM acetate, pH 5, containing 6A-methylamino-β-cyclodextrin at (a) 20 and (b) 5 mM; applied voltage: 7 kV; sample: *m*-hydroxymandelic acid (*m*-MA), mandelic acid (MA), *p*-hydroxymandelic acid (*p*-MA), 2-phenyllactic acid (2-PhL), 3-phenyllactic acid (3-PhL). Reproduced with permission of Nardi *et al.*, 1993.

tion dependence of the mobility of the analyte and of better resolution. An acrylamide-coated capillary was used for all separations.

Smith (1993) showed a separation of neutral enantiomers with an anionic carboxymethylethyl-β-CD. The separation of positional isomers (e.g. o-, p-, and m-substituted nitrobenzyl alcohols), could be carried out successfully with these charged cyclodextrins at high pH values. With charged cyclodextrins, additional separation mechanisms for enantiomers were suggested. Schmitt and Engelhardt (1993b) discussed the possibility of better stabilization of analyte-cyclodextrin complexes by hydrogen bonding of the carboxyl functions at the cyclodextrin rim. Neutral molecules like binaphthol, hexobarbital and an oxazolidinone derivative could be separated with these CDs. Because of multiple charges of the cyclodextrin anion, ion pairing mechanisms could also be postulated. This produced an altered selectivity and migration order of cationic enantiomers compared to the same CD in an uncharged state. With a change in pH leading to a deprotonation of a cyclodextrin and formation of an anion, the migration order of cationic enantiomers can be reversed.

CYCLODEXTRINS INCORPORATED IN GELS

The impressive successes of capillary gel electrophoresis with high selectivities and enormous efficiencies (up to 30 million theoretical plates) inspired attempts to separate enantiomers in gels.

Guttman et al. (1988) used cyclodextrins embedded in gels for enantiomeric separations of dansylated amino acids. This was one of the first enantiomeric separations using cyclodextrins in CE. The chiral selector was added to the

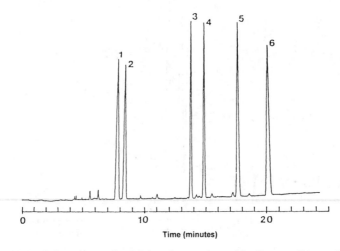

Figure 8. High efficiency separation of dansyl-D,L-amino acids. Run conditions: $L = 15$ cm, 75 μm i.d., filled with 5% T, 3.3% C crosslinked polyacrylamide in 0.1 mM Tris, 0.25 M boric acid 7 M urea, pH 8.3 containing 75 mM β-CD, $E = 300$ V/cm; sample: (1) L-Glu, (2) D-Glu, (3) L-Ser, (4) D-Ser, (5) L-Leu, (6) D-Leu; detection at 254 nm. Reproduced with permission of Guttman et al., 1988.

monomeric acrylamide solution before polymerization was started. Investigated in this system were the influences of the type of cyclodextrin, CD concentration, temperature and the addition of methanol upon analyte selectivity and migration time. A retention model was introduced based on the different equilibrium binding constants of D- and L-amino acids to the cyclodextrin, the cyclodextrin concentration, and the mobilities of the analyte-CD-complexes. The stability of gel-filled capillaries as well as the low UV-detection sensitivity below 250 nm with acrylamide gels were not discussed. A representative electropherogram is shown in Fig. 8.

Cyclodextrins have also been crosslinked with acrylamide (Cruzado and Vigh, 1992). An allyl-carbamoylated β-CD was used with copolymerizable vinyl groups. The selectivity and the efficiencies of these gels were highly dependent on the allyl-carbamoylated-CD concentration in the copolymer. With increasing concentration of the chiral selector, an increase in selectivity, accompanied by a decrease in efficiency was observed. The same detection problems as for the CDs embedded in acrylamide gels also occur in this case.

CROWN ETHERS

Recently, enantiomer separations with crown ethers were used in capillary electrophoresis. Crown ethers were already known as chiral selectors in HPLC from the work of Dotsevi *et al.* (1975). Their ability to complex ammonium compounds makes the crown ethers especially useful in the separations of primary chiral amines as well as amino acids. Another advantage of crown ethers in HPLC is their potential for chiral peptide applications (Hilton and Armstrong, 1991). The separation of chiral peptides is difficult, because the chiral center of the side chain is often too distant from the chiral centers of the selectors, so that recognition between them is not favored.

Kuhn *et al.* (1992b) successfully demonstrated the application of 18-crown-6-tetracarboxylic acid in CE. The structure of this chiral crown ether is shown in Fig. 9. Depending upon the ring size, a crown ether may selectively complex a potassium, ammonium or a protonated alkylamine. Thus, chiral alkylamines could be separated from each other via a stronger complex formation

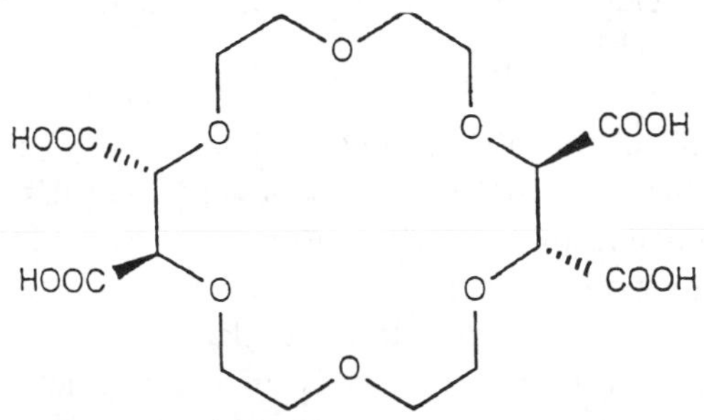

Figure 9. Structure of 18-crown-6-tetracarboxylic acid.

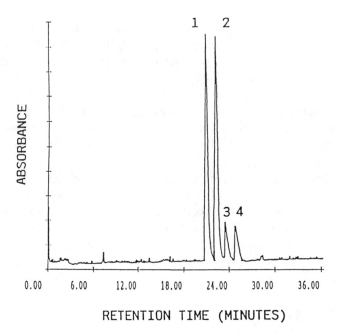

Figure 10. Separation of D,L-tryptophan and D,L-dopa. Run conditions: $L = 50/57$ cm, 75 μm i.d.; applied votage: 15 kV; detection at 254 nm; buffer: 30 mM 18-crown-6 tetracarboxylic acid in water, pH 2.2; sample: (1) L-tryptophan, (2) D-tryptophan, (3) L-dopa, (4) D-dopa. Reproduced with permission of Kuhn *et al.*, 1992b.

of one of the enantiomers with the crown ether. This is shown in Fig. 10 for D,L-tryptophan and D,L-dopa. (Note: It is not advisable to use potassium or ammonium as buffer components because they competitively bind to the crown ether). In contrast to cyclodextrin systems, an increase in crown ether concentrations in the buffer led to a better resolution with only a minimal decrease of mobilities of analytes. These data suggested the existence of diastereomeric complexes which could be separated based on their different electrophoretic mobilities.

Another report from Kuhn *et al.* (1992a) demonstrated that a single tetracarboxylate crown ether could separate enantiomers of approximately twenty underivatized amino acids. Molecules with a chiral center in the γ-position of the amino group, such as dipeptides, show good chiral recognition as well. Two mechanisms were discussed for chiral discrimination. The first was based on the four substituents of the crown ether (four carboxyl groups) forming a chiral barrier for the guest molecule by dividing the available inner space in two parts. The second recognition mechanism was described as an electrostatic interaction based on hydrogen bonding or ionic interaction between the analyte and the chiral selector. According to molecular models, good separation of dipeptides with a chiral center in the γ-position was chiefly based on chiral recognition by hydrogen bonding. Even for tripeptides, chiral recognition was achieved, despite low resolution. Nonpolar substituted amino acids could not undergo polar interactions with the crown ether. Consequently, the separation mechanism must be based primarily upon steric recognition. This was confirmed by the fact that (±)-(1-napthyl)-ethyl amine could be separated, but not its phenyl derivative. The described method could also successfully be applied for the separation of the four stereoisomers of 2-amino-3-phenylbutyric acid in one analytical run.

In the pH range used, the tetracarboxylate crown ether derivative was only slightly dissociated and migrated slowly in the opposite direction to the analyte. Together with other factors, this counter-migration could explain the high selectivities obtained. Furthermore, a separation sytem for the detection of optically clear chiral amines was described. This allows an enantiomeric separation of analytes lacking chromophores using indirect UV-detection mode without prior sample derivatization. The background electrolyte contained 6 mM benzyltrimethylammonium chloride, 5 mM tris, 15 mM crown ether, adjusted to pH 2.2 with citric acid. With this buffer, enantiomeric separation was obtained for all aliphatic amino acids discussed in the report. Furthermore, 18-crown-6-tetracarboxylic acid successfully separated chiral amino alcohols such as sympathomimetic drugs (Höhne *et al.*, 1992). This system could also be used for purity control of methoxamine, detecting 0.5% of the pharmacologically inactive enantiomer.

OPTIMIZATION OF THE SEPARATION OF ENANTIOMERS USING CYCLODEXTRINS

Cyclodextrins currently represent the most widely used chiral selector in CE. The main reasons for their success are UV-transparency at wavelengths down to 200 nm, high solubility and thus ease of inclusion in aqueous buffer systems, and the ability to form inclusion complexes with many chiral, UV-active, aromatic analytes.

Influence of the chiral selector

In the application of cyclodextrins as chiral selectros, normal cyclodextrins (α, β, γ) have a decisive influence on selectivity, not only because of differences in ring size but also because they are easily derivatized. Consider Fig. 11, showing four separations of a test mixture using four different cyclodextrins. This figure clearly illustrates the different affinities of the analyte to the cyclodextrins used in this example. The stronger this affinity, the lower the effective mobility of the analyte and the closer it runs with the electroosmotic flow marker. This does not mean that the resolution increases according to the magnitude of the affinity to the cyclodextrin. The difference in the affinities between the (+)- and the (−)-enantiomer to the chiral selector is the actual driving force for the resolution. This can be seen in Fig. 11, comparing case C (methylated β-cyclodextrin) and D (hydroxypropylated β-cyclodextrin). Although in both cases, the relative mobility for dansylated phenylalanine is almost the same, a separation was observed only using the hydroxypropylated CD.

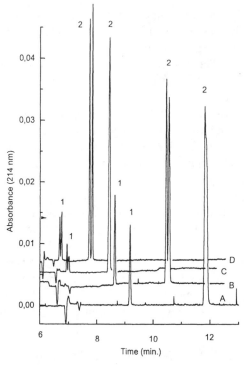

Figure 11. Separation potential of different cyclodextrins. Run conditions: $L = 50/57$ cm, $E = 350$ V/cm; buffer: 0.1 M TBE, pH 8.3; detection at 214 nm; sample: (1) D,L-hexobarbital, (2) D,L-dansylated-phenylalanine; in all cases 1.5% (w/v) of the cyclodextrin: A) α-CD, B) hydroxypropylated α-CD, C) methylated β-CD, D) hydroxypropylated β-CD.

Influence of the electroosmotic flow

Generally, a low selectivity between the D- and L-form of an analyte (α-value) causes problems in the separation of enantiomers. This low selectivity often requires long analysis times, even if an appropriate chiral selector is available for the separation. Because of the small differences in mobilities between the D- and L-form of an analyte, the presence of a strong electroosmotic flow in untreated capillaries does not allow enough time to develop a separation. The small mobility differences only permit separation of the analytes with a sufficient residence time in the capillary. The presence of electroosmotic flow shortens the residence time of an analyte in a capillary. To obtain maximum resolution in a short analysis time, coated capillaries with a strongly suppressed EOF should be used. The short run times are achieved by the use of very short capillaries (7–20 cm) in combination with high electric fields (up to 1000 V/cm). Figure 12 shows the difference between a coated and an uncoated capillary. The apparently higher efficiencies in the uncoated capillary are caused by the high velocity of the EOF, which rapidly moves the analytes through the detector. With coated capillaries better separations can be achieved in shorter times, since a shorter capillary permits the use of stronger electric

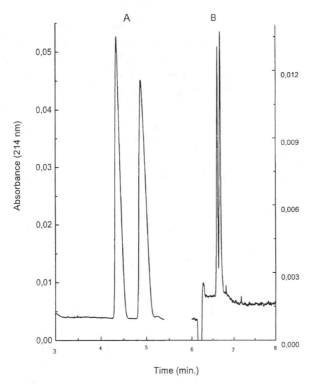

Figure 12. Enantiomeric separation of hexobarbital in a coated (A) and uncoated capillary (B). Run conditions: 0.1 M TBE pH 8.3, 1.56% (w/v) β-cyclodextrin; detection at 214 nm; injection: 1 s, 2 kV, $T = 25\,°C$. A: Coated capillary (4% T linear polyacrylamide), $E = 710$ V/cm, $L = 7/27$ cm. B: Uncoated capillary, $E = 400$ V/cm, $L = 50/57$ cm.

fields. The use of coated capillaries plays an important role in the search for an appropriate chiral selector. If a selector has a suitable selectivity for a separation problem, it can be exploited with better resolution and with a shorter analysis time in a coated capillary with suppressed EOF.

Belder and Schomburg (1992) investigated the influence of nonionic buffer additives, polyvinyl or hydroxyethyl cellulose, together with cyclodextrins as chiral selectors on the separation of chiral aromatic amines (tocainide). These additives were only used in very low concentrations ($< 0.05\%$). They caused a strong decrease in EOF and a prolonged migration time of the analytes. The lower wall adsorption of the analytes and the decrease in EOF led to a longer residence time of the analytes, and improved the peak symmetry and the resolution in the system. The use of a second, outer electric field to control the EOF did not improve the efficiencies, compared to an uncoated capillary with only one electric field.

In coated capillaries not only is the EOF suppressed, but also undesired wall adsorption between the analytes and active sites on the capillary wall (silanol groups) is decreased. One way to make a capillary coating is to polymerize acrylamide monomers with a bifunctional reagent which is chemically bound to the capillary wall. The method to produce these capillaries was first described by Hjerten (1985), and has been successfully applied in many laboratories. Capillary coatings based on linear polyacrylamide are the most popular coating materials in CE at the moment. For our purposes, a modified method for the preparation of these capillaries is described below.

Capillary pretreatment. The first step of a coating procedure is the pretreatment with alkaline hydroxides and acids. First, the capillary has to be washed with 1 M NaOH (5 h), followed by a distilled water rinse (10 min) and a 0.1 M HCl rinse (30 min). Finally, the capillary is washed with water until pH paper confirms that all acid has been removed.

Coating procedure — The first step. The water must be removed from the capillary using methanol. Afterwards, the capillary is stored in a 50% (v/v) methacryloyloxypropyl-trimethoxysilane/methanol solution for at least 12 h while the capillary ends are sealed with pieces of rubber. After this, the capillary is rinsed with methanol followed by a rinse with a mixture of methanol and water. Finally, the capillary is washed with distilled water.

Preparing the acrylamide solution. Depending upon the gel concentration used for coating, the monomer in the buffer should be adjusted to a desired concentration before the polymerization. As an example, a 4% T linear polyacrylamide gel is an appropriate concentration to polymerize in a capillary. Since polymerization is carried out in the capillary, the excess gel must be removable from the capillary by pressure if the capillary is to be used coated, not gel-filled. Monomer concentrations of more than 6% T cause problems with respect to the gel removal step. Prior to polymerization, the acrylamide solution (3 ml) should be degassed at 0.1 atm for 2 h. This is necessary because oxygen acts as an inhibitor in the free radical catalyzed polymerization reaction.

Coating procedure — The second step: Gel polymerization. The degassed solution of 4% (w/v) T-acrylamide in 0.1 M Tris/boric acid (2 mM EDTA), adjusted to pH 8.2 also containing 5 µl of a 10% (w/v) ammonium persulfate solution and 5 µl of N,N,N',N'-tetramethylethylenediamine per ml, is loaded into the capillary. The solution can be introduced with a µl syringe adapted to the capillary by Teflon tubing. Filling the capillary must be done with care to avoid introducing air bubbles. The polymerization inside the capillary requires about 12 h.

These gel capillaries, thus prepared, may be used for the separation of DNA fragments in the capillary gel electrophoresis mode. Alternatively, gels can be removed from the capillary and replaced by other polymer networks of higher or lower viscosity. For use as a coated capillary, gels can be replaced by a regular carrier electrolyte system leaving a coating along the inner capillary wall. As long as only electrolytes below pH 7 are used, this coating remains stable for several months. The stability of the coating is determined by the absence of a significant electroosmotic flow. Capillaries produced in this way show no measurable EOF for a longer time when pH values remain below 7. Their applicability at pH values above 9 is limited to a few days.

Handling of coated and gel-filled capillaries

Removing the outer polyimide coating by flame or a hot wire is no longer possible with coated or gel-filled capillaries. The polyimide coating removal should be done before the capillary pretreatment. The polyimide layer may also be removed by scratching very carefully with a razor blade, but this requires considerable skill.

Cutting a gel-filled capillary to the desired length should be done with care. To avoid pulling out part of the gel inside the column, the two pieces should be separated in a perpendicular position to each other. This guarantees a straight cut and prevents bubble formation at the ends of the columns.

Equilibrating the gel-filled capillary before using it for the first time should be performed slowly in an electric field. This can be done by a stepwise increase of the field strength to the level used normally for separations. Constant current and a constant baseline indicate the equilibration of the gel column (after aproximately 15 min).

Storing the gel columns should be done by placing the ends of the capillary into a buffer. This prevents the column from drying out, which would irreversibly destroy the capillary coating. When a polyacrylamide coating is used, the separation buffer should be removed by distilled water. The capillary can be dried by a stream of air afterwards.

Influence of pH

Generally, pH is an important optimization parameter in CE. The pH not only influences the EOF, but also determines the ionization state of functional groups. The resulting electrophoretic mobilities directly influence the separation of the analytes. When uncharged cyclodextrins are used, the enantiomers must have a charge to migrate through the 'quasi-stationary phase' of cyclodextrins. The term 'quasi-stationary' implies that a cyclodextrin only moves with the electroosmotic flow and does not exhibit any electrophoretic velocity. As in HPLC, cyclodextrins can also be referred to as a 'phase' in CE, since the separation of the analytes occurs not only because their mobilities are different, but also as a result of their interaction with a second compenent.

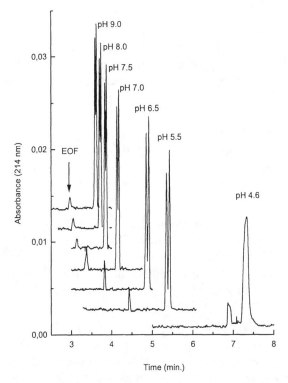

Figure 13. Influence of the pH value on the enantiomeric separation of a dihydropyridine deriva-tive. Run conditions: $L = 50/57$ cm, $E = 440$ V/cm; buffer: 20 mM phosphate with 0.4% (w/v) hydroxypropylated β-CD, different pH values as shown; detection at 214 nm.

Figure 13 shows an example of pH dependence for a separation of a racemic mixture. At a low pH, the analyte shows nearly no velocity and is thus moved by the EOF through the detector. The low mobility of the analyte reflects its low degree of ionization at the given pH value. The EOF, moreover, is very slow at this low pH. At a medium pH the analyte has a sufficient charge, and the mobility differences between the D- and L-forms are maximized. In this case, the residence time in the capillary is long enough for a good separation. The loss of separation at higher pH reflects a very fast EOF in which the analytes do not have enough time to be separated.

At pHs above 12, cyclodextrins can be deprotonated, which may also cause a change in selectivity. If the charge of the cyclodextrin and the analyte is the same, chiral recognition may be lost because of electrostatic repulsion. Additionally, high pHs can either lead to a decomposition or to a higher degree of ionization of the sample. The latter two effects may also cause a change in the selectivity of the system.

Influence of the concentration of the chiral selector

Another important part of optimization is the selection of appropriate cyclodextrin concentration in the buffer. Depending on the analyte, improved resolution, loss of resolution or even inversion of elution order with increasing CD concentration can be observed. This suggests at least two different separation mechanisms for the separation of enantiomers using cyclodextrins. Figure 14 demonstrates the influence of the CD concentration on the separation of enantiomers. In a buffer system without cyclodextrins, the enantiomers have a relatively high mobility (compared to the EOF) since they easily migrate through the buffer. Small amounts of the chiral selector added to buffer lead to a large decrease of analyte mobility as the analyte undergoes a complexation with cyclodextrin.

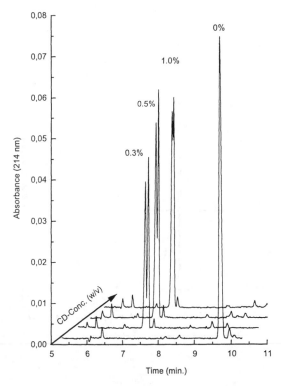

Figure 14. Influence of the hydroxypropyl β-cyclodextrin concentrations on the resolution of a dihydropyridine derivative (see structure in Fig. 13). Run conditions: $L = 50/57$ cm, $E = 350$ V/cm; buffer: 0.1 M TBE pH, 8.3, increasing concentrations of the cyclodextrin; detection at 214 nm.

By increasing the selector concentration even more, the mobility of an analyte can decline to become close to that of the EOF marker. This results in a very short residence time in the capillary and therefore in a loss of separation. Furthermore, cyclodextrins increase the viscosity of the buffer, reducing EOF and causing a longer analysis time. The same effect can be observed in capillaries with suppressed EOF where higher concentrations of cyclodextrin produce longer analysis times because of higher viscosity of the buffer.

The loss in resolution resulting from residence times that are too short can be especially observed for analytes with low electrophoretic mobility. At very high concentrations of the chiral selector, the D- and L-enantiomers are forced into an inclusion complex with the cyclodextrin. The difference in the binding constants of the analytes is reduced, resulting in the loss of resolution. If, however, the two permanent inclusion complexes of the enantiomers have different mobilities, they can still be separated. This can occur when the two diastereomeric complexes formed show different physical properties (e.g. a different charge to volume ratio). One example of such behavior is given in Fig. 15 for

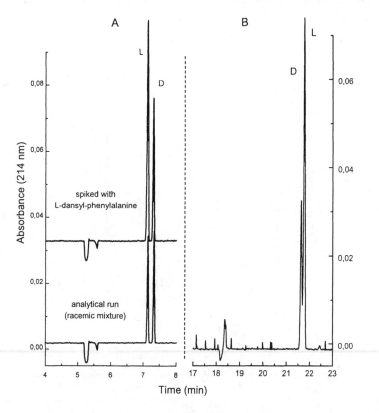

Figure 15. Reversal of the migration order with increasing of the cyclodextrin concentration. Run conditions: A: $L = 50/57$ cm, $E = 350$ V/cm, 20 mM phosphate buffer, pH 7.0, 0.5% (w/v) hydroxypropylated β-cyclodextrin. B: $L = 80/87$ cm, $E = 300$ V/cm, 20 mM phosphate buffer, pH 7.0, 15% (w/v) hydroxypropylated β-cyclodextrin; detection at 214 nm.

a separation of dansylated D,L-phenylalanine using hydroxypropylated β-CD at both low and very high concentrations. In the medium concentration range no separation is obtained. Surprisingly, the migration order of the D- and L-enantiomers reverses when going from low to high concentrations of the CD. This demonstrates that there are two separation mechanisms reponsible for the resolution, depending upon the concentration range. At a low concentration the separation depends on different binding constants of the analytes, and thus on different times they spend in the complex. At a high concentration, the separation is due to the formation of diastereomeric complexes and to different mobilities of these complexes in an electrophoretic system.

Influence of the background electrolyte on the resolution

After the choice of a chiral selector and the evaluation of optimal pH, the background electrolyte can be further optimized. As shown in Fig. 16, the mobilities of the buffer ions have a strong influence on the peak shape and efficiencies and thus also on analyte resolution. In all cases shown, the injection conditions

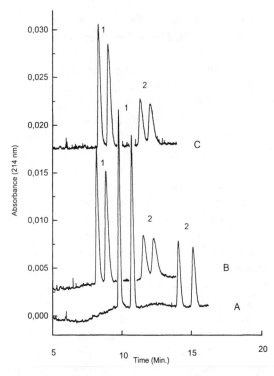

Figure 16. Influence of the background electrolyte on the peak shape and resolution. Run conditions: $L = 20/27$ cm, $E = 370$ V/cm, in all cases pressure injection: 1 s, 0.3% (w/v) hydroxypropylated β-cyclodextrin, capillary coated with 4% T linear polyacrylamide, anode at the detection side; detection at 214 nm; sample: dansyl-phenylalanine (1), dihydropyridine derivative (see structure in Fig. 13) (2). A: 10 mM citric acid/Tris, pH 6.0; B: 25 mM MES/Tris, pH 6.0; C: 20 mM phosphate buffer, pH 6.0.

as well as the pH of the buffer were unchanged. Only the composition of the electrolyte was modified in these separations. It can be seen in this particular example that the use of a citric acid buffer leads to a better resolution based on higher separation efficiencies. This also leads to higher peak signals (i.e. better detection sensitivity of the system). It is clearly demonstrated here that despite small α-values, resolution can be enhanced by improving the peak shapes and increasing the separation efficiencies. One of the contributions to peak shapes in CE is called electrodispersion and is described by the Kohlrausch Regulation Function. Bandbroadening effects are observed when the conductivities of the sample zone and the carrier electrolyte (buffer ions) differ from each other.

Kuhn *et al.* (1992b) also observed that an increase in the buffer capacity (ionic strength of the buffer) may enhance resolution in a cyclodextrin modified buffer system. This effect was explained as an increase in hydrophilicity of the buffer based on higher ionic strength, thus intensifying the hydrophobic interaction between the analyte and the cyclodextrin cavity.

Influence of buffer additives

In addition to the parameters already discussed, there are other buffer additives that have significant influence on enantiomeric separations and on chiral selector performance. The effects of additives such as organic solvents, complexing agents, detergents, etc. upon chiral recognition cannot always be predicted. As an example, the influence of urea, methanol, and SDS on migration time, peak efficiencies and resolution are shown in Fig. 17. The experiments were performed with a coated capillary (previously described in the section on EOF). A saturated solution of β-CD [1.6% (w/v)] gave the fastest migration time (Fig. 17A). Adding 7 M urea increased the solubility of the β-CD. Reflecting the lower mobility of the analyte in the presence of a higher concentration of the chiral selector as well as the higher viscosity of this buffer, the run time increased. However, the quality of resolution was not improved (Fig. 17D). The addition of methanol (Fig. 17E) had a positive influence on resolution. The migration time was only slightly higher, but the resolution was enhanced. The buffer system used in Fig. 17D together with 0.1 M SDS shortened the migration time drastically (Fig. 17B). The SDS and the analyte migrated in the same direction because they were both anionic. This synergetic effect speeded up analysis time and enhanced resolution in comparison to Fig. 17D. The addition of methanol to the buffer increased analysis time again, but without any enhancement in resolution (Fig. 17C). For the examples in Fig. 17, the increase in resolution was mostly from an increase in efficiencies in the specific running buffer. The α-values were not subject to significant changes and remained almost unchanged.

The influence of short- and long-chain tetraalkylammonium compounds on the EOF and the selectivity of β-cyclodextrin was also investigated. Quang and Khaledi (1993) discussed the separation of ephedrine derivatives and other chiral α-hydroxylamines with regular β-CD. The ammonium compounds used

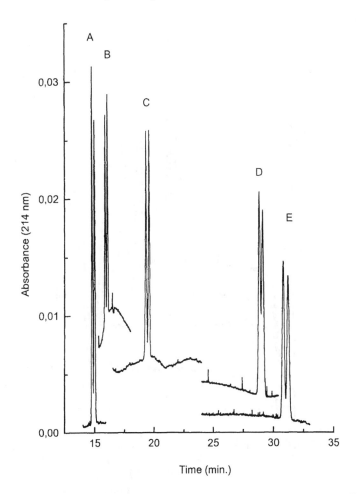

Figure 17. Influence of buffer additives on the separation of anionic enantiomers of a dihydropy-ridine derivative (see structure in Fig. 13). Run conditions: $L = 40/47$ cm, $E = 232$ V/cm (anode at the detection side), 75 μm i.d., coated capillary (4% T linear polyacrylamide); detection at 214 nm; buffer: 0.1 M TBE, pH 8.3, containing: A) 1.6% (w/v) β-cyclodextrin; B) 12% (w/v) β-cyclodextrin, 7 M urea, 0.1 M SDS; C) 12% (w/v) β-cyclodextrin, 7 M urea, 10% (v/v) methanol, 0.1 M SDS; D) 12% (w/v) β-cyclodextrin, 7 M urea; E) 12% (w/v) β-cyclodextrin, 7 M urea, 10% methanol.

were tetrabutylammonium, tetramethylammonium and hexadecyltrimethylam-monium cations. Quang found short chain tetraalkylammonium compounds to be more advantageous. These ammonium compounds could be used at higher concentrations without forming micellar systems. In comparison to the long-chain analogues, the short chain alkylammonium compounds also reversed the EOF at low pHs without competing with analytes for the hydrophobic CD cav-ity. However, disadvantages of methods utillizing short chain alkylammonium compounds are long analysis times and moderate resolution of the enantiomers (< 1.5).

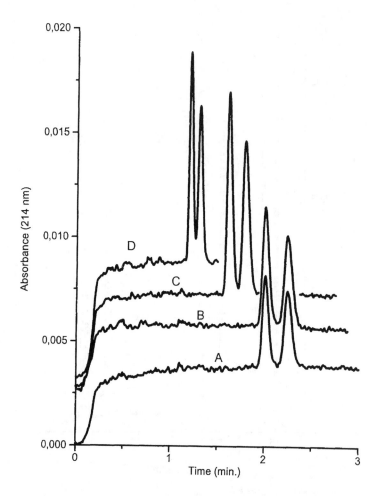

Figure 18. Optimized separation of a dihydropyridine derivative (see structure in Fig. 13). Run conditions: $L = 7/27$ cm, 75 μm i.d. coated capillary (4% T linear polyacrylamide), $T = 25\,°$C; 20 mM phosphate buffer, pH 6.0, with 0.5 (w/v) hydroxypropylated γ-cyclodextrin; detection at 214 nm; injection: A) 2 s, 3 kV; B), C) and D) 4 s, 4 kV.

In a system optimized for all important electrophoretic parameters, high resolution and very short analysis times are achievable for the separation of enantiomers. This is shown in Fig. 18 for the separation of a dihydropyridine derivative. Separation is achievable within 2 min using a hydroxypropylated γ-CD chiral selector, a short coated capillary with suppressed EOF, and a high electric field strength.

The analysis of enantiomeric impurities at trace levels can also be accomplished by CE. This is demonstrated in Fig. 19 in a dilution experiment for (±)-ephedrine. It was possible to determine (+)-ephedrine with 0.3% relative peak area compared to the main peak.

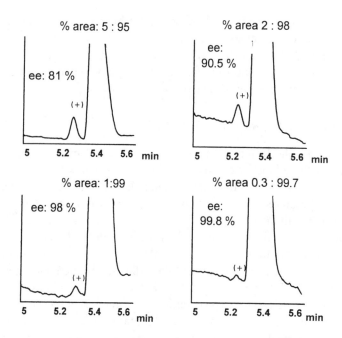

Figure 19. Determination of the enantiomeric excess of (−)-ephedrine using carboxymethylated β-CD. Run conditions: $L = 23.5/32$ cm, 75 μm i.d. fused capillary (4% T linear polyacrylamide), $E = 470$ V/cm, pressure injection 10 mbar for 3 s each; 20 mM citric acid buffer, pH 2.5, containing 2% (w/v) carboxymethylated β-CD; detection at 214 nm.

SEPARATION OF ENANTIOMERS USING MICELLAR ELECTROKINETIC CHROMATOGRAPHY

Micellar electrokinetic chromatography (MEKC) is used mostly for the analysis of neutral, uncharged species in capillary electrophoresis. It also achieves separations of ionic compounds according to different hydrophobicities of these molecules. Since hydrophobicity does not vary between any two enantiomers, a chiral selector is still required in the MEKC mode (Nishi *et al.*, 1990; Tran *et al.*, 1990; Otsuka and Terabe, 1991; Otsuka *et al.*, 1991, 1992). Chiral selectors in MEKC can be divided into three classes.

Chiral tensides (detergents). Chiral detergents can form micelles without the participation ot other nonchiral micelle-forming agents. Chiral micelle-forming detergents are also briefly summarized in Table 1.

Non-micelle-forming chiral selectors. These compounds do not exhibit a chiral interaction directly with an analyte. They behave as chiral selectors only when mixed with nonchiral surfactants. These types of selectors are incorporated into the micellar surface creating a chiral site at the outer micelle sphere.

Mixed systems. Some chiral selectors exhibit limited chiral selectivity, with optimum resolution achievable only in micellar systems. In such cases, SDS is a useful buffer additive. Other mixed systems use more than one chiral selector for the separation of enantiomers, taking advantage of synergetic effects in the selectivities of multiple chiral selectors.

CHIRAL MICELLE-FORMING DETERGENTS

Many of the micelle-forming detergents are based on amino acids. They are derivatized at the amino or carboxylic functional group with long-chain alkyl groups that form a chiral surfactant. Valine, alanine and glutamine have been described in literature as their sodium-dodecanoyl derivatives. One disadvantage connected with the use of pure selectors of this type is poor selectivity and low resolution. Only their use in mixed systems with SDS leads to acceptable efficiencies and resolution.

Otsuka and Terabe (1990) examined the influence of methanol and urea on the separation of PTH-amino acids using *N*-dodecanoyl-L-valine as a chiral selector. They observed a strong peak tailing when they used their chiral selector with methanol as a buffer additive. The addition of urea reduced the peak tailing but also destroyed the selectivity between PTH-D- and L-methionine. Even though the electropherograms shown had a relatively long run time, baseline separations were not achieved.

An improvement upon this type of selector was developed by Mazzeo *et al.* (unpublished). Further derivatization of the chiral selectors discussed above produced improved selectivity without buffer additives. The structures of these chiral surfactants are shown in Fig. 20. Another advantage of amino-acid based selectors is their usefulness in both their D- and L-forms. As a result, the migration order of the D- and the L-analyte can be reversed. This is shown in Fig. 21 for the separation of D,L-benzoin.

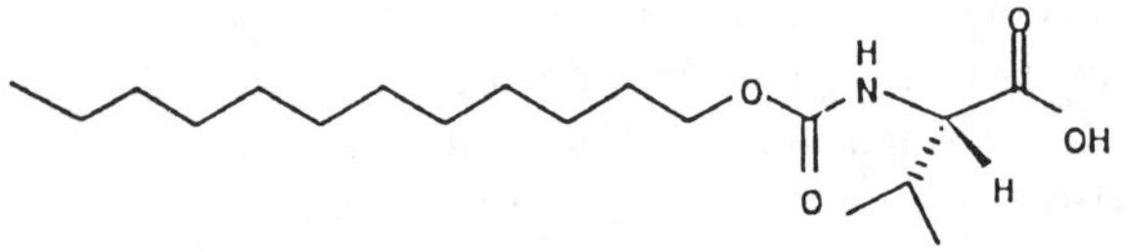

(S)-N-DODECOXYCARBONYLVALINE

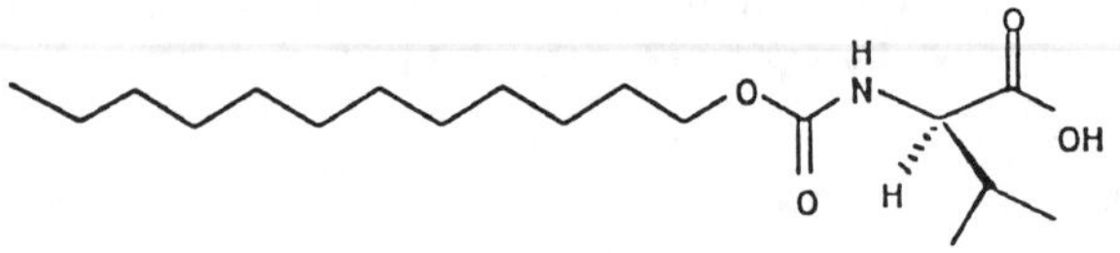

(R)-N-DODECOXYCARBONYLVALINE

Figure 20. Structure of chiral detergents in the (R)- and (S)-configuration.

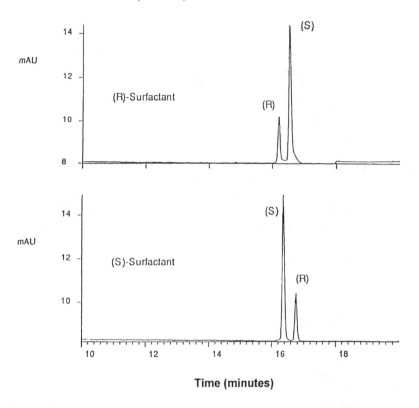

Figure 21. Chiral MEKC migration order reversal. Run conditions: $L = 60$ cm, 50 μm i.d.; detection at 254 nm; run voltage 15 kV; 50 mM phosphate buffer, pH 7.0, containing 25 mM (R) or (S)-dodecoxycarbonylvaline; sample: 3 : 1 ratio of (S)-to-(R) benzoin. Reproduced with permission of Mazzeo *et al*.

Other chiral detergents used in chiral CE are cholates and their derivatives. Their use as eluent modifiers in RP-HPLC is well established. Williams *et al.* (1990) used cholate derivatives to study their applicability as eluent modifiers for enantiomer separations. They discussed a helical structure for the micelles formed by cholates achieving the separation of binaphthyl derivatives in this system. The structure of these micelles has also been discussed in other publications (Espito *et al.*, 1987).

Terabe *et al.* (1989) employed taurodeoxycholates in pure chiral micelles for the separation of dansylated amino acids. They suggested a separation mechanism based on an aggregation of two cholate moieties. Despite unsatisfactory efficiencies, the potential of this method was established.

Nishi *et al.* (1989) employed cholates as chiral selectors in CE. A structural scheme of cholates is given in Fig. 22. It should be noted that cholates have a significantly lower aggregation number than that observed with SDS. Nishi *et al.* (1990b) used these chiral selectors for the separation of diltiazem and trimetoquinol derivatives. Small changes in the chemical structure of the cholates had a large impact on the sample separations. The pH was also an im-

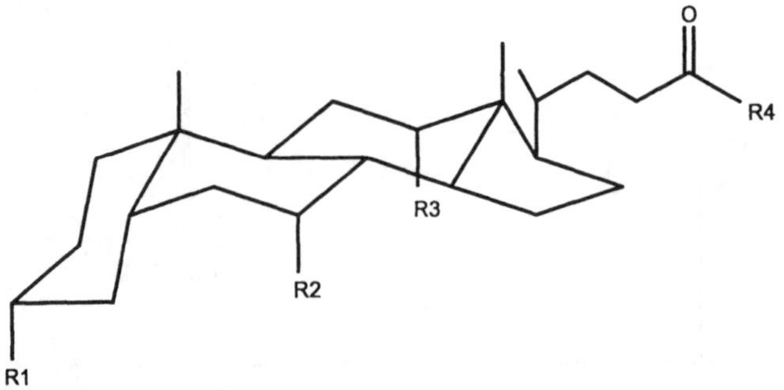

Cholate derivatives	R_1	R_2	R_3	R_4	CMC	Aggregation-number
Sodium cholate (SC)	-OH	-OH	-OH	-ONa	13	3
Sodium taurocholate (STC)	-OH	-OH	-OH	-NH-(CH$_2$-)$_2$-SO$_3$Na		
Sodium deoxycholate (SDC)	-OH	-H	-OH	-ONa	6	4
Sodium taurodeoxy-cholate (STDC)	-OH	-H	-OH	-NH-(CH$_2$-)$_2$-SO$_3$Na	9	11
Sodium dodecylsulfate (SDS)					8	63

Figure 22. Structure and substitution pattern of different cholates.

portant optimization parameter. Under neutral conditions, cholates form anions and basic analytes form cations. The resulting electrostatic attraction between the analyte and the selector was suggested as one explanation for the chiral recognition and the resulting resolution. The separation power of this method is demonstrated in Fig. 23. Generally, cholates form micelles with an aggregation number smaller than ten. These micelles possess a more or less rigid structure, consequently favoring analytes with more rigid structures. Since the separation mechanism involved is not completely understood, this field of chiral analysis remains a topic of scientific investigations.

Cole *et al.* (1990) studied the influence of organic solvents on the resolution in cholate-type systems. They analyzed negatively charged biphenyl derivatives instead of positively charged species. They found that methanol affected the cholate system in a manner similar to a SDS system. An increase in resolution with concentration of the organic modifier can be explained by the widened migration time window. It is recognized for SDS systems that the window for

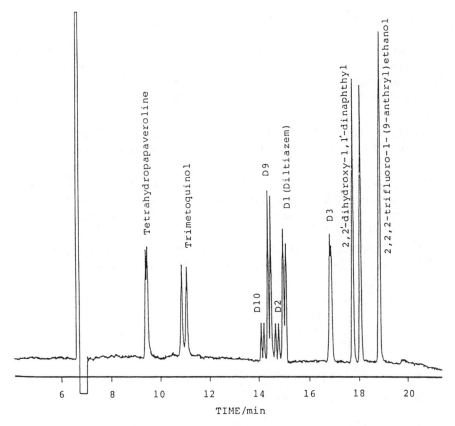

Figure 23. Chiral separation of diltiazem hydrochloride and related compounds. Run conditions: $L = 50/65$ cm, 50 μm i.d.; applied voltage: 20 kV; buffer: 50 mM sodium taurodeoxycholate in 20 mM phosphate-borate, pH 7.0; detection at 210 nm, ambient temperature. Reproduced with permission of Nishi *et al.*, 1990b.

optimal separation power lies between the migration time of the micelle (t_{mc}) and the migration time of the EOF marker (t_0). This is expressed as:

$$\alpha_{opt} = (t_{mc}/t_0)^{0.5}.$$

The number of charges on analytes also contributes significantly to resolution and chiral recognition. The best separation was achieved for a singly-charged binaphthyl derivative anion. Doubly-charged anions could no longer be resolved at pH 8. This was explained by electrostatic repulsion between the analyte and the chiral selector. A decrease of the buffer pH down to the range of the analyte's pK values led to an enhanced resolution. At that pH, the analyte was not completely dissociated and the effective charge on the analyte was reduced. Under these conditions, chiral recognition was favored, leading to a good selectivity of the system.

INCORPORATION OF A CHIRAL SELECTOR IN A MICELLE

A relatively new method in MEKC utilizes microemulsions. The method uses electrolyte systems containing a micelle-forming surfactant with additional components incorporated into the micelle to create a selectivity for a specific separation. Used alone, individual surfactants or additives would not generate any selectivity.

Aiken and Huie (1993) described a system based on microemulsions for the separation of enantiomers. Small chiral molecules, normally inactive as chiral selectors, can be incorporated into the micelle to form a 'chiral surface'. This work represents a significant contribution to the field of chiral CE, because there are many small chiral compounds available which have not yet shown any chiral activity. Microemulsion systems may lead to new aoolications with many new chiral selectors for CE enantiomeric separations. In this work, the lipophilic chiral selector (2R,3R)-di-*n*-butyltartrate was incorporated into a micellar SDS system. For the separation of (1R,2S)- and (1S,2R)-ephedrine, a high selectivity (alpha value of 2.6 as determined by the method of Terabe *et al.*, 1985a) was achieved in a short analysis time. This separation is shown in Fig. 24. Increasing the concentration of the chiral selector led to an increase

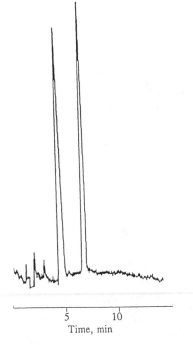

Figure 24. Chromatogram obtained from a racemic mixture of (1R,2S)- and (1S,2R)-ephedrine using a microemulsion system. Run conditions: $L = 45/75$ cm, 75 µm i.d.; applied voltage 25 kV; detection at 215 nm; buffer: 1.2% (w/w) butanol, 0.6% (w/w) SDS, 0.5% (w/w) (2R,3R)-di-*n*-butyltartrate in 15 mM Tris-hydroxyaminomethane, pH 8.1. Reproduced with permission of Aiken and Huie, 1993.

in selectivity, but unfortunately also to a significant decrease in separation efficiency. Systems with a chiral selector incorporated into a hydrophobic surface are already known from HPLC. Many systems have been described in literature using a chiral selector and a non-polar phase for the separation of enantiomers (Allenmark, 1988).

MIXED SYSTEMS

Mixed systems are often used in capillary electrophoresis when the analyte has a poor solubility in an aqueous system or when a chiral selector fails to provide suitable selectivity. In these complex systems, identification of a separation mechanism is often difficult. The mixed systems discussed here deal with micellar SDS-systems, using chiral selectors as additives. The presence of SDS micelles can be advantageous (higher selectivity) as well as disadvantageous (no selectivity) for chiral separations. Other mixed systems employ more than one chiral selector to enhance or to generate a selectivity for enantiomeric separations.

SDS-chiral detergents

Dobashi *et al.* (1989a, b) described the effects of *N*-dodecanoyl-L-valine and *N*-dodecanoyl-L-alanine for the separation of DNP-amino acid-*O*-isopropylesters. In addition to using pure micelles consisting of only one chiral surfactant, they also used a co-micellar system with SDS. Since the derivatization of amino acids often diminishes the difference in their physical properties, i.e. electrophoretic mobilities, an SDS micellar system may often be helpful to en-hance the selectivity between different species of a sample mixture. Employing mixed systems also means a dilution of the active chiral selector. This may also lead to a more homogeneous chiral surface on the micelle. As is typical in MEKC systems, organic solvents change the separation performance. System-atic studies indicate the main effect of primary organic alcohols is to change the aggregation number of the micelle.

As discussed above, it is important to work in the middle of the migration time window in MEKC to yield a maximum of selectivity. Parameters affecting this migration time window in MEKC are the type of detergent used, the concentration of the detergent, and the presence of organic solvents in the buffer.

Apart from the SDVal systems described above, Otsuka and Terabe (1990b) employed mixed nonionic, chiral micelle-forming detergents. They used mixed micelles consisting of SDS and digitonin for chiral separations. The chemical structure of digitonin is shown in Fig. 25. They found a strong pH dependence for the separation of a PTH-amino acid test mixture. At pH 7, the migration time window was not wide enough because of a low mobility of the mixed micelles relative to the EOF. Under conditions far from optimal, only a poor resolution was achieved. With a 25 mM digitonin, 50 mM SDS solution at pH 3,

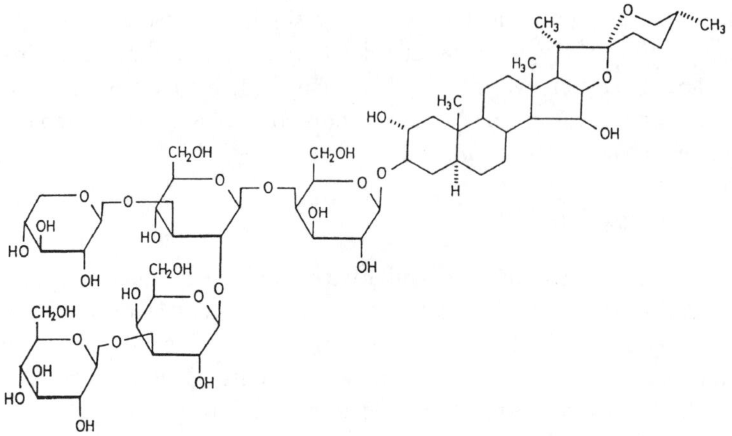

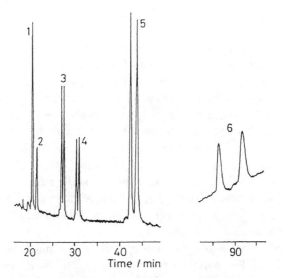

Figure 25. Structure of digitonin.

Figure 26. MEKC of six PTH derivatives of D,L-amino acids. Run conditions: $L = 49/63$ cm, 50 μm i.d.; applied voltage 20 kV; detection wavelength 260 nm, buffer: 20 mM digitonin, 50 mM SDS, pH 3; sample: tryptophan (1), norleucine (2), norvaline (3), valine (4), α-aminobutyric acid (5), alanine (6). Reproduced with permission of Otsuka and Terabe, 1990b.

EOF was reduced, the migration time window was expanded, and a separation achieved. The separation obtained with this system is shown in Fig. 26.

Similarly as the digitonin-SDS system, a pure SDVal system was also employed to study differences in selectivities between the two systems. Using the SDVal system, all PTH amino acids, previously shown in the digitonin-SDS system, were separated in a much shorter time. Unfortunately, as mentioned previously, using SDVal as a micelle-forming surfactant without any additives resulted in asymmetric peaks, low efficiencies and poor resolution. Additionally, a strong fluctuation of the baseline was observed. Assuming wall adsorption effects of the analytes, the usual procedures to recondition the capillary were not successful for this separation.

Otsuka *et al.* (1993) also reported using chiral micelle-forming detergents for the separation of racemic PTH-amino acid mixtures. They used *N*-dodecanoyl-L-glutamate systems (SDGlu systems) and mixed micelles of digitonin and sodium taurodeoxycholate. The use of urea, SDS and methanol in the SDGlu system was found to compensate for the poor peak symmetry and the poor resolution mentioned above. However, the overall differences between the SDGlu- and the SDVal-SDS system were rather small.

SDS-cyclodextrins

Cyclodextrins are not only useful as an electrolyte additive for chiral analysis, but they may also be very helpful by increasing the selectivity and solubility of many other hydrophobic compounds.

Nishi and Matsuo (1991) reported success using cyclodextrins in an SDS system for the separation of steroids and other aromatic compounds. Normally, hydrophobic compounds strongly partition into a micelle and are detected close to the migration time of the micelle. This also means that the analysis is performed far from the optimal selectivity of this system. The addition of cyclodextrins creates a third phase in this buffer system, i.e. aqueous buffer, micelles and CD. This leads to a decrease in the time which the analyte spends inside a micelle because of the competition with the CD. A faster migration time is observed in such case and the analytes are moved to a better position within the migration time window (i.e. closer to the optimal selectivity). Of course, the type of CD and the nature of the micelle-forming detergent as well as other buffer additives like organic solvents play a dominant role for the selectivity in these systems.

In another report of Nishi *et al.* (1991), the influence of the type of cyclodextrin as a third phase in an SDS system was examined. Since cyclodextrins do not dissolve in a micelle, CDs are moved by the EOF through the column. The resulting decrease in the capacity factors of a CD-modified micellar system can be very helpful for the optimization of selectivity. As already mentioned, the optimal capacity factors in the MEKC mode are around $(t_{mc}/t_0)^{0.5}$. The concentration of cyclodextrin in an SDS buffer allowed adjustment to the optimal capacity factors for that separation. In these experiments, some barbiturates, binaphthyl derivatives and anthracene derivatives were used as chiral analytes. Aside from regular cyclodextrins (α-, β-, γ-CD), DM-β-CD and TM-β-CD were also employed. Systematic studies showed that γ-cyclodextrin was better in an SDS system than other cyclodextrins. This is in contrast to HPLC data where a β-CD stationary phase was found to be the most effective selector.

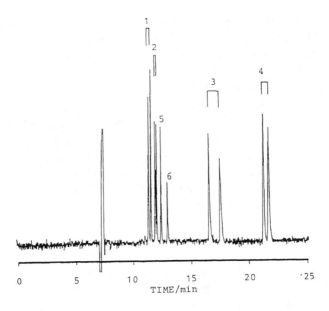

Figure 27. Chiral separation by the addition of sodium D-camphor-10-sulphonate. Run conditions: 20 mM D-camphor-10-sulphonate in 20 mM phosphate-borate buffer of pH 9.0 containing 50 mM SDS and 30 mM γ-cyclodextrin; $L = 50/65$ cm, 50 µm i.d.; detection at 220 nm; sample: sodium thiopental (1), calcium pentobarbital (2), 2,2,2-trifluoro-1-(9-anthryl)-ethanol (3), 2,2'-dihydroxy-1, 1'-dinaphthyl (4), phenobarbital (5), sodium barbital (6). Reproduced with permission of Nishi *et al.*, 1991.

The superiority of γ-CD in CE is explainable by the larger inner diameter of the cyclodextrin cavity where there is space for a surfactant molecule to enter the cavity along with the analyte. The addition of methanol to the buffer leads to a longer analysis time because of a reduced EOF. The result, however, was characterized by smaller capacity factors and decreased enantiomeric selectivity. This was explainable by the incorporation of methanol into the cyclodextrin cavity and the resulting steric hindrance of the chiral recognition between the analyte and the chiral selector. Another interesting suggestion was to use other chiral additives in the buffer that could be incorporated into the cyclodextrin. In this context, sodium D-camphor-10-sulphonate as well as L-menthol acetate were added separately to the γ-CD/SDS-buffer system. Surprisingly, this resulted in increased selectivity and resolution for the investigated analytes. This rarely observed effect was attributed to an increased chirality of the cyclodextrin when another chiral compound is incorporated. An electropherogram showing a separation in this system is given in Fig. 27. The addition of these chiral additives led to the loss of resolution for one of the analytes. The extent to which D-camphor-10-sulphonate and L-menthol acetate interact with the micelles in the systems is not discussed here. This could also be a reason for the changed chiral selectivity. Generally, an interpretation of these complex separation mechanisms in a system with many different equilibria is a difficult undertaking.

Other mixed systems were described by Terabe *et al.* (1993). They used different cyclodextrins in a micellar SDS-system for the separation of dimethyl-naphthalene isomers and other polyaromatic compounds. For dansylated amino acid enantiomers, separation was achieved employing an SDS/β-CD/γ-CD system or a buffer containing taurodeoxycholate and γ-CD.

The systems discussed above show that there may be many different combinations of the chiral selectors available. However, there are few examples where the presence of more than one chiral selector in the electrolyte improved the chiral selectivity and generated a synergetic effect. This also means that it is difficult to predict a successful combination of such chiral selectors in order to achieve a separation.

ENANTIOMERIC SEPARATION WITH OTHER CHIRAL SELECTORS

In this section, other chiral selectors, which have yet to find wide applicability in CE, will be discussed. Reasons for their restricted use are the UV-activity and low selectivity of these selectors. Also, the chiral analytes separated with these selectors can often be analyzed more efficiently with other chiral additives.

Proteins as chiral selectors for the separation of enantiomers

In HPLC, protein-based chiral stationary phases for the separation of enantiomers have long been used. Their utility is explained by the fact that protein reactions are very substrate specific, meaning that they interact only with a small number of analytes. Furthermore, proteins are built of chiral amino acyl units, giving them many chiral recognition sites for use as a chiral selector.

The first work in capillary electrophoresis with proteins as chiral selectors was published by Birnbaum and Nillson (1992). Their studies used a gel consisting of BSA as a chiral selector crosslinked with glutaraldehyde to form the polymer matrix. To ensure sensitive on-column detection (proteins show UV activity), the unpolymerized BSA-glutaraldehyde solution was only loaded in the capillary up to the detection window, after which polymerization was started. This procedure allowed the use of high concentrations of the chiral selector without a loss in detection sensitivity. They called this new method 'capillary gel affinity electrophoresis'. Possible problems arising from the stability of the rigid gels in the column and the capillary coating are not discussed by the authors.

Barker *et al.* (1992) used BSA solely as a buffer additive for the separation of enantiomers. Since both BSA and the sample are negatively charged, the mobility differences between the analyte and the chiral selector determine the extent of their interaction. The authors studied the influence of BSA concentration and pH on the separation efficiencies of this system. To prevent the proteins from adsorbing on the wall, polyethylene glycol was used both as buffer additive and dynamic modifier of the capillary wall. Because of the

UV absorbance of BSA, only a small concentration range of the chiral selector could be examined for the optimization of the separation.

Some papers deal with dextrans as buffer additives together with proteins as chiral selectors. Sun *et al.* (1993a) used a high molecular weight dextran covalently attached to BSA to achieve the separation of leucovorin. This separation was accomplished by means of capillary affinity gel electrophoresis. A liquid BSA-dextran-polymer network was introduced by pressure into a capillary coated with linear polyacrylamide. Although a polymer network was used, it could be replaced after each run like a regular liquid buffer system. They reported in another paper that BSA does not nesessarily need to be covalently bound to the dextran polymers (1993b). They separated ibuprofen and leucovorin using BSA added to a dextran solution. The function of the dextran polymer was to reduce the mobility of BSA by a sieving mechanism to allow a longer period of interaction with the analyte. Due to the UV activity of the BSA, the signal-to-noise ratio was unsatisfactory in all electropherograms shown. Practical applications for the determination of an enantiomeric excess of one enantiomer should not be advanced with this system. Furthermore, negative signals in the baseline disturbed the separation and did not allow any quantification.

Hjerten and co-workers reported the successful use of cellobiohydrolase I (Cellulase) for the separation of β-blockers in capillary electrophoresis (Valtcheva *et al.*, 1993). This was another example using acrylamide coated capillaries to prevent proteins from sticking to the wall. These capillaries had nearly no electroosmotic flow, which normally determines the migration direction of the analytes. Consequently, at the pH used, the UV-absorbing proteins (anions) migrated away from the detection window toward the injection side of the capillary (anode) while the analytes (cations) migrated in the opposite direction toward the detector. Since these proteins do not disturb the detection, high concentrations of this chiral selector in the inlet buffer vial can be used for optimization. Using this procedure it is still possible to retain sensitive detection at low wavelengths. Valtcheva *et al.* (1993) determined that only high concentrations of the chiral selector and addition of 2-propanol allowed the separation of β-blockers. This is shown in Fig. 28. Under conditions similar to HPLC for the same chiral selector, no chiral separation was possible in capillary electrophoresis.

Aside from BSA, human serum albumin was also successfully employed for the separation of kynurenine, tryptophan, lactic acid and tartaric acid derivatives (Vespalec *et al.*, 1993). Again, wall adsorption was minimized by the use of a polyacrylamide coated column. In their experiments, the authors recommended heating the protein to 60 °C for a short time to stabilize the selectivity of human serum albumin. Increasing the concentration of the protein led to an increase in resolution because of higher efficiencies although the selectivity dropped. In a direct comparison of the use of albumin selector in HPLC and CE, CE showed significantly higher separation efficiencies for the same analyte. However,

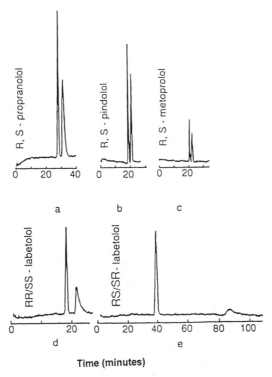

Time (minutes)

Figure 28. Separation of (R,S)-propranolol (a), (R,S)-pindolol (b), (R,S)-metoprolol (c), (RR/SS)-labetolol (d) and (RS/SR)-labetolol (e). Run conditions: $L = 8.5/11.5$ cm, 75 μm i.d. coated with linear polyacrylamide; buffer: 400 mM sodium phosphate, pH 5.1, supplemented with (a, b, c) 25% and (d, e) 30% 2-propanol, 20 mg agarose/ml buffer; capillary filled partially with 40 mg/ml cellobiohydrolase; detection at 220 nm; applied voltage 1 kV. Reproduced with permission of Valtcheva *et al.*, 1993.

the selectivities in the two systems were not directly comparable. In Fig. 29 an electropherogram of 2,3-dibenzoyl-D,L-tartaric acid is shown using human serum albumin. As mentioned above, the efficiencies in CE were much higher than in HPLC, but in this particular example they were still below what is normally attainable in CE.

Poppe and co-workers reported the use of acidic glycoproteins for the separation of enantiomers (Busch *et al.*, 1993). Orosomucoid, ovomucoid and fungal cellulose were found to be suitable as chiral selectors. BSA, a nonglycoprotein, was additionally employed for the separation of warfarin, benzoin, tryptophan, pindolol and promethazine. The addition of 1-propanol to the buffer was helpful in most cases. Higher selectivities as well as higher efficiencies could be observed with increasing concentration of the chiral selector. Very high concentrations of the 1-propanol led to a decrease in resolution. As in other examples mentioned above, sensitive detection was not possible because of the UV activity of the chiral selector. The electropherograms shown by the authors never gave a signal-to-noise ratio better than ten. Since UV-detection

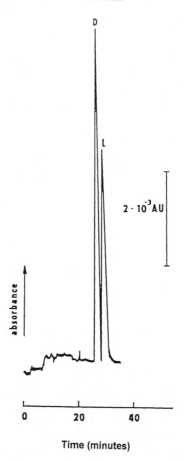

Figure 29. Separation of 2,3-dibenzoyl-D,L-tartaric acid. Run conditions: $L = 60/76$ cm, 80 μm i.d., applied voltage 9.5 kV, $T = 29–30°C$ (ambient); detection at 238 nm, 10 mM acetate-Tris, pH 8, containing 3 mg/ml human albumin, heated for 30 min at pH 9 and 60°C. Reproduced with permission of Vespalec *et al.*, 1993.

between 260 and 315 nm was used, an optimization of the concentration of the chiral selector was not possible.

In conclusion, sensitive detection using proteins is only possible when the capillary is filled with a column of buffer containing the chiral selector up to a point before the detection window. Also, problems with wall adsorption of the proteins should not be neglected at the expense of maintaining constant separation conditions in the column. This also puts some requirements on commercially available instruments, since control of complex rising procedures is necessary.

Linear sugars as chiral selectors for the separation of enantiomers

Polysaccharides belong to a widely used class of chiral selectors in HPLC. For example, acetylated celluloses immobilized on silica gel result in chiral stationary phases with high selectivity and resolution. Because of their insolubility in aqueous systems they are not easily transferable to capillary electrophoresis. Cyclic oligosaccharides such as cyclodextrins were discussed in depth earlier in this chapter. Only a few linear or branched polysaccharides available in nature have been tested as chiral selectors in capillary electrophoresis.

D'Hulst and Verbeke (1992) examined maltodextrins as potential chiral selectors for the separation of different enantiomers. This article surveyed some properties of α-(1-4)-D-glucose polymers. The separation mechanism using these linear oligosaccharides was found to be very complex. Many different parameters influenced the separation, but none was primarily responsible for chiral recognition. Buffer additives such as acetonitrile and micelle-forming detergents were investigated for their influence on the enantiomeric separation. With the addition of these components, resolution was lost in all cases. The influence of the concentration of different maltodextrins such as maltrine, maldex, glucidex and mylose was also investigated. Generally, higher concentrations of the maltodextrins enhanced the resolution. Although the structures of all samples were similar, their separation with only one maltodextrin was not successful. To determine which molecular weight range of the maltodextrins was most responsible for chiral recognition and resolution, molecular weight fractions were separated by ultracentrifugation. Oligosaccharides with molecular mass less than 3000 showed only a small decay in selectivity but the analysis time was strongly reduced. Fractions below 1000 in molecular weight no longer produced a separation. To compare the maltodextrin-based results to other chiral selectors, cyclodextrins such as β-, hydroxypropylated β- and dimethylated β-CD were used. For these selectors no separation of the chiral

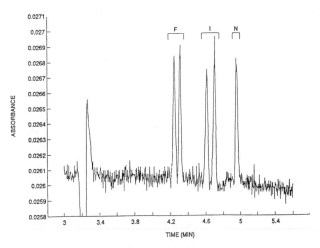

Figure 30. Electropherogram representing a chiral separation of 2-arylpropionic acid non-steroidal anti-inflammatory drugs. Run conditions: 10 mM sodium phosphate buffer, pH 7.05, 1.5% Glucidex 2, $L = 50/90$ cm; 75 μm i.d. coated capillary; applied voltage: 30 kV; detection at 214 nm. F = Flurbiprofen, I = Ibuprofen, N = Naproxen. Reproduced with permission of D'Hulst and Verbeke, 1992.

analytes was found under these conditions. It should be mentioned that not all optimization parameters for the cyclodextrins were totally exploited in this case. Nevertheless the utility of linear polysaccharides still remains questionable. None of the electropherograms presented showed a signal-to-noise ratio better than 3 to 4. Figure 30 gives an example for the separation of enantiomers using Glucidex as chiral selector.

Separation of enantiomers using electrochromatography

Some authors used HPLC-type phases in packed capillaries with elution by electroosmotic flow (Yamamoto *et al.*, 1992). Although the efficiencies seemed to be promising, no chiral stationary phases were used in *packed* capillaries for the separation of enantiomers. However, some chiral coatings were used as a chiral stationary phase in *coated* capillaries for electrochromatographic separations of enantiomers.

Meyer and Schurig (1992) were the first to demonstrate the separation of enantiomers based on chiral wall modifications of a capillary. As a chiral selector a 'Chirasil-Dex' type stationary phase from HPLC, GC and SFC was immobilized on the capillary wall. With this cyclodextrin-based coating material phenylethanol and (±)-1,1'-binaphthyl-2,2'-diyl-hydrogen phosphate were separated in these capillaries. This was remarkable since the distance from the middle of the capillary to the chiral selector on the wall had to be traversed by diffusion of the analyte. This also meant that only a few wall contacts with the chiral selector are enough for the analytes to experience a sufficient chiral discrimination.

Another example has been reported for the separation of enantiomers using a chiral capillary wall modification. Armstrong *et al.* (1993) developed an easy method for immobilization of a cyclodextrin-based chiral stationary phase in GC, SFC and CE. This method, based on polysiloxane and immobilized cyclodextrin derivatives, was used to separate mephobarbital in CE.

The stability of the chiral coatings at higher pHs was not described in any of these publications. Neither did the authors discuss the behavior of these phases under conditions of overloading the system for the determination of enantiomeric excess of one of the analytes.

Formation of diastereomeric metal complexes

The use of chiral metal complexes for the separation of enantiomers is one of the oldest chromatographic methods for the separation of enantiomers. In 1971, Davankov and Rogozhin described the synthesis of a so-called 'unsymmetric complexating' ion exchanger based on polystyrene (a chloromethylated styrene reacted with L-proline as a chiral selector). This phase showed a much higher affinity to the D-amino acids than to the L-antipodes in the presence of Cu^{2+} ions.

Gassman *et al.* (1985) were the first group to use a similar system for the separation of enantiomers in capillary electrophoresis successfully. They described the formation of diastereomeric interactions of dansylated D,L-amino acids with a Cu(II)-L-histidine complex, which was added to the background electrolyte. This chiral selector was used successfully in HPLC, facilitating an easy transfer from the HPLC experiment to capillary electrophoresis. The derivatization of amino acids with dansyl chloride and detection with laser fluorescence led to a significant increase of the detection sensitivity of the separation system. The formation and dissociation equilibria of the ternary metal complexes are responsible for the selectivity. Variations in fluorescence quenching depending on the stability of formed complexes give rise to differences in detection response for a pair of enantiomers. For quantification, it is necessary to have separate calibrations of both enantiomers.

Gozel *et al.* (1987) used the different interactions of enantiomers with a Cu(II)-aspartame complex for their separation. Aspartame, the dipeptide L-aspartyl-L-phenylalanine-methyl ester, was used with laser-induced fluorescence detection because the peptide absorbs strongly in the lower UV. In their experiments the influence of the background electrolyte, pH, and the temperature was examined along with the linearity of the detection system. The detection linearity was found to be excellent within the range of four orders of magnitude. Fluorescence quenching effects were also found to be more critical for the enantiomer more strongly bound in the complex. However, aspartame complexes showed a better resolution for dansylated amino acids in comparison to the analoguous histidine complexes.

Cohen *et al.* (1987) described the use of chiral micelle-forming selectors in chiral CE. They employed dodecyl-L-alanine, which could only be solubilized in the system with SDS. Additionally, the separation of dansylated D,L-amino acids was only successful in the presence of Cu(II)-ions using UV-detection. Therefore it could be concluded that the separation in this system is not based on a real MEKC mechanism, but more on a chiral, ion exchange supported metal chelation on a micellar surface. The 'dilution' of the active chiral selector by other detergents was also mentioned by Karger and his co-workers (Cohen *et al.*, 1987). This dilution could have allowed a better spacing between two ligand exchange centers on the micellar surface to accommodate the analytes. The selectivity was enhanced as a consequence.

The use of free tartaric acid as a buffer additive was described for the separation of some enantiomers. Fanali *et al.* (1989) reported the separation of stable enantiomeric and diastereomeric Co(III)-complexes. In their system, the selectivity for a pair of enantiomers was found to be sufficient, but in a mixture the different components of the analyte mixture did not develop high enough selectivities. Therefore cyclodextrins were added to enhance the separation. But the cyclodextrin had no influence on the chiral resolution in this case. In Fig. 31 the separation of different Co(III)-complexes using tartaric acid is shown. This is one of few examples described where small molecules are used directly as chiral selectors without being incorporated into complex structures.

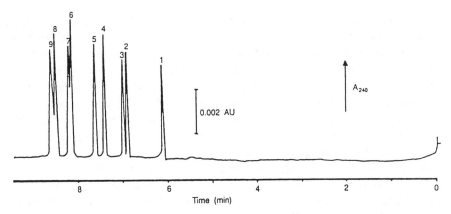

Figure 31. The CZE separation of optical isomers of cobalt complexes. Run conditions: $L =$ 35/43 cm, 100 μm i.d., applied voltage: 10 kV; detection at 240 nm, 15 mM L(+)-tartaric acid adjusted with Tris to pH 5.25, 15 mM β-CD, samples: (1) $[Co(o\text{-}phen)_3]^{3+}$, (2) $[Co(en)_2gly]^{2+}$, (3) $[Co(en)_2gly]^{2+}$, (4) $[Co(en)_2L\text{-}pro]^{2+}$, (5) $[Co(en)_2L\text{-}pro]^{2+}$, (6) $[Co(en)_2L\text{-}phe]^{2+}$, (7) $[Co(en)_2D\text{-}phe]^{2+}$, (8) $[Co(en)_2D\text{-}phe]^{2+}$, (9) $[Co(en)_2L\text{-}phe]^{2+}$. Reproduced with permission of Fanali *et al.*, 1989.

Derivatization prior to separation

Today, the derivatization of analytes is frequently performed in many analytical laboratories. It is used mostly when the properties of the analytes do not allow direct detection. With derivatization, the selectivity between the analytes may sometimes improve. These two points, selectivity of the separation sytem and detection sensitivity, are the main reasons for the derivatization of enantiomers. This is not only true for capillary electrophoresis but for any other analytical technique as well.

Ueda *et al.* (1991) focused on the derivatization of D,L-amino acids using naphthaline-2,3-dicarboxylic acid, which forms a 1-cyano-2 substituted benz[f] isoindole with cyanide. This derivatization allows the use of laser induced fluorescence detection. When achiral derivatization reagents are applied, the analyte molecules remain as enantiomers which still require a chiral selector to achieve separation. In this case a mixed system employing SDS and CDs was investigated. Ueda *et al.* found a dependence of the migration order on the structure of the D- and the L-enantiomer for β-CD in a MEKC system. Aliphatic D-amino acids migrated in front of the L-amino acids, while for aromatic amino acids, the L-enantiomer migrated first. In this paper, a γ-CD/SDS system was described as the most effective separation system for these analytes giving higher resolution than the β-CD/SDS system.

Nishi *et al.* (1990a) reported the separation of D,L-amino acids after derivatization with an optically active reagent. This kind of derivatization followed by separation is called an indirect enantiomeric separation method. The derivatization leads to the formation of diastereomers (thiourea derivatives) with different physical properties. In this case the separation was based on differential partitioning of the aromatic, heterocyclic and basic amino acids in an SDS micelle.

Of twenty-one α-amino acids, nineteen could be separated in a micellar SDS-system. The GITC derivatization was first reported for use in reverse phase HPLC of amino acid enantiomers. The GITC reagent reacted very rapidly, completely, and under mild conditions with primary and secondary amino groups of the analytes. This explains its wide acceptance in this field.

Lurie (1992) reported the separation of enantiomers in CE after derivatization with GITC. After formation of diastereomers with this reagent the separation of amphetamine, ephedrine, pseudoephedrine, norephedrine, norpseudoephedrine and methamphetamine was achieved in a 100 mM SDS system with 20% methanol. For optimization, the concentrations of both SDS as well as methanol in the system were varied and their influence on the selectivity was studied. The influence of the temperature and the separation voltage on the selectivities was also described by Lurie. Different cyclodextrins (heptakis (di-O-methyl)-β-CD) as well as a micelle-forming agent (sodium taurocholate) were compared with each other. Under these conditions, separation of the analytes was achieved in only a few cases. However, the alkaline conditions chosen in that study were not the most suitable ones for the separation of amines in a mixed system of SDS/β-CD.

CONCLUSIONS

One of the main reasons for the separation of enantiomers is the determination of the enantiomeric excess of optically active drugs. In addition to the determination of optical impurities, some work has been done in the area of method validation in chiral CE.

Altria *et al.* (1993b) validated a capillary electrophoretic method for the separation of (+)- and (−)-fluparoxan. They demonstrated that the recovery, precision, detection limit, and detector linearity fulfill the usual requirements for this kind of separation.

The ability to transfer a method from one laboratory to another is an important point for the validation of an analytical method. Altria *et al.* (1993a) reported a cross validation exercise in different laboratories and with different commercial instruments, but using the same buffer system. The experiment demonstrated that the CE method used was able to generate accurate and precise data with an acceptable reproducibility of migration time as well as of peak areas.

From these studies, we can conclude that CE is a suitable analytical method for the analysis of enantiomers for practical requirements. Another advantage of CE is fast method development. The different buffer compositions responsible for generating the selectivity can be exchanged quickly in a capillary. Long equilibration times typical of HPLC are uncommon in CE. Using fully automated CE instrumentation, an indication of the suitability of a chiral selector for a specific separation problem can be gained much faster than in HPLC, SFC or even GC. Also, chiral buffer chemicals for CE are by far less costly than the chiral stationary phases used in other analytical techniques. This is mostly due to the lower consumption of these chemicals, since smaller buffer volumes are generally required in CE.

The main disadvantage, not restricted to the separation of enantiomers in CE, is the predominantly analytical scale of this separation method. As a result of the very small sample volumes in CE (usually 5–50 nl), a high mass throughput for preparative sample collection is nearly impossible in spite of the very high mass sensitivity of the detection. Repetitive injections with sample fractionation do not seem to be sufficient for most preparative objectives. This obvious disadvantage of capillary electrophoresis can be turned into an advantage when only small amounts of sample are available and many analyses are required.

The growing number of scientific publications is a clear indicator of rapid developments in the field of capillary electrophoresis. These developments not only focus on the search for new chiral selectors, but also on the validation of these methods and their applications. There is no end in sight of new developments for the separation of enantiomers in CE. For these reasons, this review can only describe the present state of the art in an area characterized by rapid changes.

ACKNOWLEDGEMENTS

First of all, I would like to thank Thermo Separation Products, especially Dr M. Geyer for her interest and friendly support throughout this work. Special thanks belong to Professor Dr H. Engelhardt for the revision of this manuscript and his scientific contributions throughout this work.

REFERENCES

Aiken, J. H. and Huie, C. W. (1993). Use of microemulsion system to incorporate a lipophilic chiral selector in electrokinetic capillary chromatography. *Chromatographia* **35**, 448–450.

Allenmark, S. G. (1988). *Chromatographic Enantioseparation: Methods and Applications*. Ellis Horwood, Chichester.

Allenmark, S. G. (1991). *Chromatographic Enantioseparation: Methods and Applications*, 2nd edn. Ellis Horwood, Chichester.

Altria, K. D. and Simpson, C. F. (1987). High voltage capillary zone electrophoresis: Operating parameters effects on electroosmotic flow and mobilities. *Chromatographia* **24**, 527–530.

Altria, K. D., Goodall, D. M. and Rogan, M. M. (1992). Chiral separation of β-aminoalcohols by capillary electrophoresis using cyclodextrins as buffer additives. I. Effect of varying operating parameters. *Chromatographia* **34**, 19–24.

Altria, K. D., Harden, R. C., Hart, M., Hevizi, J., Hailey, P. A., Makwana, J. V. and Portsmouth, M. J. (1993a). Inter-company cross-validation exercise on capillary electrophoresis: I. Chiral analysis of clenbuterol. *J. Chromatogr.* **641**, 147–153.

Altria, K. D., Walsh, A. R. and Smith, N. W. (1993b). Validation of a capillary electrophoresis method for the enantiomeric purity testing of fluparoxan. *J. Chromatogr.* **645**, 193–196.

Armstrong, D. W., Tang, Y., Ward, T. and Nichols, M. (1993). Derivatized cyclodextrins immobilized on fused-silica capillaries for enantiomeric separations via capillary electrophoresis, gas chromatography, or supercritical fluid chromatography. *Anal. Chem.* **65**, 1114–1117.

Baeyens, W., Weiss, G., Van Der Weken, G., Van Den Bossche, W. and Dewaele, C. (1993). Analysis of pilocarpine and its trans epimer, isopilocarpine, by capillary electrophoresis. *J. Chromatogr.* **638**, 319–326.

Barker, G. E., Russo, P. and Hartwick, R. A. (1992). Chiral separation of leucovorin with bovine serum albumin using affinity capillary electrophoresis. *Anal. Chem.* **64**, 3024–3028.

Belder, D. and Schomburg, G. (1992). Modification of silica surfaces for CZE by adsorption of non-ionic hydrophilic polymers or use of radial electric fields. *HRC* **15**, 686–693.

Birnbaum, S. and Nillson, S. (1992). Protein-based capillary affinity gel electrophoresis for the separation of optical isomers. *Anal. Chem.* **64**, 2872–2874.

Busch, S., Kraak, J. C. and Poppe, H. (1993). Chiral separations by complexation with proteins in capillary zone electrophoresis. *J. Chromatogr.* **635**, 119–126.

Cellai, L., Desiderio, C., Filippetti, R. and Fanali, S. (1993). Capillary electrophoresis: Quantitation of 1-L-folinic acid in the presence of its inactive d-L-form. *Electrophoresis* **14**, 823–825.

Chen, E. T. and Pardue, H. L. (1993). Analytical applications of catalytic properties of modified cyclodextrins. *Anal. Chem.* **65**, 2563–2567.

Cohen, A. S., Paulus, A. and Karger, B. L. (1987). High performance capillary electrophoresis using open tubes and gels. *Chromatographia* **24**, 15.

Cole, R. O., Sepaniak, M. J. and Hinze, W. L. (1990). Optimization of binaphthyl enantiomer separations by capillary zone electrophoresis using mobile phases containing bile salts and organic solvent. *HRC* **13**, 579–582.

Cruzado, I. and Vigh, G. (1992). Chiral separation by capillary electrophoresis using cyclodextrin-containing gels. *J. Chromatogr.* **608**, 421–425.

Davankov, V. A. and Rogozhin, S. V. (1971). Ligand chromatography as a novel method for the investigation of mixed complexes: stereoselective effects in alpha-amino acid copper (II) complexes. *J. Chromatogr.* **60**, 280–283.

D'Hulst, A. and Verbeke, N. (1992). Chiral separation by capillary electrophoresis with oligosaccharides. *J. Chromatogr.* **608**, 275–287.

Dobashi, A., Ono, T., Hara, S. and Yamaguchi, J. (1989a). Optical resolution of enantiomers with chiral mixed micelles by electrokinetic chromatography. *Anal. Chem.* **61**, 1984–1986.

Dobashi, A., Ono, T., Hara, S. and Yamaguchi, J. (1989b). Enantioselective hydrophobic entanglement of enantiomeric solutes with chiral functionalized micelles by electrokinetic chromatography. *J. Chromatogr.* **480**, 413–420.

Dotsevi, G., Sogah, E. and Cram, D. J. (1975). Chromatographic optical resolution through chiral complexation of amino ester salts by a host covalently bound to silica gel. *J. Am. Chem. Soc.* **97**, 1259.

Drayer, D. E. (1988). In: *Drug Stereochemistry*, Wainer, I. W. and Drayer, D. E. (Eds). Marcel Dekker, Basel, pp. 209–226.

Espito, G., Giglio, E., Pavel, N. V. and Zanobi, A. (1987). Size and shape of sodium deoxycholate micellar aggregates. *J. Phys. Chem.* **91**, 356–362.

Fanali, S. (1989). Separation of optical isomers by capillary zone electrophoresis based on host-guest complexation with cyclodextrins. *J. Chromatogr.* **474**, 441–446.

Fanali, S. (1991). Use of cyclodextrins in capillary zone electrophoresis. Resolution of terbutaline and propranolol enantiomers. *J. Chromatogr.* **545**, 437–444.

Fanali, S. and Bocek, P. (1990). Enantiomer resolution by using capillary zone electrophoresis: resolution of racemic tryptophan and determination of the enantiomer composition of commercial pharmaceutical epinephrine. *Electrophoresis* **11**, 757–760.

Fanali, S., Ossicini, L., Foret, F. and Bocek, P. (1989). Resolution of optical isomers by capillary zone electrophoresis: study of enantiomeric and diastereomeric Co(III) complexes with ethylendiamine and amino acid ligands. *J. Microcolumn Separation* **1**, 190–194.

Fanali, S., Flieger, M., Steinerova, N. and Nardi, A. (1992). Use of cyclodextrins for enantioselective separation of ergot alkaloids by capillary zone electrophoresis. *Electrophoresis* **13**, 39–43.

Gassmann, E., Kuo, J. E. and Zare, R. N. (1985). Electrokinetic separation of chiral compounds. *Science* **230**, 813–815.

Gozel, P., Gassmann, E., Michelsen, H. and Zare, R. N. (1987). Electrokinetic resolution of amino acid enantiomers with copper (II)-aspartame support electrolyte. *Anal. Chem.* **59**, 44–49.

Guttman, A., Paulus, A., Cohen, A. S., Grinberg, N. and Karger, B. L. (1988). Use of complexing agents for selective separation in high-performance capillary electrophoresis. Chiral resolution via cyclodextrins incorporated within polyacrylamide gel columns. *J. Chromatogr.* **448**, 41–53.

Heumann, M. and Blaschke, G. (1993). Chiral separation of basic drugs using cyclodextrins as chiral pseudo-stationary phases in capillary electrophoresis. *J. Chromatogr.* **648**, 267–274.

Hilton, M. and Armstrong, D. W. (1991). Evaluation of a chiral crown ether LC column for the separation of racemic amines. *J. Liq. Chromatogr.* **14**, 9.

Hjerten, S. J. (1985). High-performance electrophoresis: elimination of electroendosmosis and solute adsorption. *J. Chromatogr.* **347**, 191.

Höhne, E., Krauss, G.-J. and Gübitz, G. (1992). Capillary zone electrophoresis of the enantiomers of amino alcohols based on host-guest complexation with a chiral crown ether. *HRC & CC* **15**, 698–700.

Jelinek, I., Snopek, J. and Smolkova-Keulemansova, L. (1991). Influence of counter-ion inclusion complexation on quality of cyclodextrin-supported separations in isotachophoresis. *J. Chromatogr.* **557**, 215–226.

Kuhn, R., Erni, F., Bereuter, T. and Hauser, J. (1992a). Chiral recognition and enantiomeric resolution based on host-guest complexation with crown ethers in capillary zone electrophoresis. *Anal. Chem.* **64**, 2815–2820.

Kuhn, R., Stoecklin, F. and Erni, F. (1992b). Chiral separations by host-guest complexation with cyclodextrin and crown ether in capillary zone electrophoresis. *Chromatographia* **33**, 32–36.

Li, S. and Purdy, W. C. (1992). Circular dichroism, ultraviolet, and proton nuclear magnetic resonance spectroscopic studies of the chiral recognition mechanism of β-cyclodextrin. *Anal. Chem.* **64**, 1405–1412.

Lurie, I. S. (1992). Micellar electrokinetic capillary chromatography of the enantiomers of amphetamine, methamphetamine and their hydroxyphenethylamine precursors. *J. Chromatogr.* **605**, 269–275.

Mazzeo, J., Grover, E., Swartz, M. and Peterson, J., Waters Chromatography Division of Millipore.

Meyer, S. and Schurig, V. (1992). Enantiomer separation by electrochromatography on capillaries coated with chirasil-dex. *HRC* **15**, 129–131.

Nardi, A., Ossicini, L. and Fanali, S. (1992). Use of cyclodextrins in capillary zone electrophoresis for the separation of optical isomers: Resolution of racemic tryptophan derivatives. *Chirality* **4**, 56–61.

Nardi, A., Eliseev, A., Bocek, P. and Fanali, S. (1993). Use of charged and neutral cyclodextrins in capillary zone electrophoresis: enantiomeric resolution of some 2-hydroxy acids. *J. Chromatogr.* **638**, 247–254.

Nielen, M. W. F. (1993a). Chiral separation of basic drugs using cyclodextrin-modified capillary zone electrophoresis. *Anal. Chem.* **65**, 885–893.

Nielen, M. W. F. (1993b). (Enantio-) Separation of phenoxy-acid herbicides using capillary zone electrophoresis. *J. Chromatogr.* **637**, 81–90.

Nishi, H. and Matsuo, M. (1991). Separation of corticosteroids and aromatic hydrocarbons by cyclodextrin-modified micellar electrokinetic chromatorgaphy. *J. Liq. Chromatogr.* **14**, 973–986.

Nishi, H., Fukuyama, T., Matsuo, M. and Terabe, S. (1989). Chiral separation of optical isomeric drugs using micellar electrokinetic chromatography and bile salts. *J. Microcolumn Separation* **1**, 234–241.

Nishi, H., Fukuyama, T. and Matsuo, M. (1990a). Resolution of optical isomers of 2,3,4,6-tetra-*O*-acetyl-β-D-glucopyranosyl isothiocyanate (GITC)-derivatized DL-amino acids by micellar electrokinetic chromatography. *J. Microcolumn Separation* **2**, 234–240.

Nishi, H., Fukuyama, T., Matsuo, M. and Terabe, S. (1990b). Chiral separation of diltiazem, trimetoquinol and related coumpounds by micellar electrokinetic chromatography with bile salts. *J. Chromatogr.* **515**, 233–243.

Nishi, H., Fukuyama, A. T., Matsuo, M. and Terabe, S. (1990c). Chiral separation of trimetoquinol hydrochloride and related compounds by micellar electrokinetic chromatography using sodium taurodeoxycholate solutions and application to optical purity determination. *Anal. Chim. Acta* **236**, 281–286.

Nishi, H., Fukuyama, T. and Terabe, S. (1991). Chiral separation by cyclodextrin-modified micellar electrokinetic chromatography. *J. Chromatogr.* **553**, 503–516.

Otsuka, K. and Terabe, S. (1990a). Effects of methanol and urea on optical resolution of phenylthiohydantoin-D,L-amino acids by micellar electrokinetic chromatography with sodium *N*-dodecanoyl-L-valinate. *Electrophoresis* **11**, 982–984.

Otsuka, K. and Terabe, S. (1990b). Enantiomeric resolution by micellar electrokinetic chromatography with chiral surfactants. *J. Chromatogr.* **515**, 221–226.

Otsuka, K. and Terabe, S. (1991). Chiral separation by capillary electrophoresis and electrokinetic chromatography. *Jasco Rep.* **33** (1), 1–5.

Otsuka, K., Kawahara, J., Takekawa, K. and Terabe, S. (1991). Chiral separation by micellar electrokinetic chromatography with sodium *N*-dodecanoyl-L-valinate. *J. Chromatogr.* **559**, 209–214.

Otsuka, K., Kashihara, M., Kawaguchi, Y., Koike, R. and Hisamitsu, T. (1992). Optical resolution by capillary electrophoresis. *Kuromatogurafi* **13**, 317–318.

Otsuka, K., Kashihara, M., Kawaguchi, Y., Koike, R., Hisamitsu, T. and Terabe, S. (1993). Optical resolution by high-performance capillary electrophoresis. Micellar electrokinetic chromatography with sodium *N*-dodecanoyl-L-glutamate and digitonin. *J. Chromatogr.* **652**, 253–257.

Pasteur, L. (1853). Transformation des acides tartriques en acide racemique. Decouverte de l'acide tartrique inactif. Nouvelle methode de separation de l'acide racemique en acides tartriques droit et gauche. *C.R. Hebd. Seanc. Acad. Sci.* (Paris) **37**, 162.

Penn, S. G., Goodall, D. M. and Loran, J. S. (1993). Differential binding of tioconazole enantiomers to hydroxypropyl-β-CD studied by capillary electrophoresis. *J. Chromatogr.* **636**, 149–152.

Peterson, T. E. (1993). Separation of drug stereoisomers by capillary electrophoresis with cyclodextrins. *J. Chromatogr.* **630**, 353–361.

Peterson, T. E. and Trowbridge, D. (1992). Quantitation of L-ephedrine and determination of the D-L-ephedrine enantiomer ratio in a pharmaceutical formulation by capillary electrophoresis. *J. Chromatogr.* **603**, 298–301.

Quang, C. and Khaledi, M. G. (1993). Improved chiral separation of basic compounds in capillary electrophoresis using β-cyclodextrin and tetraalkylammonium reagents. *Anal. Chem.* **65**, 3354–3358.

Rawjee, Y. Y., Staerk, D. U. and Vigh, G. (1993a). Capillary electrophoretic chiral separations with cyclodextrin additives. I. Acids: chiral selectivity as a function of pH and the concentration of β-CD for fenoprofen and ibuprofen. *J. Chromatogr.* **635**, 291–306.

Rawjee, Y. Y., Williams, R. L. and Vigh, G. (1993b). Capillary electrophoretic chiral separations using β-cyclodextrin as resolving agent. II. Bases: chiral selectivity as a function of pH and the concentration of β-cyclodextrin. *J. Chromatogr.* **652**, 223–245.

Schmitt, T. and Engelhardt, H. (1993a). Derivatized cyclodextrins for the separation of enantiomers in capillary electrophoresis. *HRC* **16**, 35–39.

Schmitt, T. and Engelhardt, H. (1993b). Charged and uncharged cyclodextrins as chiral selectors in capillary electrophoresis. *Chromatographia* **37**, 475–481.

Schurig, V. and Jung, M. (1990). In: *Recent Advances in Chiral Separations*, Stevenson, D. and Wilson, I. D. (Eds). Plenum Press, New York.

Schutzner, W. and Fanali, S. (1992). Enantiomers resolution in capillary zone electrophoresis by using cyclodextrins. *Electrophoresis* **13**, 687–690.

Schutzner, W., Fanali, S., Rizzi, A. and Kenndler, E. (1993). Separation of diastereomeric derivatives of enantiomers by capillary zone electrophoresis with a polymer network — use of polyvinylpyrrolidone as buffer additives. *J. Chromatogr.* **639**, 375–378.

Shibukawa, A., Lloyd, D. K. and Wainer, I. W. (1993). Simultaneous chiral separation of leucovorin and its major metabolite 5-methyl-tetrahydrofolate by capillary electrophoresis using cyclodextrins as chiral selectors: Estimation of the formation constant and mobility of the solute-cyclodextrin complexes. *Chromatographia* **35**, 419–429.

Smith, N. W. (1993). Separation of positional isomers and enantiomers using capillary zone electrophoresis with neutral and charged cyclodextrins. *J. Chromatogr.* **625**, 259–262.

Snopek, J., Soini, H., Novotny, M., Smolkova-Keulemansova, E. and Jelinek, I. (1991). Selected applications of cyclodextrin selectors in capillary electrophoresis. *J. Chromatogr.* **559**, 215–222.

Soini, H., Riekkola, M. L. and Novotny, M. V. (1992). Chiral separation of basic drugs and quantitation of bupivacine enantiomers in serum by capillary electrophoresis with modified cyclodextrin buffers. *J. Chromatogr.* **608**, 265–274.

Sun, P., Barker, G. E., Hartwick, R. A., Grinberg, N. and Kaliszan, R. (1993a). Chiral separations using an immobilized protein-dextran polymer network in affinity capillary electrophoresis. *J. Chromatogr.* **652**, 247–252.

Sun, P., Wu, N., Barker, G. and Hartwick, R. A. (1993b). Chiral separations using dextran and bovine serum albumin as run buffer additives in affinity capillary electrophoresis. *J. Chromatogr.* **648**, 475–480.

Swartz, M. E. (1991). Method development and selectivity control for small molecules pharmaceutical separations by capillary electrophoresis. *J. Liq. Chromatogr.* **14**, 923–938.

Szejtli, J. (1982). *Cyclodextrins and their Inclusion Complexes.* Akademia Kiado, Budapest, pp. 162–178.

Terabe, S. (1989). Electrokinetic chromatography: an interface between electrophoresis and chromatography. *Trends Anal. Chem* **8**, 129–134.

Terabe, S., Otsuka, K. and Ando, T. (1985a). Electrokinetic chromatography with micellar solution and open-tubular capillary. *Anal. Chem.* **57**, 834.

Terabe, S., Ozaki, H., Otsuka, K. and Ando, T. (1985b). Electrokinetic chromatography with 2-O-carboxymethyl-β-cyclodextrin as a moving 'stationary' phase. *J. Chromatogr.* **332**, 211–217.

Terabe, S., Shibata, M. and Miyashita, Y. (1989). Chiral separation by electrokinetic chromatography with bile salt micelles. *J. Chromatogr.* **480**, 403–411.

Terabe, S., Miyashita, Y., Shibata, O., Barnhart, E. R., Alexander, I. R., Patterson, D. G., Karger, B. L., Hosoya, K. and Tanaka, N. (1990). Separation of highly hydrophobic compounds by cyclodextrin-modified micellar electrokinetic chromatography. *J. Chromatogr.* **516**, 23–31.

Terabe, S., Miyashita, Y., Ishihama, Y. and Shibata, O. (1993). Cyclodextrin-modified micellar electrokinetic chromatography: separation of hydrophobic and enantiomeric compounds. *J. Chromatogr.* **636**, 47–55.

Tran, A. D., Blanc, T. and Leopold, E. J. (1990). Free solution capillary electrophoresis and micellar electrokinetic resolution of amino acid enantiomers with L- and D-Marfey's reagent. *J. Chromatogr.* **516**, 241.

Ueda, T., Kitamura, F., Mitchell, R., Metcalf, T., Kuwana, T. and Nakamoto, A. (1991). Chiral separation of naphthalene-2,3-dicarboxaldehyde-labeled amino acid enantiomers by cyclodextrin-modified micellar electrokinetic chromatography with laser-induced fluorescence detection. *Anal. Chem.* **63**, 2979–2981.

Valtcheva, L., Mohammad, J., Petterson, G. and Hjerten, S. (1993). Chiral separation of β-blockers by high-performance capillary electrophoresis based on non-immobilized cellulase as enantioselective protein. *J. Chromatogr.* **638**, 263–268.

Vespalec, R., Sustacek, V. and Bocek, P. (1993). Prospects of dissolved albumin as a chiral selector in capillary zone electrophoresis. *J. Chromatogr.* **638**, 255–263.

Williams, R. W., Sheng, F. Z. and Hinze, W. L. (1990). Micellar bile salt mobile phases for the liquid chromatography separation of routine compounds and optical, geometrical and structural isomers. *J. Chromatogr. Sci.* **28**, 292.

Wren, S. A. C. (1993). Theory of chiral separation in capillary electrophoresis. *J. Chromatogr.* **636**, 57–62.

Wren, S. A. C. and Rowe, R. C. (1992a). Theoretical aspects of chiral separation in capillary electrophoresis. I. Initial evelution of a model. *J. Chromatogr.* **603**, 235–241.

Wren, S. A. C. and Rowe, R. C. (1992b). Theoretical aspects of chiral separation in capillary electrophoresis. II. The role of organic solvent. *J. Chromatogr.* **609**, 363–367.

Wren, S. A. C. and Rowe, R. C. (1993). Theoretical aspects of chiral separation in capillary electrophoresis. III. Application to beta-blockers. *J. Chromatogr.* **635**, 113–118.

Yamamoto, H., Baumann, J. U. and Erni, F. (1992). Electrokinetic reversed-phase chromatography with packed capillaries. *J. Chromatogr.* **593**, 313–319.

Yoneda, H. and Miura, T. (1970). Complete resolution of the racemic triethylenediaminecobalt (III) complex into its optical antipodes by means of electrophoresis. *Bull. Chem. Soc. Jpn* **43**, 574.

APPENDIX 1 – ABBREVIATIONS

α-coefficient	$\alpha = k_2/k_1$ where $k = \{(t_r - t_0)/[1 + (t_0/t_{mc})]\}t_0$ t_r – retention time of solute, t_0 – ret. time of water, t_{mc} – ret. time of micelles, see also (Terabe *et al.*, 1985)
BSA	bovine serum albumin
CBI-amino acids	1-cyano-2-substituted-benz(f)isoindole-amino acids
CD	cyclodextrin
CE	capillary electrophoresis
CZE	capillary zone electrophoresis
DM-β-CD	heptakis (di-O-methyl)-β-CD
DNP-amino acids	dinitrophenol-amino acids
DNS-amino acids	dansylated amino acids
E	electric field strength (volts/cm)
EOF	electroosmotic flow
GITC-derivatives	2,3,4,6-tetra-O-acetyl-β-D-glucopyranosyl isothiocyanate-derivatives
GC	gas chromatography
HPLC	high performance liquid chromatography
i.d.	internal diameter of the capillary
ITP	isotachophoresis
MEKC	micellar electrokinetic chromatography
MES	2-[N-morpholino] ethanesulfonic acid
MW	molecular weight
PTH-Amino acids	phenylthiohydantoin amino acids
SDAla	sodium dodecanoyl-alanine
SDGlu	sodium dodecanoyl-glutamic acid
SDS	sodium dodecyl sulfate
SDVal	sodium dodecanoyl-valine

% T	Total content of monomer in a gel (in w/v %)
TBE	0.1 M Tris, 0.1 M boric acid, 2 mM EDTA;
	Tris = Tris(hydroxymethyl)-aminomethane,
	EDTA = Ethylenediaminetetraacetic acid
TM-β-CD	heptakis (tri-O-methyl)-β-CD
Tris	Tris-(hydroxymethyl)-aminomethane

APPENDIX 2 – CHIRAL NOMENCLATURE

Allenisomers	Chirality based on a chiral axis, i.e. conjugated, odd number of double bonds as in cumulens
Asymmetric	Carbon atom with four different substituents
Asymmetry	Totally lacking elements of symmetry
Atropisomers	Based on hindered rotation, e.g. biphenyl derivatives
Chirality, chiral	(Greek: handed); molecules which exist as mirror images
D or L	Absolute configuration of a molecule according to the D- or L-form of glyceraldehyde as a reference
Diastereomers	Stereoisomers which are not mirror images
EE: enantiomeric-excess	Percentage by which one enantiomer is in excess in a mixture of the two
Enantiomers	Stereoisomers whose molecules are non-superimposable mirror images of each other
Enantioselectivity	Degree to which a chemical synthesis produces more of one of the enantiomers relative to the other
Epimers	Molecules which differ in the configuration of one of several chiral centers
Configuration	Specific orientation of atoms or substitutions in a molecule
Meso-compounds	Molecules with two or more chiral centers but with a plane of symmetry, i.e. meso-tartaric acid or erythritol
Optical activity	A pure enantiomer rotates the plane of polarized light to a certain degree to the observer's right or left
Optical antipodes	Less frequently used expression for enantiomers
Optical isomers	Molecules which show optical activity; this term is occasionally used for enantiomers
Optical purity	Determination of the purity of an enantiomer in a mixture of two by measuring the angle of optical rotation of polarized light
(+) or (−)	Rotation of plane of polarized light to the right (+) or to the left (−)
Pro-chirality	Property whereby molecules can be transformed by a single chemical reaction into enantiomers; substituents on corresponding prochiral centers are called 'enantiotops'

R or S	Description of absolute configuration around an asymmetric carbon atom according to the rule of Cahn, Ingold and Prelog
Racemate, racemic mixture	Equimolar mixtures of (+) and (−) rotating enantiomers; the angle of rotation of polarized light is zero since both enantiomers have the same optical rotation but in the opposite direction
Racemization	Conversion of pure enantiomers into a racemic mixture
Stereoisomers	Different spatial orientation of atoms in a molecule with an otherwise identical carbon atom chain

HPCE OF ANIONS AND CATIONS

Progress in HPLC-HPCE, Vol. 5, pp. 445–471
H. Parvez *et al.* (Eds)
© VSP 1997.

Negative absorbance: Application to the detection of inorganic anions and cations in high performance capillary electrophoresis

PIERRE ROLAND-GOSSELIN,[1] PIERRE CAUDY,[1] HASAN PARVEZ[2,3]
and SIMONE PARVEZ[3]

[1] *Thermo Separation Products, France, Hightec Sud, 12 Avenue des Tropiques,
Z. A. de Courtabeuf, BP 141, 91944 Les Ulis Cedex, France*
[2] *CNRS, 91400 Orsay, France*
[3] *Neuroendocrinologie et Neuropharmacologie du Développement, Université de
Reims-UFR Sciences, BP 347, 51062 Reims Cedex, France*

INTRODUCTION

The commercial availability of capillary zone electrophoresis (CZE) has generated particular interest in the analytical field over the past four years. The
possibility of achieving separations using physical phenomena other than those
mostly used in gas or liquid chromatography has permitted exploration and
realization of new methods of separation which were impossible to set up tech-
nically before. The increase in number of papers published recently is evidence
of the possibilities of capillary electrophoresis for the detection of inorganic or
organic species in the field of environmental science or biology. The aim of
the present chapter is to elaborate the significance of this new analytical development. After considering the difficulties encountered in the use of negative
absorbance by literature review, an attempt will be made to demonstrate these
problems in practical applications of inorganic and organic separations.

NEGATIVE ABSORBANCE

Introduction

The method called negative absorbance has been known for some time in
HPLC (Small and Miller, 1982). Though this method makes UV detection

universal, its application in HPLC still lacks development. Specific detectors have been developed in parallel, bringing a better detection selectivity and increased sensitivity. Conductivity detectors, well adapted for inorganic ion detection, allow us to achieve acceptable detection limits in the range of µg/l, especially when the method of conductivity suppression developed by Dionex is utilized. Amperometric/electrochemical detectors can supplement conductivity detection for the detection of HS– and CN– (Carpentier, 1986; Parvez *et al.*, 1987; Rosset *et al.*, 1991). The application of these two types of detectors in capillary electrophoresis (CE) has been described in several publications but, till this date, none of these detectors is commercially available (Avdalovic *et al.*, 1993; Hauser *et al.*, 1994; Nann and Pretsch, 1994). Contrary to electrochemical or conductivity detection, the adaptation of UV detection to high performance capillary electrophoresis (HPCE) has been quite easy, since all commercial instruments are equipped with UV detection as standard.

Negative absorbance

This technique relies upon the use of a working electrolyte showing a very high absorbance at the working wavelength. When different chemical species showing low or zero absorbance are coupled with the working electrolyte, the detector measures the decrease in absorbance at the same wavelength. Figure 1 illustrates the principle of this measure. According to Yeung (1989), the

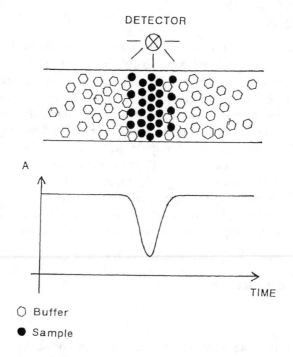

Figure 1. Indirect UV detection.

minimal detectable concentration obeys equation (1):

$$C_{\lim} = \frac{C_m}{TR \times DR},$$

(1)

where C_m is the concentration of the chemical species absorbing in the electrolyte; TR is a transfer ratio and DR is the dynamic reserve.

The transfer ratio is a number of ionic equivalents of the buffer which are displaced by each equivalent of the solute. If the absorbing ion is chromate (CrO_4^{2-}), the TR will be 1 for sulfate and 0.5 for fluoride. This shows that for divalent solutes, the sensitivity will be twice that of monovalent solutes when the absorbing ion is divalent. Starting from Kohlrausch theory, Nielen (1991) has discussed the possibility of using a universal calibration as practiced in ion chromatography. The plotting of a calibration curve for each species will eventually allow measurement of all species having the same charge. In reality, though such a concept is valid theoretically, this process of calibration is not used in practice because other factors such as the conductivity of the solute-electrolyte contribute to an significant imprecision. The term DR of equation (1) represents the dynamic reserve for evaluation of the ratio between signal intensity over background noise. In negative absorbance, the term DR follows equation (2) (Jandik and Bonn, 1993):

$$DR = \left(\frac{\varepsilon L C_m}{AN} \right),$$

(2)

where ε is the molar coefficient of absorbance, L is the optical length, C_m is the concentration of the visual ion in the electrolyte and AN is the background noise. According to (2), we can see that the DR depends both upon the quality of the detector and visualising ion used. The experimenter should choose a visualising ion having a mobility close to that of the solute and with a significant molar absorbance coefficient.

In addition, one should utilize a concentration as high as possible without increasing either the Joule effect or the background noise. Substituting for the term DR from (2) into (1) we get

$$C_{\lim} = \frac{AN}{\varepsilon \times L} \times \frac{1}{TR}.$$

(3)

The minimal detectable concentration is independent of the concentration of the visualising ion. If we consider that the actual detectors in use produce a background noise AN in the range of 10^{-5} AU and we monitor at an ε of 10^3 with an optical length of 75 μm, the minimal detectable concentration can be calculated as:

$$C_{\lim} = \frac{1}{TR} \times \frac{10^{-5}}{10^3 \times 75 \times 10^{-4}} = \frac{1.33 \times 10^{-6}}{TR} \text{M}.$$

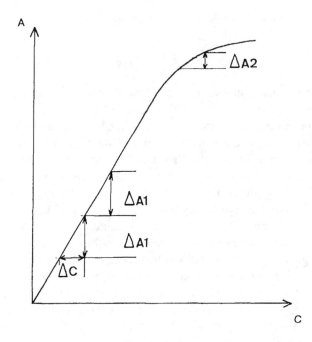

Figure 2. Curve concentration $A = f(C)$. Reproduced with permission.

This type of calculation gives us an idea about the detection limit that we can achieve but in no way reflects the absolute detection limit which we can obtain practically. As a matter of fact, it depends on the value of *TR* but also on the concentration effect at the time the current is applied, and on the symmetry of the peaks showing the ratio of signal to noise. Regardless of the fact that the value of the concentration limit is independent of the visualising ion concentration, it remains in the limit of linearity as defined by the Beer–Lambert law. According to Fig. 2, the trace of curve $A = f(C)$ shows that an identical decrease ΔC of the visualising ion does not induce a similar decrease in absorbance when measured in the non-linear region of the curve.

ION SEPARATION

Selection of visualising ion

On the criterion of its mobility. The mobility of the visualising ion compared to that of the ions to be detected is the most important criterion to be considered for the choice of the visualising ion. As a matter of fact, as shown in Fig. 3,

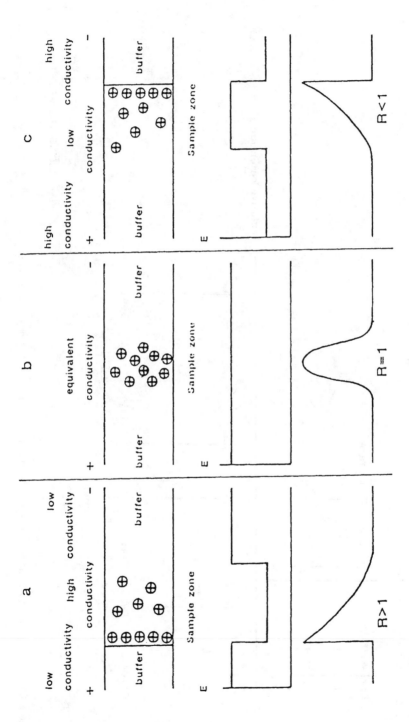

Figure 3. Electrodispersion due to mismatched sample and buffer conductivities. Reproduced with permission.

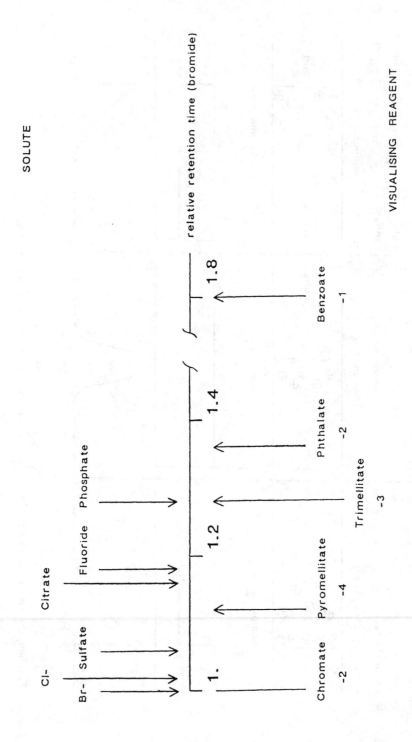

Figure 4. Mobility of the visualising ion and of more frequently analyzed ions. Reproduced with permission.

Table 1.

Visualising ions and their extinction coefficients

Visualising ions	Molar extinction coefficient, $1\ mol^{-1}\ cm^{-1}$
Chromate	3180
Pyromellitate	7062
Trimellitate	7147
Phthalate	1357
Benzoate	809

From Cousins *et al.*, 1994.

the peak asymmetry can be determined using the ratio of the mobility of the visualising ion present in the buffer to that of the solute (Mikkers *et al.*, 1979). This ratio (R) can lead to three different situations: $R > 1$, $R = 1$ and $R < 1$.

The importance of the asymmetry of the peak will be greater with increase in absolute value of R. When $R > 1$, the solute peak will appear near the end. With R close to 1, the peak will be symmetrical and if $R < 1$, the solute will be near the beginning. Figure 4 shows the mobility of the visualising ion as well as that of the more frequently analyzed ions. Therefore, it may be seen that the peaks of bromide, chloride, and sulfate will be symmetrical whereas those of fluoride and phosphate will be asymmetric if the visualising ion is chromate or pyromellitic acid.

On the criterion of its ε. From equation (3) we can deduce that the choice of visualising ion is directly linked to the value of ε. Therefore, it is necessary to choose an eluent, the molecular absorbtion coefficient of which is in the range of 10^3, as is the case with all the visualising ions presented in Table 1.

On the criterion of wavelength detection. The choice of the working wavelength is based on two different criteria:

a) One should be at the maximal absorbance wavelength of the visualising ion.

b) This wavelength should be as different as possible from those of the solutes to be measured: some species such as bromide, nitrite and nitrate absorb at 210 nm. If one wishes to obtain an appreciable detection limit for these ions, the selected wavelength must reflect the minimal possible absorbance in order to obtain maximal absorbance difference between those of the visualising ions and of the solutes (Fig. 5).

On the criterion of transfer ratio, TR. We have seen previously that the choice of the visualising ion is directly linked to the value of *TR*: in order to keep the whole electrolyte neutral, a molecule of visualising ion should be replaced by

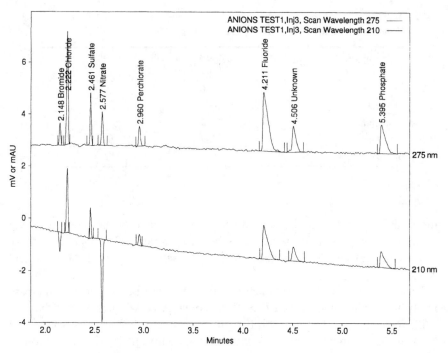

Figure 5. Influence of the wavelength on the response. At 210 nm bromide and nitrate have own absorbance. The detection limit for nitrate is better at 210 nm. Conditions. Capillary: Fused silica, 75 μm, 44 cm, 37 cm. Carrier electrolyte: 5 mM $K_2Cr_2O_7$ – 0.5 mM hexamethonium hydroxide – NaOH (pH 8). Voltage: −25 kV.

a molecule of solute having equal valence. It has been shown that the variation of the concentration Δ[B] of visualising ion B induced by the concentration [A] of solute A responds to the following equation:

$$\Delta[\text{B}] = -\frac{z_A}{z_B} \times \frac{\mu_B[\mu_A + \mu_C]}{\mu_A[\mu_B + \mu_C]} \times [\text{A}], \qquad (4)$$

where z is the valence of solute A, as well as of visualising ion B, μ is the electrophoretic mobility of solute A, visualising ion B and counter ion C. This equation shows that the replacement of one equivalent by solute can only be expected if the two species demonstrate a similar mobility. The experimental determination of the *TR* parameter is reported in Table 2 according to Cousins *et al.* (1994). The values collected in Table 2 show that the transfer ratio, *TR*, increases when the mobility of the visualising ion is close to that of the solute and therefore confirms the theory.

Conclusions regarding visualising ion. The best visualising ion is the one which allows the best compromise between the sensitivity and separation efficiency. Separation of anions performed by several authors using chromate,

Table 2.
Variation of transfer ratio *TR* in relation to visualising ions

Solute	Determination mode	Benzoate	Phthalate	Trimellitate	Pyromellitate	Chromate
			Visualizing ions			
Chloride	a	0.184	0.205	0.325	0.366	0.608
"	b	1.000	0.500	0.333	0.250	0.500
"	c	0.651	0.423			0.521
Fluoride	a	0.421	0.233	0.319	0.356	0.718
"	b	1.000	0.500	0.333	0.250	0.500
"	c	0.748	0.486			0.599
Sulfate	a	0.897	0.575	0.612	0.625	1.124
"	b	2.000	1.000	0.667	0.500	1.000
"	c	1.277	0.831			1.023
Phosphate	a	1.065	0.463	0.585	0.584	1.218
"	b	2.000	1.000	0.667	0.500	1.000
"	c	1.476	0.960			1.182
Citrate	a	1.629	0.614	0.931	0.593	1.756
"	b	3.000	1.500	1.000	0.750	1.500
"	c	2.019	1.313			1.617

a) Experimental; b) Calculated on the basis of an equivalent to equivalent exchange; c) Calculated according to Nielen (1991). Reproduced with permission.

pyromellitate or trimellitate leads to different conclusions from one publication to the other. Some authors prefer pyromellitate whereas others chromate. From our point of view, we shall withhold our conclusions at this stage, as other factors not yet considered might interfere in such a choice.

Electroosmotic flux inversion

Problems on inverse electroosmotic flux (EOF). The separation of negatively charged species is accompanied by the particular problem that migration of these species is always in the opposite direction to the EOF. Therefore, the time of analysis is long and leads to an increase in peak width, with an adverse effect on the sensitivity. It is then necessary to control the direction of the electroosmotic flux, to divert it towards the same direction as anionic solutes. Reijenga *et al.* (1983) were the first to demonstrate that the use of cationic tensioactives allows reduction of the inversion of the EOF as a function of the concentration used.

Mechanism of flux inversion. The tensioactives utilized possess a short or long alkyl chain onto which is grafted an ammonium group at one or both ends.

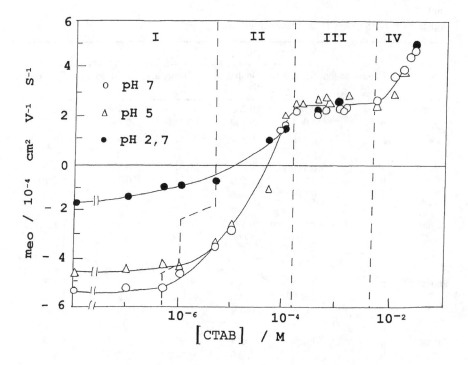

Figure 6. Relation between electroosmotic flux and [CTAB]. Reproduced with permission.

Since the studies of Fuerstenau (1956) and Somasundaran *et al.* (1964), many authors have carried out investigations to explain the mechanism of interaction between ammonium and silanol groups (Janini *et al.*, 1993; Stathakis and Cassidy, 1994). Recently Kaneta *et al.* (1993) have proposed a mechanism which is in good agreement with the observations made by other authors (Jones and Jandik, 1991; Stathakis and Cassidy, 1994). It is known that EOF increases with the pH. The variations of the concentration of CTAB at different pH values are shown in Fig. 6 (Kaneta *et al.*, 1993). These authors have distinguished four zones in the curves obtained.

In zone 1, EOF depends upon electrolyte pH. The dissociation constant of the silanol groups is $10^{-5.3}$ in aqueous solution (Schwer and Kenndler, 1991). The level of dissociation of silanol groups varies greatly between pH 3 and pH 7. One can postulate that, at very low cationic tensioactive concentrations, their absorption on the capillary wall is due only to electrostatic interactions between silanols and ammonium.

In zone 2, the electroosmotic mobility varies considerably and one can observe inversion of the EOF. The electroosmotic mobility depends again on the pH. Fuerstenau (1956) has shown that in this range of concentration, hemi-aggregates were appearing. In these conditions, the adsorption of surfactant results from electrostatic and van der Waals effects. The excess of positive charges at the surface of these hemi-aggregates is the origin of the inversion of EOF.

In zones 3 and 4, the value of EOF is independent of pH value. The cause of this phenomenon is related to the fact that all silanol groups are ionized in SiO^- and are linked by van der Waals type forces rather than covalent or coulombic ones to ammonium groups.

Influence of alkyl chain length. The observations of Buchberger and Haddad (1992) on the influence of alkyl chain length have shown that the mean migration time of anions studied doubles when the alkyl chain passes from C16 to C12. They also demonstrated that cationic tensioactives also play an ion exchanger role. The relative migration time of certain anions such as fluoride and chlorate are considerably modified. If some authors report similar observations, no study has been made on the mechanism to check if mono- and di-ammoniums interact in a similar manner (Morin *et al.*, 1994). The use of di-ammonium originates from the observation of precipitate formation between chromate and mono-ammoniums in certain cases (CTAB, TTAB; Harrold *et al.*, 1993).

Influence of counter ion. Alkyl ammoniums are commercialized along with a counter ion which is generally a chloride or a bromide. Unfortunately, such ions induce aberrations on the base line as well as an absorbance decrease at the migration interval of bromide or chloride. This phenomenon evidently perturbs the precision and sensitivity of the measurements. Therefore, it is recommended to pass the ammonium used through anionic exchange resin in order to have only OH^- as a counter ion, since these ions will not disturb measurements because of their higher mobility as compared to those of the species determined.

Influence of ionic strength of the electrolyte

This specificity of the electrolyte is important for the two following reasons. First of all, the intensity of the current for a given working tension is proportional to the ionic strength of the electrolyte. The higher the current intensity will be, the greater will be the need to eliminate the heat produced. According to its value, this heat production will be more or less difficult to eliminate. If the increase in temperature is not correctly controlled, it produces an increase in background noise. During negative absorbance, the increment in background noise is higher at higher absorbance. It is therefore imperative to minimize the ionic strength of the electrolyte. However, as the separation efficiency is directly proportional to the ionic strength of the electrolyte the efficiency of separation decreases in parallel with the decrease in ionic strength (Fig. 7).

The peak width in capillary electrophoresis is proportional to the concentration ratio of solute to electrolyte (Reijenga *et al.*, 1983). It is therefore necessary

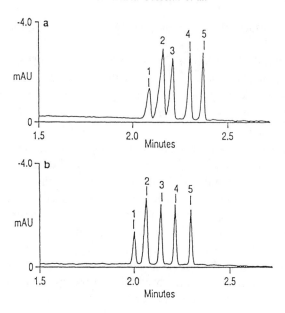

Figure 7. Influence of ionic force on separation efficiency. Reproduced with permission.

to work with equivalent concentrations of solute and electrolyte which in our case are mainly formed by visualizing ion. The maintenance of high efficiency for a given separation obliges us to optimize the visualizing ion concentration without increasing significantly the ionic strength of the electrolyte.

Influence of pH value

The pH of the electrolyte induces several effects on the separation.

1) It interacts with the degree of ionization of the visualizing ion. If we utilize pyromellitate, its four functional acidic groups will be ionized in relation to the pH value. Therefore, it will have an effect on *TR*.

2) The mobility will vary according to the ionization of the visualizing ion: the smaller its ionization, the lower will be its mobility. The same visualizing ion, therefore, will be applicable to solutes having completely different mobilities.

3) The mobility of certain solutes, such as carbonate, phosphate and acetate, varies as a function of their values of pH and ionization. This is, on the one hand, an advantage and, on the other, an inconvenience. The advantage is that it enables us to achieve a better separation by changing the pH whereas the inconvenience is the extreme sensitivity of these species to pH variations.

4) Stability of pH value. The application of the Henderson-Hasselbach equation demonstrates the variation of the pH of a buffer caused by water hydrolysis when a current is applied (Weinberger and Schlabach, 1991). We also know that Lewis acid-base reactions do take place with the silica of the flasks. According to these two observations, it can be deduced that it is imperative to

Table 3.

pK_a values of some amine buffers

Amines	pK_a (H_2O)
Hydroxylamine	6.0
Triethanolamine	7.8
Diethanolamine	9.0
Trimethylamine	9.8
Triethylamine	10.7

use a buffer that can have both buffering effects to minimize the above observations as well as better mobilities of the solutes. Therefore, to achieve this, it seems interesting to modify the electrolyte to the required value of pH by an agent which is neither a strong acid nor a strong base since its utilization will have two consequences. First, it will augment the total ionic strength of the electrolyte and secondly, if the chosen pH is 8, NaOH will have no buffering effect at this pH, especially when one wishes to utilize chromate, pyromellitate or others as visualizing ions. The possibility of using an amine as a buffering agent and a modifier of pH is also interesting when its proper pH$_a$ values represent that of different strong acids or bases. The second advantage linked to the use of these amines is that they do not augment the ionic force of the electrolyte. The third advantage achieved by the use of these amines is that they do not introduce new counter ions into the medium since they are available in the form of free bases. Table 3 summarizes some of the buffering amines along with their pH$_a$ values (Benz and Fritz, 1994).

Influence of solvation

When we add an organic solvent in an electrolyte, it is observed that the migration time of the anion increases. This can be related to the fact that the intensity of the current is lower in an electrolyte added with an organic solvent. The interaction between organic solvent and silica decreases the number of free sites accessible to quaternary ammonium ions. It is therefore necessary to use lower ammonium concentration to obtain the inversion of electroosmotic flux and maintain a similar flux. The same organic solvent also has some effects on solvation of anions, since variation of the migration time of some of them (fluoride, nitrite and nitrate) is observed when added with some organic solvent.

Influence of the current

All the electrolytes used by different authors have the common characteristic of a weak ionic force, which allows us to work in the maximum range of the intrument (30 kV) without provoking a significant Joule effect.

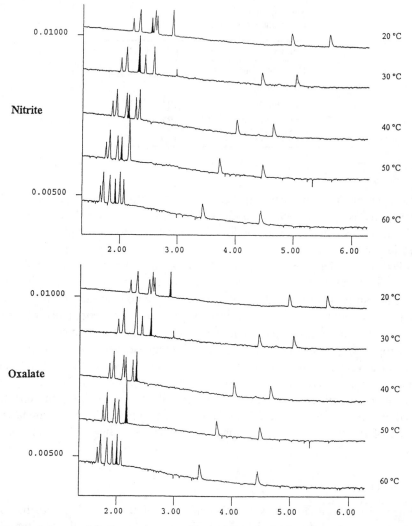

Figure 8. Influence of the temperature on migration time of anions (nitrite and oxalate). Reproduced from Geyer (1993).

Effect of temperature

Figure 8 illustrates the influence of temperature on analysis time, peak resolution as well as the order of separation of anions. Here, we can distinguish the importance of an efficient temperature regulation system.

Effect of capillary length

Figures 9 and 10 show that the capillary length plays an important role for the appearance of the first peak but has little influence on the quality of separation of the peaks.

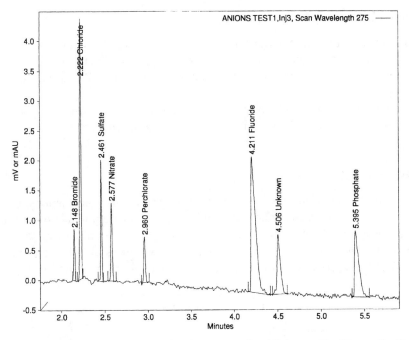

Figure 9. Influence of the capillary length on separation efficiency. Conditions. Capillary: Fused silica, 75 μm, 44 cm, 37 cm. Other conditions as in Fig. 5.

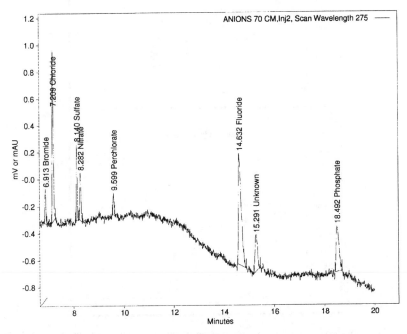

Figure 10. Influence of the capillary length on separation efficiency. Conditions. Capillary: Fused silica, 75 μm, 70 cm, 63 cm. Other conditions as in Fig. 9.

Conclusions on separation of inorganic anions

To summarise this presentation about separation of inorganic anions we can say:
 – that the best visualizing agents are chromate, pyromellitate and trimellitate,
 – that the most commonly used alkyl ammoniums are TTAB or hexamethonium,
 – that it is indispensable to have an electrolyte with a certain ionic force,
 – and that it is necessary to have an amine as a pH stabilizer.

THE ORGANIC ACIDS

Since the organic acids have a lower mobility than the inorganic anions, it will
be necessary to utilize a visualizing ion having a similar amplitude of mobility
to achieve separation in good conditions. It will also be necessary to apply
the same rules applicable to inorganic anions as well as all other parameters
of separation. Some examples of separations have been presented in Figs 11

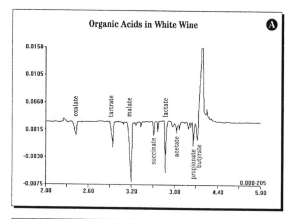

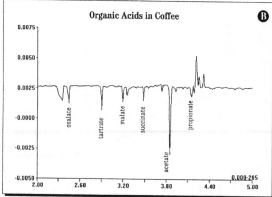

Figure 11. Separation of organic acids by high performance capillary electrophoresis. A: Tartrate
and malate are the major organic acids in white wine. B: Organic acids in coffee, diluted 1:10 in
water.

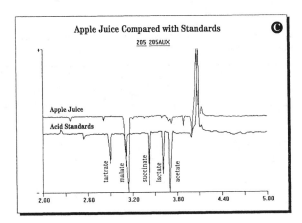

Figure 11 (continued). C: Apple juice diluted 1:10 in water is shown. The electropherogram is overlaid with an electropherogram of the standards tartrate, malate, succinate, lactate and acetate. Malate is identified as a major component of apple juice.

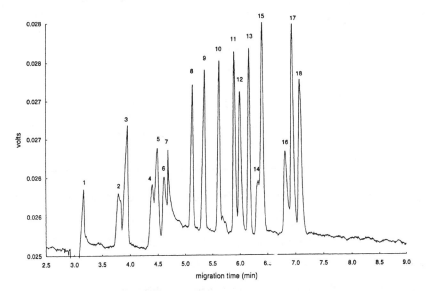

Figure 12. Capillary electropherogram of eighteen-component organic acid mixture. 4-Hydroxybenzoate buffer at pH 4.75 at 25 °C, with 2.5% (v/v) OFM. Peaks: 1 = oxalic acid (0.2); 2 = formic acid (0.1); 3 = fumaric acid (0.1); 4 = pyruvic acid (0.15); 5 = maleic acid (0.1); 6 = L-malic acid (0.1); 7 = citric acid (0.25); 8 = D-lactic acid (0.1); 9 = succinic acid (0.1); 10 = DL-aspartic acid (0.1); 11 = D-glucuronic acid (0.1); 12 = D-gluconic acid (0.1); 13 = acetic acid (0.1); 14 = L-ascorbic acid (0.1); 15 = shikimic acid (0.1); 16 = gallic acid (0.2); 17 = propionic acid (0.1); 18 = n-butyric acid (0.1 mM). Reproduced with permission of Devêvre *et al.* (1994).

and 12. It is generally accepted that organic acids migrate in the order of their pK_a values within their class (mono-, di- or tribasic acids) and this phenomenon is nearly stable in reality. Inversions in the elution pattern are observed which are certainly caused by differences in solvation of the solutes. This notion is

Table 4.

List of organic acids with experimental values of their pK_a values, the charge/mass ratio and their mobilities. Reproduced with permission

Organic acid	pK_{a1}	pK_{a2}	Charge/mass (pH 4.75, T = 25°)	Mobility $10^{-4} cm^2 s^{-1} V^{-1}$	Migration order
Oxalic	1.27	4.27	0.0222	8.26	1
Formic	3.75		0.0217	6.91	2
Fumaric	2.03	4.54	0.0173	6.61	3
Pyruvic	2.49		0.0113	5.94	4
Maleic	1.97	6.24	0.0086	5.82	5
L-Malic	3.40	5.13	0.0074	5.67	6
Citric	3.13	4.76 6.39 (pK_{a3})	0.0052	5.58	7
D-Lactic	3.83		0.0111	5.09	8
Succinic	4.21	5.64	0.0084	4.88	9
DL-Aspartic	1.99	3.90	0.0150	4.65	10
D-Glucuronic		—		4.44	11
D-Gluconic	3.60		0.0050	4.36	12
Acetic	4.76		0.0166	4.24	13
L-Ascorbic	4.17		0.0056	4.14	14
Shikimic	4.21		0.0057	4.09	15
Gallic	4.41		—	3.84	16
Propionic	4.87		0	3.77	17
n-Butyric	4.83		0	3.69	18

not taken into consideration when calculating the ratio valence/mass. Table 4 gives some examples of organic acids (Devêvre *et al.*, 1994).

THE CATIONS

The cations, similar to anions, require negative absorption since none of them demonstrates significant UV absorbance. Contrary to anions, the direction of their displacement in an electric field is always in the same path as the EOF; therefore, it will not be necessary to invert the EOF. The mobilities of compounds of the alkalis and alkali earth metals are relatively close to each other. Since electrostatic interactions can be observed between these cations and silanols, it raises the question if complexation of these elements is required to undergo reproducible separation, detection and assay.

Separation of alkalis and alkali earths

In regard to the separation of alkalis and alkali earths such as ammonium, sodium, potassium, barium, magnesium and calcium, two analytical techniques

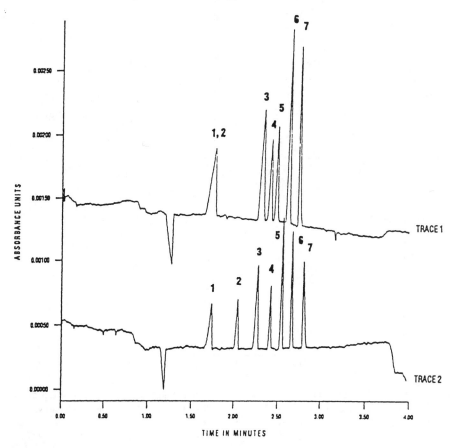

Figure 13. Influence of crown ether in buffer solution on separation of cations in high performance capillary electrophoresis.

[a] 1 – ammonium; 2 – potassium; 3 – sodium; 4 – calcium; 5 – magnesium; 6 – lithium; 7 – barium.

[b] Conditions. Capillary: Fused silica, 50 μm, 44 cm, 37 cm. Carrier electrolyte: 4 mM copper (II) sulphate, 4 mM formic acid, with (Trace 1) and without (Trace 2), 4 mM 18-crown-6. Voltage: +20 kV. Injection: 2 s electrokinetic, 10 kV or hydrodynamic. Detection: 215 nm, indirect. Temperature: 15 °C. Sample concentration: 10 ppm ammonium, 10 ppm potassium, 0.5 ppm sodium, 5 ppm calcium, 0.1 ppm magnesium, 0.2 ppm lithium and 20 ppm barium as the chloride salts in water, stored in plastic vials. Instrument: SpectraPHORESIS® 1000.

[c] This pair of electropherograms shows the effect of 18-crown-6 in separating potassium from ammonium. Both electropherograms show the same sample of chloride salts. The lower Trace 2 was run with the cation buffer containing 4 mM 18-crown-6, while a buffer was prepared without the crown ether additive for the upper Trace 1. Potassium is sequestered by the crown ether, thus changing its effective mobility, but ammonium does not complex with the crown ether. Without the additive, the two small cations comigrate.

are employed, since potassium and ammonium have the same electrophoretic mobility. The first analytical strategy employs a primed ether to form an inclusion complex with potassium (Figs 13 and 17). The other alternative concerns pH modification so that the ammonium-ammoniac equilibrium is in

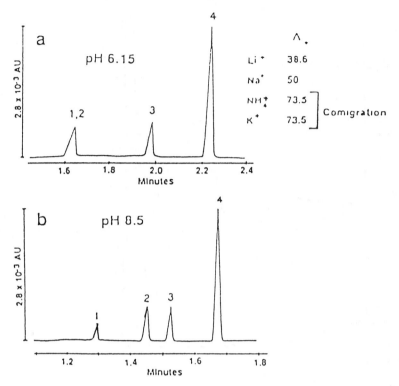

Figure 14. HPCE separation of alkali cations using ammonium and indirect UV detection at 214 nm and at applied voltage of +25 kV. MES (5 mM) was added to carrier electrolyte by adjusting the pH to 6.15. Reproduced with permission.

the favour of the base (Jandik and Bonn, 1993; Fig. 14). The use of an electrolyte at a more basic pH induces an increase in EOF and consequently results in a shorter analysis interval with loss of resolution. The use of citrate to improve resolution and separation by forming cation complexes has been tried (Fig. 15: Weston *et al.*, 1992). The mobility of the formed complexes is inferior to that of the corresponding cations (Fig. 16; Jandik *et al.*, 1992). Some examples of separations of ions from aqueous solutions are presented in Fig. 17.

Separation of metals

The application of HPLC to assay metals after formation of complexes that absorb in the UV region provides a very weak response in conductivity detection. The metals, among themselves, have similar sizes and valencies. Therefore, their mobilities are very close to each other. Hence, it is necessary to utilize complex formation which allows us to separate these metals. Because there is little understanding of how the complexation of these metals takes place in solution, the number of publications in this domain are fewer than on the separation of anions.

Two different approaches have been adopted.

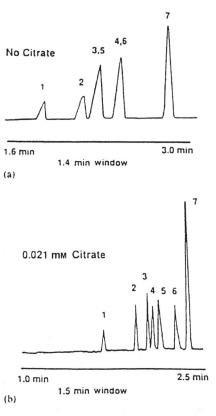

Figure 15. HPCE separation of two cations. A: carrier electrolyte: 5 mM CATI, pH 5.2, capillary 75 μm × 60 × 52 cm fused silica. Inverse UV detection at 214 nm. Separation voltage +25 kV. 1 (potassium), 2 (barium), 3 (strontium), 4 (sodium), 5 (calcium), 6 (magnesium), 7 (lithium) in concentration of 0.2 to 0.5 ppm. Reproduced with permission.

Mobility of Cations in the Capillary

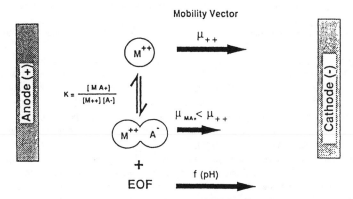

Figure 16. Modification of ionic mobility by complexation. The mobility of cation M is reduced by complexation with complexing ligand A. Reproduced with permission.

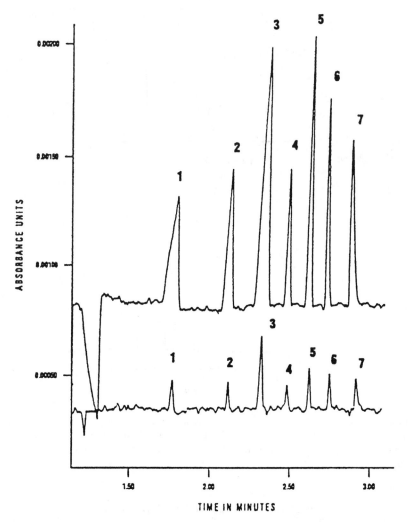

Figure 17. Separation of a standard mixture of cations: comparison of different injection modes. Reproduced with permission.

[a]1 – ammonium; 2 – potassium; 3 – sodium; 4 – calcium; 5 – magnesium; 6 – lithium; 7 – barium.

[b]Conditions. Capillary: Fused silica, 50 μm, 44 cm, 37 cm. Carrier electrolyte: 4 mM copper (II) sulphate, 4 mM formic acid, 4 mM 18-crown-6, the TSP Cation Buffer Kit P/N 4659-010. Voltage: +20 kV. Injection: 2 s electrokinetic at 10 kV (upper trace), or hydrodynamic for 2 s (lower trace). Detection: 215 nm, indirect. Temperature: 15°C. Sample concentration: 2.5 ppm ammonium, 2.5 ppm potassium, 0.125 ppm sodium, 1.25 ppm calcium, 0.125 ppm magnesium, 0.05 ppm lithium and 5 ppm barium as the chloride salts in water, stored in plastic vials. Instrument: SpectraPHORESIS® 1000.

[c]Cations in low ppm concentrations are easily separated using the TSP Cation Buffer Kit. The figure shows the focusing effect obtained by using electrokinetic injection. Peaks are much larger, with concomitantly lower detection limits. For quantitation, however, hydrodynamic injection is recommended because it offers much wider range of dynamic response. The electrokinetic injections usually show a good dynamic response at low concentrations, while at higher levels the response becomes essentially independent of analyte concentration.

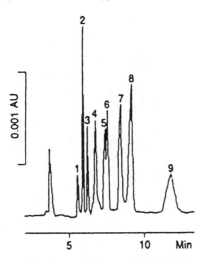

Figure 18. Micellar CE separation of metal chelates of PAR. Capillary, $42/50$ cm $\times$ 75 mm i.d.; carrier electrolyte, 10 mM ammonium phosphate buffer containing 75 mM SDS and 1×10^{-4} M PAR (pH 8.0); voltage, 15 kV; injection, 30 s (hydrostatic); detection, 254 nm. Metals (mol/l): $1 = $ Cr(III) (2.4×10^{-4}); $2 = $ Co(II) (6×10^{-5}); $3 = $ Cu(II) (8×10^{-5}); $4 = $ Pb(II) (8×10^{-5}); $5 = $ Ni(II) (8×10^{-5}); $6 = $ Fe(II) (8×10^{-5}); $7 = $ Zn(II) (1.6×10^{-4}); $8 = $ Fe(III) (8×10^{-5}); $9 = $ Cd(II) (2.4×10^{-4}). The first-migrating peak belongs to acetone. Reproduced with permission.

Table 5.

Detection limit (mol/l) of PAR/metal complexes

Metal	Detection wavelength (nm)	
	254 nm	500 nm
Co(II)	8×10^{-7}	1×10^{-7}
Cu(II)	2×10^{-6}	6×10^{-7}
Fe(II)	4×10^{-6}	9×10^{-7}
Ni(II)	3×10^{-6}	1×10^{-6}
Zn(II)	2.4×10^{-5}	8×10^{-7}

Reproduced with permission of Timerbaev *et al.* (1994).

The first involves forming a complex with a ligand demonstrating a strong absorbtion in the UV region before injection (Saitoh *et al.*, 1989; Timerbaev *et al.*, 1993). The separation is performed afterwards by micellular electrophoresis employing direct detection. An example of separation utilizing PAR as ligand is provided in Fig. 18 (Timerbaev *et al.*, 1993). The first advantage of this method is linked to the non-utilisation of a visualizing ion having a mobility close to that of the solutes, since we separate the whole complex formed after reaction with the chromophore. The second advantage is a lower detection limit compared to that which we obtain generally in inverse absorption. Some examples can be seen in Table 5 which lists detection limits obtained by metal-PAR complexation. The detection limit in inverse absorption is around 10^{-6} mol/l.

P. Roland-Gosselin et al.

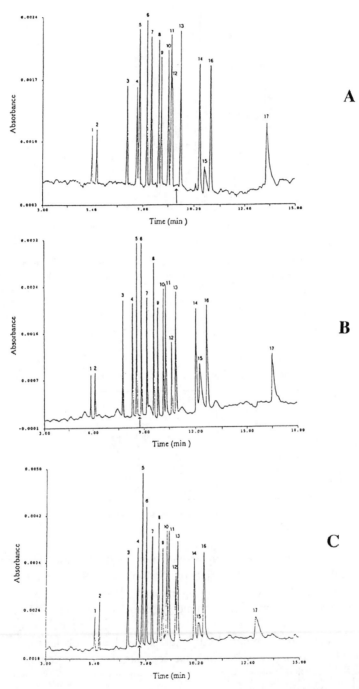

Figure 19. A: HPCE separation of 17 metal ions using 10 mM benzylamine-16 mM glycolic acid at pH 4. The upward arrow indicates the migration position of benzylamine. B: same but separated using 10 mM pyridine-12 mM glycolic acid at pH 4. C: same but separated using 10 mM imidazole-13 mM glycolic acid. Reproduced with permission.

An improvement of the basic technique was undertaken by Regan *et al.* (1994). They realized an on-line preconcentration by introducing a small quantity of PAR reagent just after introducing the sample in the capillary. It is possible to obtain complexation of the sample with reagent in the capillary playing with the differences in valencies and EOF. Detection limits obtained by this method are of the same order as these described previously. The main interest of this method is linked to a single on-line manipulation without a supplementary protocol.

The second method involves the use of a visualising ion and a complexing agent. Several visualising agents were investigated. It seems that imidazole represents a mobility compatible with metal separation. α-Hydroxyisobutric acid (HIBA) is often employed as a complexing agent since it is a weak acid and the metal ions complex partially. In addition, the separation time is not too long. The migration time is greater if the stability of the complex formed is higher, as illustrated in some examples found in the published literature (Fig. 19).

Regardless of the fact that, today, the principle of the separation mechanism is well understood, the separation of standard mixtures is still achieved with relative difficulty. Therefore, for the above analyses, no privileged method of separation, such as that for anions, is actually available. The nature of the samples also poses problems. It is very often required to undertake analysis in a highly acidic or basic medium, leading to enormous pH differences between the sample and the electrolyte. Such situations greatly influence the quality of separation. Nevertheless, we should highlight a very original method of separation which employs cyano complexation for separation of an anion mixture (Buchberger *et al.*, 1993).

CONCLUSIONS

We have tried to demonstrate that utilisation of high performance capillary electrophoresis offers several possibilities for separation of inorganic ions. This field of ion separation is thoroughly investigated today and allows rapid analysis. However, the separation of cations such as metals still remains difficult but capillary electrophoresis is a relatively new analytical method and in future years greater accomplishments and improvements are certainly going to come.

ACKNOWLEDGMENTS

The authors express their sincere gratitude to different authors or publishers to allow the reproduction of some of the figures in the text.

REFERENCES

Avdalovic, N., Pohl, Ch. A., Rocklin, R. D. and Stillian, J. R. (1993). Determination of cations and anions by capillary electrophoresis combined with suppressed conductivity detection. *Anal. Chem.* **65**, 1470–1475.

Bachmann, K., Boden, J. and Haumann, I. (1992). Indirect fluorimetric detection of alkali and alkaline earth metal ions in capillary zone electrophoresis with cerium (III) as carrier electrolyte. J. Chromatogr. 626, 259–265.

Benz, N. J. and Fritz, J. S. (1994). Studies on the determination of inorganic anions by capillary electrophoresis. J. Chromatogr. 671, 437–443.

Buchberger, N. and Haddad, P. R. (1992). Effect of carrier electrolyte composition on separation selectivity in capillary zone electrophoresis of low molecular-mass ions. J. Chromatogr. 608, 59–64.

Buchberger, N., Semenova, O. P. and Timerbaev, A. R. (1993). Metal ion capillary zone electrophoresis with direct UV detection: Separation of metal cyanide complexes. J. High Resol. Chromatogr. 6, 153–156.

Carpentier, G. (1986). Etude du comportement d'un détecteur electrochimique, voltampérométrique. Conception originale d'un détection voltampérométrique différentielle en opposition couplée à un HPLC pour des application en neurobiologie. Doctor of Sciences Thesis, Paris University.

Charlot, G. (1966). Les méthodes de la Chimie Analytique. Masson, Paris.

Cousins, S. M., Haddad, P. R. and Buchberger, W. (1994). Evaluation of carrier electrolytes for capillary zone electrophoresis of low molecular-mass anions with indirect UV detection. J. Chromatogr. (A) 671, 397–402.

Devêvre, O., Putra, D. P., Botton, B. and Garbaye, J. (1994). Sensitive and selective method for the separation of organic acids by capillary zone electrophoresis. J. Chromatogr. (A) 679, 349–357.

Fuerstenau, D. W. (1956). Streaming potential studies on quartz in solutions of aminium acetates in relation to the formation of hemimicelles at the quartz-solution interface. J. Phys. Chem. 60, 981–985.

Geyer, M. (1993). Product applications and information (Thermo Separation Products): influence of capillary temperature on anions with PNA buffer system.

Harrold, M. P., Wojtusik, M. J., Riviello, J. and Henson, P. (1993). Parameters influencing separation and detection of anions by capillary electrophoresis. J. Chromatogr. 640, 463–471.

Hauser, P. C., Renner, N. D. and Hong, A. P. C. (1994). Anion detection in capillary electrophoresis with ion-selective microelectrodes. Anal. Chim. Acta 295, 181–186.

Jandik, P. and Bonn, G. (1993). Capillary Electrophoresis of Small Molecules. VCH Press, New York.

Jandik, P., Jones, W. R., Weston, A. and Brown, P. R. (1992). Electrophoretic capillary ion analysis: Origins, principles and applications. L. C. G. C. Int. 5, 20–25.

Janini, G. M., Chan, K. C., Muschik, G. M. and Issaq, H. J. (1994). Analysis of nitrate and nitrite in water and urine by capillary zone electrophoresis. J. Chromatogr. (B) 657, 419–423.

Jones, W. R. and Jandik, P. (1991). Controlled changes of selectivity in the separation of ions by capillary electrophoresis. J. Chromatogr. 546, 445–458.

Kaneta, T., Tanaka, S. and Taga, M. (1993). Effect of cetyltrimethylammonium chloride on electroosmotic and electrophoretic mobilities in capillary zone electrophoresis. J. Chromatogr. (A) 653, 313–319.

Mikkers, F. E. P., Everaerts, F. M. and Verheggen, Th. P. E. M. (1979). J. Chromatogr. 169, 1–10.

Morin, P., François, C. and Dreux, M. (1994). Séparation de cations et d'anions inorganiques par électrophorèse capillaire ionique avec une détection spectrométrique UV indirecte. Analusis 22, 178–189.

Nann, A. and Pretsch, E. (1994). Potentiometric detection of anions separated by capillary electrophoresis using an ion-selective microelectrode. J. Chromatogr. (A) 676, 437–442.

Nielen, M. W. F. (1991). Quantitative aspects of indirect UV detection in capillary zone electrophoresis. J. Chromatogr. 588, 321–326.

Parvez, H., Bastart-Malsot, M., Parvez, S., Nagatsu, T. and Carpentier, G. (1987). Electrochemical Detection in Medicine and Chemistry. VNU Science Press, Utrecht.

Regan, F. B., Meaney, M. P. and Lunte, S. M. (1994). Determination of metal ions by capillary electrophoresis using on-column complexation with 4-(2-pyridylazo)resorcinol following trace enrichment by peak stacking. *J. Chromatogr.* (*B*) **657**, 409–417.

Reijenga, J. C., Aben, G. V. A., Verheggen, Th. P. E. M. and Everaerts, F. M. (1983). Effect of electroosmosis on detection in isotachophoresis. *J. Chromatogr.* **260**, 241–254.

Rosset, R., Caude, M. and Jardy, A. (1991). *Chromatographie en Phase Liquide et Supercritique.* Masson, Paris.

Saitoh, T., Hoshino, H. and Yotsuyanagi, T. (1989). Separation of 4-(2-pyridylazo)resorcinolato metal chelates by micellar electrokinetic capillary chromatography. *J. Chromatogr.* **469**, 175–181.

Schwer, C. and Kenndler, E. (1991). Electrophoresis in fused-silica capillaries: The influence of organic solvents on the electroosmotic velocity and the potential. *Anal. Chem.* **63**, 1801–1807.

Small, H. and Miller, Th. E. Jr. (1982). Indirect photometric chromatography. *Anal. Chem.* **54**, 462–469.

Somasundaran, P., Healy, T. W. and Fuerstenau, D. W. (1964). Surfactant adsorption at the solid–liquid interface: Dependence of mechanism on chain length. *J. Phys. Chem.* **68**, 3562–3566.

Stathakis, C. and Cassidy, R. (1994). Cationic polymers for selectivity control in the capillary electrophoretic separation of inorganic ions. *Anal. Chem.* **66**, 2110–2115.

Timerbaev, A. R., Buchberger, W., Semenova, O. P. and Bonn, G. K. (1993). Metal ion capillary zone electrophoresis with direct UV detection: Determination of transition metals using an 8-hydroxyquinoline-5-sulphonic acid chelating system. *J. Chromatogr.* **630**, 379–389.

Timerbaev, A. R., Semenova, O. P., Jandik, P. and Bonn, G. K. (1994). Metal ion capillary electrophoresis with direct UV detection: effect of a charged surfactant on the migration behaviour of metal chelates. *J. Chromatogr.* **671**, 419–427.

Weinberger, S. R. and Schlabach, T. (1991). Poster PM 36, *HPCE 1991*. San Diego, CA.

Weston, A., Brown, Ph., Jandik, P., Jones, W. R. and Heckenberg, A. L. (1992). Factors affecting the separation of inorganic metal cations by capillary electrophoresis. *J. Chromatogr.* **593**, 289–295.

Yeung, E. S. (1989). Indirect detection methods: looking for what is not there. *Acc. Chem. Res.* **22**, 125–130.

Author index

Subject index